Soil Respiration and the Environment

Soil Respiration and the Environment

Yiqi Luo and Xuhui Zhou

AMSTERDAM • BOSTON • HEIDELBERG
LONDON • NEW YORK • OXFORD
PARIS • SAN DIEGO • SAN FRANCISCO
SINGAPORE • SYDNEY • TOKYO
Academic Press is an imprint of Elsevier

Academic Press is an imprint of Elsevier
30 Corporate Drive, Suite 400, Burlington, MA 01803, USA
525 B Street, Suite 1900, San Diego, California 92101-4495, USA
84 Theobald's Road, London WC1X 8RR, UK

This book is printed on acid-free paper. ♾

Library of Congress Cataloging-in-Publication Data

British Library Cataloguing-in-Publication Data
A catalogue record for this book is available from the British Library.

ISBN 13: 978-0-12-088782-8
ISBN 10: 0-12-088782-7

For information on all Academic Press publications
visit our Web site at www.books.elsevier.com

Working together to grow
libraries in developing countries

www.elsevier.com | www.bookaid.org | www.sabre.org

ELSEVIER   BOOK AID
              International   Sabre Foundation

Table of Contents

Preface

Soil respiration is an ecosystem process that releases carbon dioxide from soil via root respiration, microbial decomposition of litter and soil organic matter, and fauna respiration. Research on soil respiration has been remarkably active in the past decade partly because it is among the least understood subjects in ecosystem ecology and partly because it represents the second largest flux of carbon cycling between the atmosphere and terrestrial ecosystems. As one key process of ecosystems, soil respiration is related to ecosystem productivity, soil fertility, and regional and global carbon cycles. Since the global carbon cycle regulates climate change, soil respiration also becomes relevant to climate change, carbon trading, and environmental policy. In short, soil respiration is nowadays a multidisciplinary subject that is of concern not only to ecologists, soil scientists, microbiologists, and agronomists but also to atmospheric scientists, biogeochemists, carbon traders, and policy-makers. To date, no book has been published to synthesize extant information on soil respiration in spite of its importance in many disciplines. We write this book to fill this void and to stimulate broad interests in this subject among students, scientists, environmental managers, and policy makers from different disciplines,

The active research in the past decade has substantially advanced our understanding but, meanwhile, created much confusion with considerable repetitive work in the research community. Much of the confusion and repetition stems from the lack of a systematic organization of knowledge on fundamental processes of soil respiration. It was our initial motivation to lay down the foundation of the soil respiration sciences and to clarify some of the confusion. Toward that goal, we make an attempt to progressively introduce and rigorously define concepts and basic processes. We also try to structure the book in such a way that all the major up-to-dated research findings can be logically summarized. The book is accordingly divided into four sections—context, mechanisms, regulation, and approaches—and ten chapters. Chapters 1 and 2 offer a contextual view of the soil respiration science and lay down its relationships with a variety of issues in carbon research. Chapters 3 and 4 describe fundamental processes of CO_2 production and transport. Chapters 5–7 present regulatory mechanisms of soil respiration, including controlling factors, spatial and temporal variations, and responses to natural and human-made perturbations. Chapters 8—10 illustrate research approaches to measurement of soil respiration, partitioning to various components, and modeling. It is our hope that this book helps clarify confusion and identify knowledge gaps where research may be most productive.

We write the book for undergraduate and graduate students, professors and researchers in areas of ecology, soil science, biogeochemistry, earth system science, atmosphere, climate molders, microbiology, agronomy, plant physiology, global change biology, and environmental sciences. The book introduces concepts and processes in a logical way so that students and laymen who do not have much background in this area are can read the book without too much difficulty. The book has also summarized the contemporary research findings with extensive references. Scientists who are actively working on soil respiration should find this book as a useful reference book for their research. We also recognize that the field of soil respiration research is evolving very quickly. Even within the time span from the manuscript submission to the publication of this book, many important papers have been published. Inevitably, many good papers may have been left out. We are sorry if we miss your work in this book but welcome you to write us emails and send us the postal mails with your important publications. We will try our best to incorporate your work into the new version of the book in the future.

This book is first dedicated to our fellow researchers. Their devotion to and passionate on the soil respiration science are the impetus of advances in our understanding on this subject. Their imagination and creativity result in, for example, diverse ideas, experimental evidence from different angles, and measurements by distinct methods. Their rigorous logic helps critique results, identify new issues to be addressed, and generate new ideas to be tested. Their meticulous methodology checks measurement and modeling results once and again, enhancing the robustness of our knowledge. Their collective effort helps establish the soil respiration science and, more importantly, bring it into a focal research area in the earth system science. We hope that this book will stimulate further interest in this fascinating subject and promote high-quality scientific contribution.

We also dedicate this book to our families. Our parents taught us to work hard no matter what we are doing, which becomes the lifetime gift to us. The hardship of lives in our childhoods makes us appreciate what we have every-day. We thank our spouses for their understanding of our career choices and for their support to our effort on book writing. They have sacrificed countless hours of family activities to make time for us to work on the book. Our children brought us tremendous fun to our busy lives. In particular, Jessica Y. Luo has read the first two chapters and offered suggestions to improve reader-ship of the book.

Yiqi Luo is also grateful for students and post-doctoral fellows in his laboratory who have worked with him to develop ideas, test various hypotheses, and contribute to discussion in the research community via publications and participation in international meetings.

Finally, we are indebted to many colleagues and authors who have sent us reprints of their papers and manuscripts. We are grateful to Eric A. Davidson, Joseph M. Craine, Dafeng Hui, Changhui Peng, Weixing Cheng, and Kiona Ogle for their time to read the manuscript and for many helpful suggestions and criticisms they have offered. We also thank Kelly D. Sonnack and Meg Day of Academic Press/Elsevier for their patience and encouragement for this project, Cate Barr for providing a cover design and Deborah Fogel for help in editing manuscripts. Yiqi Luo thanks Dr. Lars Hedin for hosting his sabbatical leave at Princeton University where the manuscript was finalized. Yiqi Luo also acknowledges the financial support from US Department of Energy and National Science Foundation, which has helped maintain his active research in the past decade.

Yiqi Luo and Xuhui Zhou
Norman, Oklahoma
April 12, 2006

Context

Introduction and Overview

Soil respiration is a crucial piece of the puzzle that is the earth's system. To understand how the earth's system functions, we need to figure out the role that soil respiration plays in regulating atmospheric CO_2 concentration and climate dynamics. Will global warming instigate a positive feedback loop between the global carbon cycle and climate system that would, in turn, aggravate climatic warming? How critical is soil respiration in regulating this positive feedback? To answer these questions, we have to understand the processes involved in soil respiration, examine how these processes respond to environmental change, and account for their spatial and temporal variability.

Since climate change is one of the main challenges facing humanity, quantification of soil respiration is no longer just a tedious academic issue. It is also relevant to farmers, foresters, and government officials. Can respiratory carbon emission and/or photosynthetic carbon uptake be manipulated to maximize carbon storage so that farmers and foresters can earn cash awards in global carbon-trading markets? To effectively manipulate respiratory carbon emission from terrestrial ecosystems, we need to identify the major factors that control soil respiration. Even if we can manipulate respiratory processes, how could signatory countries to the Kyoto treaty verify carbon sinks in the biosphere to claim their credits during the intergovernmental negotiations? All these issues make it necessary for us to invent reliable methods to measure soil respiration accurately in croplands, forest areas, and other regions. Can the managed carbon sinks last long enough to mitigate greenhouse gas emission effectively in the future? How will soil respiration respond to natural and human-made perturbations? To answer all these questions, it is necessary to develop a predictive understanding of soil respiration, aiming toward a mechanistic modeling of soil respiration. It is evident from all these examples that studying soil respiration is not only desirable for

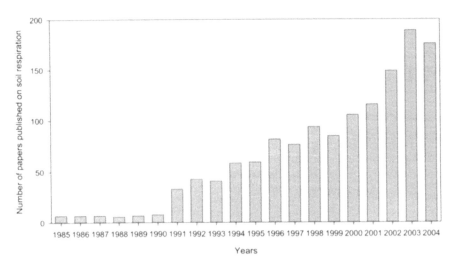

FIGURE 1.1 Number of papers published on soil respiration since 1985. The number was obtained from a search for the key terms "soil respiration," "soil CO_2 efflux", and "belowground respiration" in the Web of Science database.

purely academic reasons but also crucial in the commercial and political arenas.

Due to the recent societal need to mitigate climate change and the scientific aspiration to understand soil respiration itself, the research community has been very active in studying soil respiration. During the past 15 years, the number of papers published on soil respiration has linearly increased and reached nearly 200 papers in 2003–2004, compared with about 10 papers in 1985–1990 (Fig. 1.1). The active research also partially reflects the fact that soil respiration remains least understood among ecosystem carbon processes, despite its central role in the global carbon cycle and climate change. This book lays down the fundamentals of soil respiration while synthesizing the recent literature in this field.

1.1. DEFINITION AND INTRODUCTION

The word respiration, derived from the Latin prefix *re-* (back, again) and root word *spirare* (to breathe), literally means breathing again and again. It is thus used to describe the process of gas exchange between organism and environment. Physiologically, respiration is a series of metabolic processes that break down (or catabolize) organic molecules to liberate energy, water, and carbon dioxide (CO_2) in a cell. All living organisms—plants, animals, and micro-organisms alike—share similar pathways of respiration to obtain the energy that fuels life while releasing CO_2. Respiration is often studied in relation to energy supply at the biochemical and cellular levels as a major component of bioenergetics. However, bioenergetics in soils is not well developed (Dilly 2005), and soil respiration is studied predominantly in relation to CO_2 and O_2 exchanges. In this book the word respiration is used mainly to describe CO_2 production rather than energy supply.

For the purposes of this book, soil respiration is defined as the production of carbon dioxide by organisms and the plant parts in soil. These organisms are soil microbes and fauna, and the plant parts are roots and rhizomes in the soil. Additionally, soil is often defined as a mixture of dead organic matter, air, water, and weathered rock that supports plant growth (Buscot 2005). Some authors (e.g., Killham 1994) also include living organisms in the definition of soil, treating roots, soil microbes, and soil fauna as part of soil. Therefore, it makes sense to talk about soil that can breathe. Soil respiration means that the living biomass of soil respires CO_2, while soil organisms gain energy from catabolizing organic matter to support life.

Soil respiration is sometimes called belowground respiration, in contrast with aboveground respiration. The latter refers to respiratory CO_2 production by the plant parts above the soil surface. Although the definition of soil usually does not include dead plant materials at the soil surface that have not been well decomposed, CO_2 production via litter decomposition in the litter layers is generally included in soil respiration (or belowground respiration) in many publications and, for the sake of simplicity, in this book as well.

Technically, the rate of CO_2 production in the soil (i.e., the soil respiration rate) cannot be directly measured in the field. Measurements are often made at the soil surface to quantify a rate of CO_2 efflux from the soil to the atmosphere. The instantaneous rate of soil CO_2 efflux is controlled not only by the rate of soil respiration but also by the transport of CO_2 along the soil profile and at the soil surface (see Chapter 4). The CO_2 transport is influenced by the strength of the CO_2 concentration gradient between the soil and the atmosphere, soil porosity, wind speed, and other factors. At a steady state, the CO_2 efflux rate at the soil surface equals the rate of CO_2 production in soil. In this case, soil CO_2 efflux is practically equivalent to soil respiration, and the two terms are thus interchangeable.

However, there are several situations in which CO_2 production may not be at a steady state with CO_2 transport. For example, soil degassing occurs during rainfall or irrigation, driving CO_2 stored in the soil air space out of the soil. After rainfall or irrigation, CO_2 produced by soil organisms is partially stored in the soil to rebuild the CO_2 concentration gradient. Carbonic acid reaction and microbial methanogensis could each produce or consume

CO_2, depending on conditions that influence reaction equilibriums (see Chapter 3). Thus, the CO_2 released at the soil surface could be generated by carbonic acid reactions during rock weathering, particularly in arid lands where carbonic reaction is very strong. On the other hand, the CO_2 produced by soil living tissues could be absorbed by microbes during methanogenic processes. However, the amount of CO_2 produced and/or consumed by carbonation and methanogenesis is generally trivial in comparison with soil respiration, except in very dry lands. The non-steady-state CO_2 efflux at the soil surface occurs mostly during rainfall or irrigation after long periods of drought (Liu et al. 2002a, Xu et al. 2004). In absence of major perturbation, the rate of CO_2 production in soil is indistinguishable from the rate of CO_2 efflux at the soil surface on a daily or longer time-scale (Hui and Luo 2004). Thus, the term soil respiration is practically interchangeable with soil surface CO_2 efflux on a long-term scale. However, soil CO_2 efflux rates measured at shorter time-scales may not be equivalent to the rate of soil respiration.

Soil respiration usually accounts for the majority of ecosystem respiration, which is the sum of soil respiration and respiration of aboveground parts of plants (see Chapter 2). Some methods can directly measure ecosystem respiration, from which soil respiration is estimated indirectly (see Chapter 8). Thus, the soil and ecosystem respirations are closely related. Although this book focuses on soil respiration, it often describes ecosystem respiration as well.

As a preview, Figure 1.2 shows a typical time course of CO_2 efflux rates from soil. The time course, which was measured at the soil surface in a tall-grass prairie of Oklahoma, displays a distinct seasonal pattern of high soil respiration during summer and low respiration in winter. The seasonal pattern is roughly repeated in subsequent years. Nonetheless, there are observable variations from year to year. For example, the summer peak of soil respiration reaches nearly $6\,\mu\mathrm{mol\,m^{-2}\,s^{-1}}$ in 2002 and is less than $4\,\mu\mathrm{mol}$ $\mathrm{m^{-2}\,s^{-1}}$ in 2001. The winter low is nearly $0\,\mu\mathrm{mol\,m^{-2}\,s^{-1}}$ in 2002 but 0.3–$0.5\,\mu\mathrm{mol\,m^{-2}\,s^{-1}}$ in other years. In most years, there are dips in the measured soil respiration during the late summer and early autumn, but in 2004 the seasonal pattern is relatively smooth. This kind of year-to-year variation exemplifies the term "interannual variability."

Similar seasonal patterns have also been observed in northern semiarid grasslands (Frank et al. 2002), forests (Salvage and Davidson 2001, Epron et al. 2004, King et al. 2004), and croplands (Beyer 1991). For example, soil respiration varies from nearly $0\,\mu\mathrm{mol\,m^{-2}\,s^{-1}}$ in the winter to about $10\,\mu\mathrm{mol\,m^{-2}\,s^{-1}}$ in the summer over one year in the Duke Forest, North Carolina (King et al. 2004). This seasonal pattern repeats from 1997 to 2002, and interannual variation is apparent with different peaks in summer and valleys in winter.

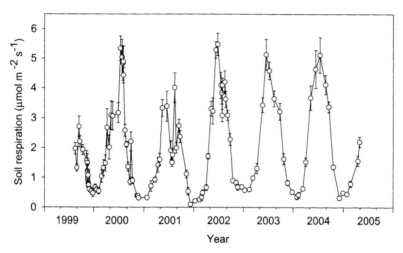

FIGURE 1.2 Measured rate of soil CO_2 efflux in a tallgrass prairie of Oklahoma, USA from 1999 to 2005. Open circles represent data points, and bars indicate the one standard error below and above the data points. Data are only for the measured soil CO_2 efflux in the control treatment in a warming and clipping experiment and adopted from Luo *et al.* (2001), Wan *et al.* (2005), and Zhou *et al.* (2006).

From the observed soil respiration patterns, we can ask many questions. For example, what causes such seasonal and interannual variations? Why does soil respiration vary from one site to another? How can we scale up the plot-level measurements to estimate total carbon losses on regional and global scales? Can we derive general mechanisms from the observed patterns and then predict future changes in soil respiration? What percentage of the lost carbon is from root respiration? How much is the carbon released by soil respiration directly from the recent photosynthesis? This book will address these questions, among others, as it lays down the basic principles of soil respiration. Before turning to these issues, however, let's first review the history of research on soil respiration.

1.2. HISTORY OF RESEARCH

Research on soil respiration has an impressively long history (Fig. 1.3) and can be dated back to papers by Wollny (1831), Boussingault and Levy (1853), and Möller (1879). The earliest studies of soil respiration were intended to characterize soil metabolism. Twentieth-century research on soil respiration can be divided into roughly four major periods. During the first few decades of the century, research on soil respiration was conducted primarily in the laboratory with agricultural soil. Soil respiration was used to evaluate soil fertility and biological activities in soil. Chemical fertilizers, invented in the late 19th century, were applied to crops to stimulate growth and considerably enhanced agricultural productivity as a result. At that time, researchers emphasized understanding the soil properties that influence crop production. Soil respiration was used as an index of soil fertility for agricultural production (Russell and Appleyard 1915), because in a field study, fertilization of agricultural crops generally increases soil respiration rates (Lundegårdh 1927). Some laboratory studies, however, showed that nutrient release was not proportional to the carbon release during mineralization (Waksman and Starkey 1924, Pinck et al. 1950).

During that period, some primitive methods for the measurement of soil respiration were developed. Stoklasa and Ernest (1905) passed CO_2-free air over soil samples contained in a flask and measured the amount of CO_2 released from the soil samples using the alkali absorption method. Lundegårdh (1927) recognized that measured CO_2 efflux from soil samples in the laboratory might not be representative of that from intact soils in the field, where, he argued, diffusion was a chief process controlling efflux of CO_2. He was probably the first scientist to make in situ measurements of rates of CO_2 efflux from field soil by covering the soil surface with a chamber for a period of time. Then he took air samples with brass tubes from the chamber, as well as from air spaces in the soil at three different depths. The air samples were passed through alkali solutions for measurements of soil respiration. Humfeld (1930) modified Lundegårdh's method and passed air through the chamber with inlet and outlet ports to collect the CO_2-enriched air in an alkali absorption train. The alkali absorption chamber method, first introduced by Lundegårdh (1921), modified by Humfeld (1930) and others, and widely used in the following decades, places static alkali solution within the chamber followed by titration of chloric acid.

By this time the major factors that influence soil respiration had been identified. Greaves and Carter (1920) were among the first to document a consistent relationship between soil water content and microbial activity. Turpin (1920) reviewed soil respiration and concluded that the primary source of CO_2 efflux from soils was attributable to bacterial decomposition. Lundegårdh (1927) pointed out that soil diffusion was important in controlling the efflux of CO_2. Smith and Brown (1933) indicated that the rate of diffusion of CO_2 through the soil correlated with CO_2 production. Lebedjantzev (1924) observed that air drying of soil samples increased fertility (such as NH_4-N, amide-N, and phosphorus) of a variety of soils and decreased the number of microorganisms in pot experiments.

Few publications on soil respiration can be identified during the relatively inactive research period from the late 1930s to the early 1950s, possibly due to the worldwide social turbulence of that period. From the late 1950s to the 1970s, research activity on soil respiration resumed (Fig. 1.3), mainly from an ecological perspective, as scientists tried to understand heterotrophic processes in the soils of native ecosystems (Lieth and Ouellette 1962, Witkamp 1966, Raguotis 1967, Schulze 1967, Reiners 1968, Kucera and Kirkham 1971). During that period, research advanced the science of soil respiration in many respects, including (1) methods of measurement, (2) controlling factors, (3) partitioning into components, (4) relationships with other ecosystem carbon processes, and (5) synthesis and scaling to global estimation.

Many studies were devoted to careful evaluation of the various factors that affect the accuracy of the alkali absorption method (Walter 1952, Howard 1966, Kirita and Hozumi 1966, Kirita 1971, Chapman 1971, 1979, Anderson 1973, Gupta and Singh 1977). The accuracy of the method was found to vary with factors such as the amount and strength of alkali used, the area of covered soil, the chamber height above the ground, the depth of the chamber inserted into soil, the surface area and the height of the alkali container within the chamber, the duration of measurement, and the rates of soil CO_2 efflux. Minderman and Vulto (1973) suggested the use of fine-grained soda lime instead of alkali solution to absorb CO_2.

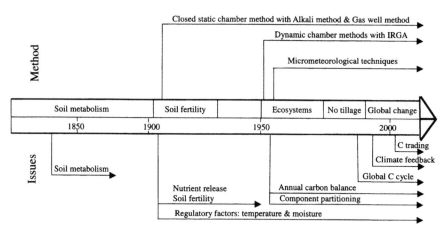

FIGURE 1.3   Schematic illustration of the history of soil respiration research since the 1830s. Within the main axis are major themes in different eras of research. There is little research activity from late 1930s to early 1950s. Above the axis is method development for measurement of soil respiration. Below the axis are major issues that have been addressed by and/or motivate soil respiration research during the different eras.

One major technical advance was made in the 1950s: infrared gas analyzer (IRGA) was used for the measurement of soil respiration. Haber (1958) first used IRGA to calibrate the alkali absorption method. Golley et al. (1962) were among the first to make field measurements of soil respiration on the peat floor of a mangrove forest using IRGA. Reiners (1968) examined how gas flow rates influenced IRGA measurement of CO_2 evolution, while Kanemasu et al. (1974) studied effects of air "suction" and "pressure" on IRGA measurements of soil respiration. Measured CO_2 efflux with the suction chamber was one order of magnitude higher than with the pressure chamber. The suction chamber drew CO_2 from the soil outside the chamber and/or in deep layers via mass flow. Edwards and Solins (1973) designed an open flow system with the chamber linked to IRGA to measure soil respiration continuously. Edwards (1974) used movable chambers that were lowered onto the forest floor during measurements and lifted between measurements. The movable chambers allowed natural drying of the soil and litterfall onto the measurement surface. The IRGA measurements of soil CO_2 efflux were then compared with those using the alkali absorption method (Kirita and Hozumi 1966). Many studies found that the alkali method underestimated soil CO_2 efflux compared with the IRGA measurements (Haber 1958, Witkamp 1966, Kucera and Kirkham 1971). Other studies did not detect any significant differences between the two methods (e.g., Ino and Monsi 1969).

The gas-well method first used by Lundegårdh (1927) to estimate soil respiration from a CO_2 concentration gradient along soil profiles was fully developed by de Jong et al. (1979). Meanwhile, a variety of micrometeorological methods, such as Bowen ratio and eddy flux, have been developed to measure gas exchanges within and above the plant canopy (Monteith 1962, Monteith et al. 1964), from which soil respiration was indirectly estimated.

From the late 1950s to the 1970s, knowledge of factors that regulate soil respiration was greatly enriched. Bunt and Rovira (1954) studied soil respiration in a temperature range of 10 to 70°C. They found that O_2 uptake and CO_2 release increased with temperature up to 50°C, above which it declined. Many studies demonstrated that soil respiration correlated exponentially with temperature (Wiant 1967, Kucera and Kirkhma 1971, Medina and Zelwer 1972). Drobnik (1962) estimated Q_{10}, that is, a quotient indicating the temperature sensitivity of soil respiration (see Chapter 5), to be 1.6 to 2.0 in response to temperatures ranging from 8 to 28°C. Wiant (1967) estimated Q_{10} to be approximately 2 for temperatures from 20 to 40°C. Soil moisture was also identified as important in influencing soil respiration. A laboratory study suggested that microbial respiration decreased when soil moisture was below 40% or above 80% of the field-holding capacity (Ino and Monsi 1969). Soil temperature and moisture combined could account for up to 90% of the variation of soil respiration measured in the field (Reiners 1968).

Birch and his colleague (Birch and Friend 1956, Birch 1958) conducted a notable study demonstrating that when a soil was dried and rewetted, decomposition of its organic matter was enhanced, leading to a flush of CO_2 production. They explained that the drying-wetting effect was not related to microbial stimulation or microbial death but rather caused by liberation of rapidly decomposable material from the clay. The clay protected the organic materials from microbial attacks under consistently moist conditions.

During that period, components of soil respiration were clearly identified into two major categories: autotrophic and heterotrophic respiration. The autotrophic components are the metabolic respiration of live root, associated mycorrhiza, and symbiotic N fixing nodules. The heterotrophic respiration is from microbial decomposition of root exudates in rhizosphere, aboveground and belowground litter, and soil organic matter (SOM). Coleman (1973b) measured total respiration of intact soil cores and individual components of roots, litter, and soil. Contribution to the total soil respiration was 8 to 17% from roots, 6 to 16% from litter, and 67 to 80% from soil microbes in a successional grassland. Edwards and Sollins (1973) partitioned total soil respiration from a forest into 35% from roots, 48% from litter, and 17% from soil. Richards (1974) found it difficult to partition soil respiration among different soil fauna, fungi, and bacteria.

Field measurements over the whole growing seasons made it possible to scale up individual measurements to estimate annual carbon efflux. Kucera and Kirkham (1971) estimated annual soil CO_2 efflux to be $452\,g\,C\,m^{-2}\,yr^{-1}$ in a tallgrass prairie by applying a temperature-respiration regression to continuous temperature records. Coleman (1973a) scaled up monthly averages of soil respiration in a grassland and estimated annual soil CO_2 efflux to be 357 to $421\,g\,C\,m^{-2}\,yr^{-1}$. Estimated annual soil CO_2 releases were about $1000\,g\,C\,m^{-2}\,yr^{-1}$ in many forests (Edwards and Sollins 1973, Garrett and Cox 1973).

Estimated annual efflux from soil respiration was often compared with annual carbon influx via aboveground litterfall, although the two processes are not completely comparable. Reiners (1968) showed that total soil respiratory carbon release was three times higher than litter carbon input. Edwards and Sollins (1973) found that litter decomposition accounted for only one-fifth of annual soil respiration. Anderson (1973) showed that annual soil respiration released 2.5 times as much carbon in annual litterfall. However, several studies demonstrated that carbon released by soil respiration was equivalent to that input from litterfall (Colemen 1973a, Witkamp and Frank 1969).

The accumulation of studies during that period offered opportunities to synthesize and compile results from many ecosystems. Singh and Gupta (1977) produced a major synthesis on the carbon processes of litter decomposition, soil respiration, root respiration, microbial respiration, faunal respiration, and SOM dynamics. Schlesinger (1977) reviewed many studies on soil respiration in the literature in order to develop latitudinal patterns of soil respiration worldwide and estimate a global total of carbon released via soil respiration.

Bunnell *et al.* (1977) and Minderman (1968) suggested that decomposition could best be represented by the summation of the exponential decay curves for all major chemical constituents, including sugars, cellulose, hemicellulose, lignin, waxes, and phenols. Henin *et al.* (1959) appeared to have been the first to propose a model that explicitly relates the two exponential rates to fresh plant carbon and "humified" carbon.

Long-term no-till plots were first established at the International Institute of Tropical Agriculture, Ibadan, in 1971 and continued through 1987 (Lal 2004). In the 1980s the agricultural practice of no tillage stimulated research on soil properties. Soil respiration was often used to indicate biological activities in soil with different tillage treatments (Anderson 1982). For example, Linn and Doran (1984) studied how no tillage affected soil water–filled pore space and its relationships with CO_2 and N_2O production. The level of soil aeration using microbial respiration rates of aerobic heterotrophs was also examined for compaction problems in a no-tillage management system (Linn and Doran 1984, Wilson *et al.* 1985, Neilson and Pepper 1990).

Since the 1990s, research on soil respiration has been driven primarily by global change. While climate research has its own long history (Weart 2003), the ecology research community, stimulated by the International Geosphere Biosphere Program (IGBP) and by a U.S. National Research Council (NRC) report (NRC 1986), has been involved in global change research in the past two decades and has studied ecosystem-level responses to climate change since the early 1990s (Mooney *et al.* 1991). In particular, the paper by Tan *et al.* (1990) played a critical role in attracting researchers' attention to the land biosphere. Their analysis of atmospheric CO_2 data suggested that land biosphere may absorb a large portion of the emitted carbon from anthropogenic sources. Three reports by the Intergovermental Panel on Climate Change (IPCC, 1990, 1995, 2001) and Schimel (1995) provided a global perspective on the carbon cycle in terrestrial ecosystems. Cox *et al.* (2000) linked a carbon cycle model with a global circulation model and highlighted the importance of the temperature sensitivity of respiration in future climatic predictions. That study continues to stimulate great interest in the temperature sensitivity of soil respiration among the research community.

Advances in measurement techniques have also stimulated modern, active research on soil respiration. Portable IRGAs have been widely used to measure soil surface CO_2 fluxes since the early 1990s (Norman *et al.* 1992). The IRGA method requires relatively less technique training than the traditional alkali or soda-lime absorption methods, but it provides quicker measurements of soil surface CO_2 effluxes. Meanwhile, many companies have retooled IRGA sensors and developed various chambers specifically for the measurement of soil CO_2 effluxes (see Chapter 8 and Appendix) facilitating research on soil respiration.

1.3. OVERVIEW OF THE BOOK

This book, which comprises 10 chapters, is dedicated to providing an understanding of various aspects of soil respiration. Chapters 1 and 2 provide a context of soil respiration science. Chapters 3 and 4 describe fundamental processes of CO_2 production and CO_2 transport. Chapters 5 through 7 present regulatory mechanisms of soil respiration, including controlling factors, spatial and temporal variations, and responses to natural and human-made perturbations. Chapters 8 through 10 discuss research approaches to measurement of soil respiration, partitioning to various components, and modeling.

Following the introduction and brief history of research on soil respiration covered in this chapter, Chapter 2 places soil respiration in the context of ecosystem carbon balance, nutrient cycling, regional and global carbon cycling, climate change, and carbon storage and trading. Soil respiration releases a large portion of carbon fixed by photosynthesis and strongly regulates net ecosystem productivity. Carbon dioxide released via microbial decomposition of litter and SOM is accompanied by either immobilization or mineralization of nutrients and is thus related to soil nutrient dynamics. Soil respiration plays a critical role in regulating global and regional carbon cycles. Its temperature sensitivity is a key issue in modeling feedback between global carbon cycling and climate change in response to anthropogenic warming. Although it is not the direct mechanism underlying land carbon storage, soil respiration is relevant to understanding carbon sequestration and global carbon trading markets.

Chapter 3 focuses on the processes of CO_2 production, including the fundamental biochemistry of respiratory processes, root respiration, microbial respiration in rhizosphere, and microbial decomposition of litter and SOM. The primary biochemical process of CO_2 production is the tricarboxylic acid (TCA) cycle. Root respiration in an ecosystem is determined by root biomass growth and the specific rates of root respiration. Microbial respiration occurs while root exudates are broken down, litter is decomposed, and SOM oxidated. Microorganism communities that use root exudates, litter, and SOM as substrates differ greatly and are briefly described in this chapter.

Chapter 4 describes processes of CO_2 transport along vertical profiles within the soil, at the soil surface, in the canopy, and in the planetary boundary layer. Soil CO_2 transport is driven primarily by gradients of CO_2

concentration along soil vertical profiles and determined by diffusion and mass flow processes. The CO_2 release at the soil surface depends on CO_2 gradients and is strongly affected by wind gusts, turbulences, and atmospheric pressure fluctuation. The CO_2 transport in the canopy and planetary boundary layer may not be directly relevant to soil respiration per se but is influenced by and often used to estimate soil respiration indirectly.

Soil respiration is affected by many factors, such as substrate supply, temperature, moisture, oxygen, nitrogen, soil texture, and pH value. Chapter 5 focuses on how individual factors regulate component processes of soil respiration and attempts to show that many of the factors influence multiple processes in various magnitudes and at different directions, leading to variable responses and complex patterns of soil respiration. The interactive effects of multiple factors on soil respiration are very complex and poorly understood.

Chapter 6 presents spatial and temporal patterns of soil respiration. It discusses temporal variations in soil respiration at multiple time-scales—from diurnal and weekly to seasonal, interannual, and decadal and centennial. Spatial patterns emerge at the stand level, landscape and regional scales, and across biomes. The chapter comparatively presents soil respiration among ecosystem types and examines general relationships of soil respiration to ecosystem productivities, prevailing environmental variables, and soil characteristics. This chapter also examines how soil respiration varies along latitudinal, altitudinal, and topographical gradients.

Chapter 7 describes changes in soil respiration in response to a variety of perturbations, such as elevated CO_2, global warming, changes in precipitation frequency and intensity, disturbances and manipulation of substrate supply, nitrogen deposition and fertilization, and agricultural cultivation. Generally speaking, soil respiration increases when substrate availability increases, such as under elevated CO_2 and litter addition. Soil respiration decreases if substrate supply is reduced under disturbances of fire, burning, forest cutting, cutting and grazing in grasslands, and litter removal. Agricultural cultivation usually stimulates soil respiration in the short term because of soil disturbances but results in a long-term decrease in soil carbon content. Climatic warming also causes short-term stimulation of soil respiration and may induce long-term acclimation. Responses of soil respiration to changes in precipitation and nitrogen addition are highly variable.

Chapter 8 introduces a variety of methods for measurement of soil respiration. The most commonly used are chamber methods, which include the closed dynamic-chamber method, the open dynamic-chamber method, and the closed static-chamber method. Soil respiration can also be estimated from air samples from different depths of soil using the gas-well method. This chapter describes the basic principles behind those methods, discusses chamber designs and deployment, and assesses the accuracy and potential issues of those methods. It also briefly describes a few indirect methods for estimation of soil respiration.

The partitioning of soil respiration is critical for developing predictive understanding of soil respiration. Chapter 9 introduces three groups of methods—experimental manipulations, isotope tracers, and indirect inference analysis—for partitioning. The experimental methods manipulate the substrate supply to different pathways of soil respiration and separate components of soil respiration. The isotope methods take advantages of isotope signals of C_3 and C_4 plants and soils, CO_2 experiments that fumigate CO_2 with different isotope values, bomb ^{14}C that enriched ^{14}C in the atmosphere in 1950s and 1960s, and labeling experiments. The inference methods are to estimate component contributions through regression extrapolation and deconvolution analysis. This chapter summarizes estimates of contributions of each source component to the total soil respiration.

Chapter 10 provides a general description of models and modeling studies of soil respiration. In general, the modeling studies are based on three types of models: empirical models, CO_2 production models, and CO_2 production and transport models. The empirical models are derived primarily from regression analysis of soil respiration with temperature, moisture, and some surrogate quantities of substrate availability. The production models usually incorporate carbon processes of photosynthesis, partitioning, and decomposition of litter and SOM. The production-transport models consider transport processes of soil CO_2 along a soil profile from the production sites to soil surface. This chapter examines modeling studies according to different spatial and temporal scales and discusses model development and evaluation.

Importance and Roles of Soil Respiration

Soil respiration is a subject that is of concern not only to ecologists but also to scientists who study atmospheric dynamics and earth system functioning. As an integral part of the ecosystem carbon cycle, soil respiration is related to various components of ecosystem production. Soil respiration is also intimately associated with nutrient processes such as decomposition and mineralization. Moreover, soil respiration plays a critical role in regulating atmospheric CO_2 concentration and climate dynamics in the earth system. Thus, it becomes relevant to the mitigation of climate change and the implementation of international climate treaties in terms of carbon storage and trading. This chapter relates soil respiration to ecosystem carbon balance and production, nutrient cycling, regional and global carbon cycling, climate change, and carbon storage and trading.

2.1. SOIL RESPIRATION AND ECOSYSTEM CARBON BALANCE

The carbon cycle in an ecosystem usually initiates when plants fix CO_2 from the air and convert it to organic carbon compounds through photosynthesis (Fig. 2.1). Some of the organic carbon compounds are used to grow plant tissues. Some are broken down to supply the plants with energy. During this

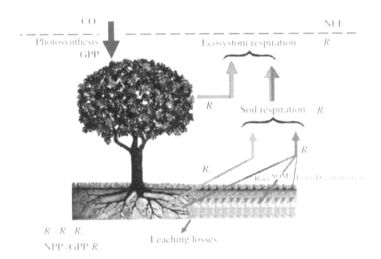

FIGURE 2.1   Schematic diagram of ecosystem carbon processes. Abbreviation see text.

process, CO_2 is released back into the atmosphere through plant respiration. The grown tissues include leaves, stems (e.g., wood for trees), and roots. Leaves and fine roots usually live for several months up to a few years before death, whereas woody tissues may grow for hundreds of years in forests. Dead plant materials (i.e., litter) are decomposed by microorganisms to provide energy for microbial biomass growth and other activities. At the same time, CO_2 is released back into the atmosphere through microbial respiration. The live microbial biomass is mixed with organic residuals of dead plants and dead microbes to form soil organic matter (SOM). SOM can store carbon in soil for hundreds and thousands of years before it is broken down to CO_2 through respiration by microbes.

Through the carbon cycle, CO_2 is produced by both plant respiration (R_p) and microbial respiration (R_m) that occurs during decomposition of litter and SOM. R_p is often called autotrophic respiration and can be separated into aboveground plant respiration (R_a) and belowground plant respiration (R_b). (The belowground plant respiration is often equivalent to root respiration.) Microbial respiration (R_m) during the decomposition of litter and SOM is called heterotrophic respiration. The efflux rate measured at the soil surface (R_s) is the sum of root respiration and microbial respiration:

$$R_s = R_b + R_m \tag{2.1}$$

The CO_2 efflux measured at the soil surface can be considered as soil respiration when CO_2 production and transport are at a steady state (see

Chapter 1). Thus, ecosystem respiration (R_e), the total CO_2 emission from an ecosystem, can be estimated by:

$$R_e = R_a + R_s \qquad (2.2)$$

The relationship of R_s with R_e, as seen in equation 2.2, is well illustrated by data collected from an aspen-dominated mixed hardwood forest in Michigan from 1999 to 2003 (Curtis et al. 2005). On average, over the five years R_s accounts for 71% of R_e, while leaves and aboveground live wood combined (R_a) contribute the rest of R_e (Table 2.1). The relative contribution of R_s to R_e varies considerably in a year. R_s contributes nearly 100% of R_e for most of the winter; the contribution drops to about 60% during the period of fast leaf expansion and then gradually increases during the growing season as soil warms, reaching about 75% at the time of leaf abscission in the autumn (Curtis et al. 2005). Typically, R_s contributes 30–80% of R_e in forests.

Soil respiration is not only an important component of ecosystem respiration but also closely related to ecosystem production such as gross primary production (GPP), net primary production (NPP), and net ecosystem production (NEP). GPP is annual carbon assimilation by photosynthesis ignoring photorespiration. In the Michigan forest, for example, soil respiration is approximately 63% of GPP (Table 2.1). NEP is GPP minus R_e and also related to soil respiration by:

$$NEP = GPP - R_a - R_s \qquad (2.3a)$$

or R_s is related to NEP though NPP, which is GPP minus autotrophic plant respiration, by:

$$\begin{aligned} NEP &= NPP - R_m \\ &= NPP + R_b - R_s \end{aligned} \qquad (2.3b)$$

TABLE 2.1 Various components of ecosystem carbon fluxes in a mixed hardwood forest from 1999 to 2003

Year	R_s	R_e	R_s/R_e	GPP*	R_s/GPP	NPP	NEP
1999	1116	1538	0.73	1637	0.68	656	99
2000	987	1396	0.71	1580	0.62	678	184
2001	1005	1412	0.71	1615	0.62	704	203
2002	946	1404	0.67	1549	0.61	618	145
2003	960	1375	0.70	1545	0.62	650	170
Mean	1003	1425	0.71	1585	0.63	661	160

Note: GPP was estimated by a biometrical approach that sums up different components. The biometrically estimated GPP was higher than that estimated by eddy-flux measurements by nearly 30%. Units are g C m^{-2} yr^{-2}. Modified with permission from New Phytologist: Curtis et al. (2005)

Equation 2.3 is a quantitative basis of the biometrical approach to estimation of net carbon storage in an ecosystem (i.e., *NEP*). *NPP* can be estimated by measuring yearly increments in plant biomass. R_a is often estimated from measured respiration rates of aboveground plant parts (i.e., leaves and live wood in forest). R_b is estimated either from measured respiration rates of roots or indirectly from R_s through partitioning techniques (see Chapter 9). With measured soil respiration, *NEP* can be estimated from Equation 2.3. In the Michigan hardwood forest, the estimated *NEP* by the biometrical method ranged from 100 to $200\,g\,C\,m^{-2}\,yr^{-1}$ (Table 2.1) (Curtis *et al.* 2005).

Another rate of flux in the ecosystem carbon cycle that can be relatively easily measured, especially in forests, is aboveground litterfall. For a long time scientists have sought a relationship between measured litterfall and soil respiration (e.g., Reiners 1968). By synthesizing experimental results from many forests in different regions with various types and ages of forests, Raich and Naderhoffer (1989) generalized the relationship (Fig. 2.2) as:

$$R_s = aL_a + b \tag{2.4}$$

where L_a is aboveground litterfall and a and b are coefficients. Both R_s and L_a are expressed in units of $g\,C\,m^{-2}\,yr^{-1}$. The regression coefficient a is usually about 3 (Raich and Naderhoffer 1989, Davidson *et al.* 2002a), suggesting that carbon release from soil respiration is nearly three times the carbon input

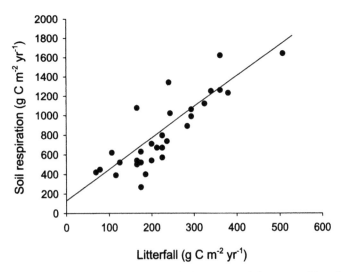

FIGURE 2.2   Correlation of soil respiration with the amount of aboveground litterfall across many forest ecosystems (Redrawn with permission from Ecology: Raich and Naderhoffer 1989).

TABLE 2.2   Annual carbon fluxes for mid-rotation loblolly pine plantations

Component	Control	Irrigated	Fertilized	Fertilized and Irrigated
Soil CO_2 release	1263	1489	1293	1576
Root respiration	663	745	942	1062
Microbial respiration	600	744	351	514
NPP	500	635	1020	1235
NEP	−100	−109	669	721

Note: Units are $g\,C\,m^{-2}\,yr^{-1}$. Modified with permission from Canadian Journal of Forest Research Maier and Kress (2000).

from aboveground litter. Indeed, soil respiration releases carbon from sources of root litter, root exudates, and root respiration in addition to the aboveground litterfall. The correlation was poor, however, among years at a single site (Davidson *et al.* 2002a).

The relationships of R_s with other fluxes can also be used to examine responses of an ecosystem to perturbations. Table 2.2, for example, presents annual carbon fluxes in mid-rotation loblolly pine plantations as affected by fertilization and irrigation (Maier and Kress 2000). Annual R_s is mainly affected by irrigation, ranging from 1263 to $1576\,g\,C\,m^{-2}\,yr^{-1}$ among the four treatments. Belowground root respiration (R_b) is much more responsive to fertilization than to irrigation, whereas R_m is considerably depressed by fertilization. As a consequence, the relative contribution of R_b to R_s increases from 52% under control to 73% under fertilization. Fertilization substantially increased *NPP*, resulting in net carbon storage in the forest. *NEP* is negative by $100\,g\,C\,m^{-2}\,yr^{-1}$ without fertilization and becomes positive to $700\,g\,C\,m^{-2}\,yr^{-1}$ with fertilization (Table 2.2).

2.2. SOIL RESPIRATION AND NUTRIENT CYCLING

A major component of soil respiration is from microbial decomposition of litter and SOM that releases CO_2, meanwhile immobilizing or mineralizing nutrients (Coleman *et al.* 2004). During the initial phases of decomposition, nitrogen that is mineralized from litter substrate is simultaneously immobilized by microbes for their own growth, leading to an increased nitrogen concentration in the mixture of litter substrate and microbes. Since the litter substrate and microbes are not easily separated, in practice the mixture is also called litter. The nitrogen concentration of decomposing litter usually increases, while the absolute amount of nitrogen in the litter may or may not increase during the decomposition. The absolute amount of nitrogen increases when nitrogen from exogenous sources in soil or from fixation is incorporated into microbial biomass growth. The release of carbon combined with nitrogen immobilization during the litter decomposition gradually decreases carbon-nitrogen ratio (C:N) until mineralized nitrogen from litter substrate is greater than required for microbial growth. After that point, litter decomposition leads to a net release of nitrogen. Similarly, phosphorus and sulfur may also increase in absolute amounts during initial phases of decomposition.

Decomposition of SOM usually results in net releases of nitrogen, since C:N of SOM is generally smaller than 20, much closer to C:N of microbes than litter (Paul and Clark 1996). Degradation of proteins and nucleic acids in SOM releases nitrogen in a mineral form (i.e., NH_4^+). The mineralized nitrogen from SOM is partly immobilized for growth of microorganisms and partly added to the mineral nitrogen pool in soil.

Due to the coupled carbon and nitrogen mineralization during microbial decomposition of litter and SOM, the rate of nitrogen mineralization often correlates with microbial respiration. For example, Zak et al. (1993) studied carbon and nitrogen releases from labile organic matter within the forest floor and mineral soil of Jack pine, red pine, balsam fir, sugar maple, and quaking aspen forests in Michigan. Carbon released from microbial decomposition was correlated with mineralized nitrogen (N_{min}) by $R_m = 15.9 N_{min} + 27.4$ with $r = 0.853$ and $n = 154$ for litter and $R_m = 7.1 N_{min} + 159.9$ with $r = 0.616$ and $n = 154$ for SOM from a laboratory incubation. Similar relationships between net carbon and nitrogen mineralization were found in organic substrates with low C:N ratios (Gilmore et al. 1985, Moorhead et al. 1987, Ruess and Seagle 1994, Eriksen and Jensen 2001). Across different types of soils from three communities in an Alaskan boreal forest, rates of soil respiration were associated with rates of microbial turnover and nitrogen mineralization in a laboratory incubation study (Vance and Chapin 2001). In the field research, nitrogen mineralization may not be well correlated with soil respiration due to the nitrogen immobilization.

2.3. SOIL RESPIRATION AND REGIONAL AND GLOBAL CARBON CYCLING

Soil respiration plays a critical role in the regulation of carbon cycling on regional and global scales. The carbon cycle on the global scale involves exchanges of CO_2 among the land biosphere, the atmosphere, oceans, and the earth's crust (Fig. 2.3). Each year, photosynthesis of land plants takes up approximately $120 Pg (10^{15} g) C yr^{-1}$ from the atmosphere. A similar amount of carbon is released back to the atmosphere through ecosystem respiration. Oceans absorb nearly $92 Pg C yr^{-1}$ from the atmosphere and release

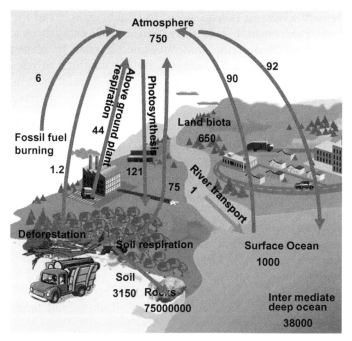

FIGURE 2.3 The global carbon cycle. Pools in Pg (= 10^{15} g) C and fluxes in $PgCyr^{-1}$ as indicated by arrows.

90.6 $PgCyr^{-1}$ back to the atmosphere through physiochemical exchanges of CO_2 at the air-sea surface and through photosynthesis and respiration of marine organisms.

The global soils contain as high as 3150 Pg C, including 450 Pg C in wetlands, 400 Pg C in permanently frozen soils, and 2300 Pg C in other ecosystems (Sabine *et al.* 2003). The latter 2300 Pg C can be further divided into 1500 Pg C in the top soils to the depth of 1 meter and 800 Pg C in the deeper soil layers to the depth of 3 meters according to distribution profiles of soil carbon along depths (Jobbágy and Jackson 2000). Plants contain 650 Pg C, slightly smaller than the carbon pool size in the atmosphere (750 Pg C). The sum of soil and plant carbon contents is 3800 Pg C, five times the size of the atmospheric pool.

The burning of fossil fuels by humans presently adds about 6 $PgCyr^{-1}$ to the atmosphere. Land clearing, deforestation, and fire release an additional 1.2 $PgCyr^{-1}$ to the atmosphere. The amount of CO_2 added to the atmosphere by human activities may seem very small in comparison with the rates of fluxes through natural processes such as photosynthesis and respiration. But it takes only a small change to upset the balance of the global carbon cycle. Of the total anthropogenic emission, a little over half remains in the atmosphere, while the rest is sequestered in land biosphere and the oceans. Modeling and experimental studies suggest that land ecosystems sequester approximately one-third of the anthropogenic emission in plant and soil pools (Schimel *et al.* 2001). As human activities continue to release CO_2, atmospheric CO_2 concentration is expected to keep increasing. Whether the terrestrial carbon sinks are sustainable, however, is highly uncertain.

To understand how the global carbon cycle responds to human perturbation and climate change, we have to understand different aspects of carbon processes, including soil respiration. Soil respiration accounts for a large portion of the total biosphere respiration and is the second largest flux from terrestrial ecosystems. A number of studies have compiled data from field measurements and scaled them up to estimate the global respiratory flux of CO_2 from soils. Schlesinger (1977) estimated global flux at a rate of approximately $75\,Pg\,C\,yr^{-1}$, roughly 2.5 times larger than the input of fresh litter to the soil surface. Raich and Schlesinger (1992) compiled available data from the literature and estimated global flux to be $68\,Pg\,C\,yr^{-1}$ from soils. Global soil respiration consists of $50\,Pg\,C\,yr^{-1}$ from decomposition of litter and SOM, and $18\,Pg\,C\,yr^{-1}$ from live roots and mycorrhizae. Using a global model, Raich and Potter (1995) updated the estimate of global soil respiration to $77\,Pg\,C\,yr^{-1}$.

At the global scale, soil respiration releases carbon at a rate that is more than one order of magnitude larger than the anthropogenic emission. The soil pool from which soil respiration releases carbon is about four times the atmospheric pool. Thus, a small change in soil respiration can seriously alter the balance of atmosphere CO_2 concentration. To predict changes in the carbon cycle in response to global change, soil respiration has to be carefully studied.

Soil respiration is very sensitive to environmental changes. The sensitivity of soil respiration to changes in temperature, for example, is a critical parameter in the regulation of the global carbon balance. Results from seasonal measurements usually yield a relationship that the rate of soil respiration increases with temperature (Raich and Schlesinger 1992). In light of this relationship, global warming is expected to stimulate soil respiration and diminish the sink strength of terrestrial ecosystems.

Because of its crucial role in regulating the global carbon cycle, the temperature sensitivity of soil respiration has been extensively studied, using both experimental and modeling approaches. Giardina and Ryan (2000) and Liski *et al.* (1999) found that decomposition of old, recalcitrant SOM or organic carbon in mineral soils is less sensitive to temperature than labile carbon. Luo *et al.* (2001a) conducted a warming experiment in a natural grassland and revealed a phenomenon of acclimation whereby the sensitivity of soil respiration to warming decreases after the ecosystem is exposed to experimental warming for a certain time. Thus, short-term data may not capture long-term characteristics of respiratory responses to rising temperature. Such results from those and many other studies challenge a common assumption in global models that respiratory carbon release from decomposing organic matter increases with global warming. However, recent soil incubation studies showed that the temperature sensitivity of the decomposition of SOM does not change with soil depth, sampling method, and incubation time (Fang *et al.* 2005). Using a three-pool model, Knörr *et al.* (2005) analyzed soil incubation data and claimed that the temperature sensitivity of slow carbon pools is even higher than that of the faster pools. We need data from well-designed, long-term experiments to resolve the issue of how soil respiration varies with long-term changes in temperature.

The differences in temperature sensitivity of soil respiration nonetheless have global and regional implications. Grace and Rayment (2000) used simple models to illustrate that forest carbon sink diminishes if respiration rises with long-term increases in temperature. When respiration is insensitive to longer-term temperature changes, the forest ecosystems become increasingly effective at sequestering carbon as atmospheric CO_2 continues to increase. Thus, the assumption made about the temperature sensitivity of soil respiration has a profound effect on long-term projections of the global and regional carbon cycles and climate change.

The temperature sensitivity of soil respiration may also be a key factor in determining regional carbon balance. Results from a network of CO_2 flux sites across forests in Europe show that respiration increases, but photosynthesis does not vary along the latitudinal band from Iceland to Italy (Valentini *et al.* 2000). Tropical regions have large pools of SOM with relatively rapid turnover times. Carbon fluxes in the tropical regions are also larger than those in temperate and northern forests. Global warming potentially stimulates great losses of soil carbon in the tropics (Trumbore *et al.* 1996). Boreal forests and tundra have the largest store of labile organic matter and the greatest predicted rise in temperature. Organic carbon accumulated in the soil over previous, colder periods is now decomposing and being released through soil respiration as the soil warms in response to climate change. Thus, understanding soil respiration in different regions is critical in predicting regional and global carbon cycles.

2.4. SOIL RESPIRATION AND CLIMATE CHANGE

Soil respiration becomes relevant to climate change because the CO_2 released from soil respiration is one of the greenhouse gases. The greenhouse gases permit incoming solar radiation to reach the surface of the earth but restrict the outward flux of infrared radiation. They absorb and reradiate the outgoing infrared radiation, effectively storing some of the heat in the atmosphere. In this way, greenhouse gases trap heat within the atmosphere, resulting in climate warming near the earth's surface.

The increased concentration of greenhouse gases in the atmosphere enhances the absorption and emission of infrared radiation. The atmosphere's opacity increases so that the altitude from which the earth's radiation is effectively emitted into space becomes higher. Because the temperature at higher altitudes is lower, less energy is emitted, causing a positive radiative forcing (IPCC 2001). If the amount of CO_2 is doubled instantaneously, with everything else remaining the same, the outgoing infrared radiation would decrease by about $4\,W\,m^{-2}$. That is, the radiative forcing corresponding to a doubling of the CO_2 concentration is $4\,W\,m^{-2}$. To counteract this imbalance, the temperature of the surface-troposphere system would have to increase by 1.2°C (with an accuracy of ±10%), in the absence of other changes. In reality, complex feedbacks in the climate system (e.g., via clouds and their interactions with radiation) are predicted to amplify the temperature increase to 1.5 to 4.5°C (IPCC 2001).

In addition to feedback loops within the climate system, the atmosphere interacts with the biosphere through climate-carbon cycle loops. The terrestrial ecosystems presently absorb approximately $2\,Pg\,C\,yr^{-1}$, primarily resulting from fertilization effects of rising atmospheric CO_2 concentration and N deposition on plants. As atmospheric CO_2 concentration continues to increase at the "business-as-usual" emission scenario (IS92a) (IPCC 1992), the land biosphere will take up an average of $7.5\,Pg\,C\,yr^{-1}$ by the end of the 21st century without the coupled climate-carbon cycle feedbacks (IPCC 2001).

Rising CO_2 concentration in the atmosphere enhances greenhouse effects, likely resulting in global warming. The global warming could substantially stimulate respiration, resulting in more release of CO_2 to the atmosphere to trap heat. Thus, the climate system and the global carbon cycle form a positive feedback loop to reinforce each other (Friedlingstein et al. 2003). Based on temperature sensitivity with a fixed Q_{10} value (e.g., 2.0) across the globe, global warming by 2°C would increase additional carbon release from soil respiration by more than $10\,Pg\,C\,yr^{-1}$, which is larger than the current anthropogenic carbon emission. The additional carbon release aggravates anthropogenic warming.

To examine the positive feedback loop between the climatic system and global carbon cycle, Cox et al. (2000) carried out three simulations. The first simulation set the atmospheric CO_2 concentration in the model as in the IS92a scenario without climate warming. The model projects that soils in the land ecosystems absorb a net of nearly $400\,Pg\,C$ from 2000 to 2100 (Fig. 2.4). The second simulation examines climate warming and its effects on the global

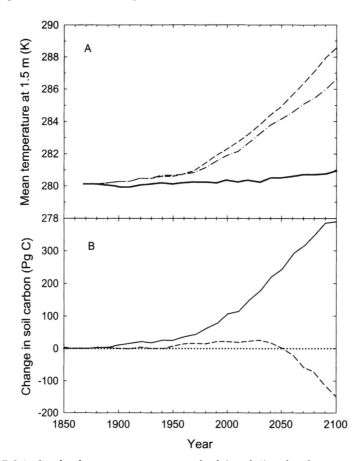

FIGURE 2.4 Simulated mean temperature over land (panel A) and carbon storage in soil (panel B) as affected by rising atmospheric CO_2 concentration, climate warming, or both. The dashed line indicates simulated land surface temperature (panel A) and soil carbon storage (panel B) by the fully coupled carbon cycle climate model, the dot-dashed line in panel A indicates the simulated temperature by a standard global circulation model of climate change with prescribed CO_2 concentration (IS92a) and fixed vegetation, and the solid line indicates simulated temperature (panel A) and carbon storage (panel B) by a model that neglects direct CO_2-induced climate change. The slight warming in the latter is due to CO_2-induced changes in stomatal conductance and vegetation distribution (Redrawn with permission from Nature: Cox et al. 2000).

carbon cycle without the climate-carbon cycle feedback. Rising atmospheric CO_2 concentration, as predefined by the IS92a scenario, induces a 5.5°C warming over land. The climate warming stimulates plant and microbial respiration. The land ecosystems become a source of 60 Pg C to the atmosphere over the 21st century. In the third simulation, the climate model is coupled with the carbon cycle model. The simulation by the coupled model projected an atmospheric CO_2 concentration of 980 ppm in 2100, rather than the 700 ppm as in IS92a. The land ecosystems release 170 Pg C in the simulation, with the coupled model due to stimulated respiration. The global temperature was projected to increase by 8.0°C over land, 2.5°C greater than the simulation of the climate model not coupled to the carbon cycle model. The dramatic increase in global temperature is largely due to stimulated respiration and oxidation of organic matter in warmer soils. Similar positive feedbacks between climate warming and global carbon cycling are demonstrated in simulations by Dufresne *et al.* (2002). Thus, soil respiration is a critical process that is involved in the positive feedback between climate change and the global carbon cycle. An understanding of responses of soil respiration to global warming is now urgently needed in order to evaluate uncertainty in global climate change projections.

2.5. SOIL RESPIRATION AND CARBON STORAGE AND TRADING

Climate change is not merely a scientific issue but also one of the main challenges facing humanity. To address this challenge, business opportunities have been created for carbon trading in a global market. The market provides incentives for reducing atmospheric CO_2 by those countries seeking to meet their obligations under the framework of the Kyoto Protocol as well as by voluntary national or regional jurisdictions outside the Kyoto Protocol. The Kyoto Protocol, formally known as the United Nations Framework Convention on Climate Change (UNFCCC), was forged in Kyoto, Japan, in December 1997. It has been ratified by most of the world's developed countries and took effect as an international treaty in February 2005. Under the treaty, the participating countries (i.e., the developed and/or market-oriented ones) are legally bound to reduce their greenhouse gas emissions by 2008–2012 to 5% below their levels in 1990 (Sanz *et al.* 2004).

Global change markets have existed for carbon trading since 2002. The markets traded approximately US$10 million worth of emission allowances in European Union countries in 2002 and will trade as much as US$1 billion per year in allowances by 2010 (Johnson and Heinen 2004). This emerging carbon market is potentially quite substantial (estimated at US$10 billions per year) and introduces a clear financial value for the capture and mitigation of CO_2 emissions in land ecosystems.

Under the Kyoto Protocol, management of natural terrestrial carbon sinks can earn a direct cash award in the carbon mitigation market. The natural sinks reside primarily in expanded forest stocks and increased soil sinks, which can be managed to increase sink strength and reduce atmospheric CO_2. The emission-trading market provides the opportunity for farmers and foresters to profit by selling emission credits to those parties looking to partially offset their CO_2 reduction obligation. The buying parties may find it less costly to outsource part of their emission mitigation commitment in the natural sinks than to take other measures to reduce emissions. This market-trading practice provides the selling parties with new financial incentives for environmentally friendly land management and forest rehabilitation.

Forests cover about $42 \times 10^{12} m^2$ globally. Forest carbon storage can be achieved in three principal ways: (1) improving the management of currently forested areas, (2) expanding the currently unforested area via afforestation and agroforestry, and (3) reducing the rate of deforestation. All the management options alter the balance between carbon fluxes into the forest ecosystems (i.e., photosynthesis) and fluxes out of the forests via plant and microbial respiration and biomass harvests, resulting in increased carbon stocks in tree biomass, litter mass, soil SOM, and wood products. Potential forest sequestration could approach $1 PgC yr^{-1}$. But more realistic estimates of achievable sequestration are approximately $0.17 PgC yr^{-1}$ from improved management of existing forests and $0.2 PgC yr^{-1}$ from afforestation on formerly wooded and degraded lands (Watson et al. 2000). Financial costs are modest to high (US$3 to $120 per ton of carbon) in so-called Annex I countries (i.e., industrialized countries or those that are undergoing the process of transition to a market economy) but often low (US$0.2 to $29 per ton of carbon) elsewhere. Management measures to improve carbon storage in forestry include prolonging rotations, changing tree species, continuous-cover forestry, fire control, combined water storage with peat swamp afforestation, fertilization, thinning regimes, and mixed species rotation. Once management improvements saturate forest carbon sinks, forest ecosystems achieve a steady state, so that any further net carbon storage is unlikely to occur.

Cessation of deforestation is another major method of promoting carbon storage in forest ecosystems. Currently, land use changes result in a net release of $1.2 PgC yr^{-1}$ (Fig. 2.3). Deforestation, mainly in the tropics, accounts for a large portion of the net release. While complete cessation of deforestation is unrealistic for a variety of social and economic reasons, it offers the single most effective potential solution to mitigate climate change by forest ecosystems. Agroforestry has been widely practiced in the Punjab and India, where crops grow under a canopy of trees. The combinations of trees, crops, and forages in agroforestry may promote carbon sequestration and the sustainable use of other resources.

The other major natural sink in terrestrial ecosystems can be realized mainly through the recapturing of some portion of carbon released from cultivation in world soils. Natural soils retain carbon in stable microaggregates for up to hundreds and thousands of years unless environmental conditions are changed and the stable soil structure is damaged. Cultivation practices such as plowing break soil aggregates, expose originally protected organic matter in soils to microbial attacks, and thus accelerate decomposition and respiratory carbon losses to the atmosphere. Soils degraded by cultivation are more susceptible to accelerated erosion, which carries carbon to rivers and oceans, where it is partially released into the atmosphere by outgassing (Richey *et al.* 2002). After conversion of natural to agricultural ecosystems, organic carbon in soils has been depleted by as much as 60% in temperate regions and 75% or more in the tropics. Some soils have lost as much as 2000 to $8000\,g\,C\,m^{-2}$. Land clearance by humans for agricultural activities began 8000 years ago in Eurasia (McNeill and Winiwarter 2004) and became substantial enough to cause preindustrial CO_2 anomalies in the atmosphere 2000 years ago. Ruddiman (2003) estimated that land conversion during the preindustrial era may cause carbon loss at a rate of $0.04\,Pg\,C\,yr^{-1}$ for 7800 years and that the total carbon emission from terrestrial ecosystems is $320\,Pg\,C$, including carbon losses from plant and soil pools. The global cumulative loss of carbon from terrestrial ecosystems is estimated to be 136 to $160\,Pg\,C$ over the past 200 years. Carbon loss from soils is approximately $78\,Pg\,C$, including $52\,Pg\,C$ by soil respiration and $26\,Pg\,C$ by soil erosion (Lal 2004), with 2.0 (±1.4) $Pg\,C\,yr^{-1}$ in the 1980s and 1990s alone (Houghton 2002). In comparison, carbon emission from fossil fuel combustion was $270\,Pg\,C$ between 1850 and 1998 and approximately $5\,Pg\,C$ in the 1990s. Land use change transformed land covers of temperate regions before about 1950 to the tropics in recent decades (Achard *et al.* 2002, DeFries *et al.* 2002, Houghton 2003), resulting in substantial CO_2 effluxes from soils in every continent except Antarctica (DeFries *et al.* 1999).

The potential carbon sink capacity in soils through ecosystem management approximately equals the cumulative historical carbon loss. The attainable soil sink is 50 to 66% of the potential capacity. The optimistic rate of soil carbon sequestration is at 0.6 to $1.2\,Pg\,C\,yr^{-1}$ (Lal 2003) and a more likely rate at 0.3 to $0.5\,Pg\,C\,yr^{-1}$ (Sauerbeck 2001). Carbon sequestration at the optimistic rate would restore most of the lost carbon within 50 to 100 years. Thus, carbon sequestration in soils potentially offsets fossil fuel emissions by 0.4 to $1.2\,Pg\,C\,yr^{-1}$, or 5 to 15% of the global fossil fuel emissions.

Based on the principles of increasing plant carbon inputs, slowing soil carbon decomposition rates, or both, soil carbon can be built through a variety of agronomic management techniques (Fig. 2.5). Carbon inputs can be enhanced by growing higher biomass crops, by leaving more crop biomass to decompose in situ, by increasing belowground *NPP*, and by growing cover crops during portions of the year. Decomposition rates can be slowed by reducing tillage and by growing crops with low residue quality. No tillage

Cropland soils: 1350 Mha
[0.4 to 0.8 Pg C yr⁻¹]
❖ Conservation tillage (10-100)
❖ Cover crops (5-25)
❖ Manuring and INM (5-15)
❖ Diverse cropping systems (5-25)
❖ Mixed farming (5-20)
❖ Agroforestry (10-20)
Acid savanna soils, 250 Mha in South
America, have a high potential.

Range lands and grass lands:
[0.01 to 0.3 Pg C yr⁻¹]
3.7 billion ha in semi-arid and sub-
humid regions
❖ Grazing management (5-15)
❖ Improved species (5-10)
❖ Fire management (5-10)
❖ Nutrient management

Potential of carbon
sequestration in
world soils
[(0.4 – 1.2 Pg C yr⁻¹]

Restoration of degraded and
desertified soils: 1.1 billion ha
[0.2 to 0.4 Pg C yr⁻¹]
❖ Erosion control by water (10-20)
❖ Erosion control by wind (5-10)
❖ Afforestation on marginal lands
(5-30)
❖ Water conservation/harvesting
(10-20)

Irrigated soils: 275 Mha
[0.01 to 0.03 Pg C yr⁻¹]
❖ Using drip/sub-irrigation
❖ Providing drainage (10-20)
❖ Controlling salinity (6-20)
❖ Enhancing water use efficiency/water
conservation (10-20)

FIGURE 2.5 Soil C sequestration potential in cropland, grazing/range land, degraded/deserti-
fied lands, and irrigated soils. Rates of C sequestration given in parentheses are in $gCm^{-2}yr^{-1}$.
These are not additive and low under on-farm conditions (Redrawn with permission from
Nature: Lal 2004 with references to original papers for the listed rates).

implants seeds without turning the soil with a plow and reduces the loss of
SOM. The low-quality residue contains organic carbon that is more difficult
for microbes to decompose. Thus, soil restoration and woodland regeneration,
no-till farming, cover crops, nutrient management, manuring and sludge
application, improved grazing, water conservation and harvesting, efficient
irrigation, agroforestry practices, and growing energy crops on spare lands
are recommended management practices (RMPs) to increase the soil carbon
sequestration (Silver *et al.* 2000, Nordt *et al.* 2001, West and Marland 2002,
Lal 2004). Those management practices add high amounts of biomass to the
soil, cause minimal soil disturbance, conserve soil and water, improve soil
structure, enhance activity and species diversity of soil fauna, and strengthen
mechanisms of elemental cycling (Fig. 2.5).

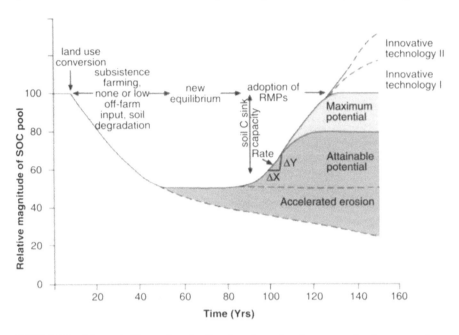

FIGURE 2.6 Schematic illustration of soil carbon dynamics after conversion from natural to agricultural ecosystems and subsequent recovery using recommended management practices (RMPs). The maximum potential equals the magnitude of historical carbon loss (adapted with permission from supplemental material of Lal 2004).

The capacity of soil carbon sequestration varies with time (Lal 2004). The rate of soil carbon sequestration through land managements usually follows a gradual decline. It reaches a maximum in the first 5 to 20 years after land conservation and continues until SOM attains a new equilibrium (Fig. 2.6). The rates of soil carbon sequestration in agricultural and restored ecosystems range from 0 to $15\,g\,C\,m^{-2}\,yr^{-1}$ in dry and warm regions (Armstrong *et al.* 2003) and 10 to $100\,g\,C\,m^{-2}\,yr^{-1}$ in humid and cool climates (West and Post 2002). These rates may continue for 20 to 50 years with the continuous uses of recommended management practices and then decline as the soil carbon content reaches a steady state. The global carbon-trading markets can be a major incentive in promoting the management practices that increase carbon storage in soils. To implement the carbon-trading markets, on the other hand, we have to develop the ability to measure photosynthesis, respiration, and short-term (three- to five-year) changes in SOM pool for verification of carbon credits.

PART II

Mechanisms

Processes of CO₂ Production in Soil

Soil respiration involves several processes, including CO_2 production in the soil and CO_2 transport from the soil to the atmosphere. This chapter describes the CO_2 production processes whereas the CO_2 transport processes are presented in Chapter 4.

Soil respiration releases gaseous CO_2 molecules that are produced by roots, soil microbes, and soil fauna within soil and litter layers. The CO_2 produced by the living tissues is a by-product of metabolisms that yield energy and/or carbon intermediates needed for the maintenance, growth, ion uptake and reproduction of organisms. According to sources of carbohydrate substrate supply, CO_2 production in the soil can be attributed to root respiration, microbial respiration in rhizosphere by consuming labile carbohydrate exudates from roots, decomposition of litter, and oxidation of SOM (Fig. 3.1). Soil fauna may contribute a nontrivial proportion of respiratory fluxes in an ecosystem, but as the portion of CO_2 production by soil fauna has not been well quantified, this chapter does not describe the respiration of soil fauna in detail.

At the biochemical level, CO_2 production by all the living tissues shares common processes that are primarily through the tricarboxylic acid (TCA) cycle (the citric acid cycle, also known as the Krebs cycle) in the

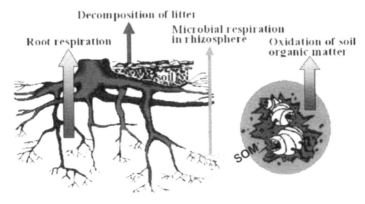

Sources of soil respiration

Decomposition of litter

Microbial respiration
in rhizosphere

Root respiration

Oxidation of soil
organic matter

Soil

SOM

FIGURE 3.1 Schematic representation of CO$_2$ production processes in soil. Those processes are root respiration, rhizosphere respiration, litter decomposition, and oxidation of SOM (Modified with permission from Arlene Mendoza-Moran).

aerobic condition and fermentation of glucose in the anaerobic conditions. Although biochemical metabolisms in soil result mainly in CO$_2$ production, there are other processes in the soil that either consume or produce CO$_2$ such as methanogenesis, phototrophs or carbonic reactions. This chapter first describes the biochemistry of respiratory processes and then outlines each of the CO$_2$ production processes according to the supply sources of carbon substrates.

3.1. BIOCHEMISTRY OF CO$_2$ PRODUCTION PROCESSES

CO$_2$ can be produced through several biochemical pathways, the most common being the TCA cycle. Other CO$_2$ production processes include the fermentation of glucose to organic acids and methanotroph to oxidize methane. The fermentation happens in anaerobic environments such as wetlands, waterlogged areas, and anaerobic microsites within soil particles, whereas the TCA cycle and methanotroph occur in aerobic conditions. Although the carbonation reaction is a geochemistry topic, since it may produce or consume CO$_2$ in soil, it is also described briefly in this section.

TRICARBOXYLIC ACID (TCA) CYCLE

Under aerobic conditions in the presence of oxygen, respiration generates energy by oxidizing sugars. The overall chemical reaction for the oxidation of glucose (or other carbohydrates) to carbon dioxide can be described as:

$$C_6H_{12}O_6 + 6O_2 \rightarrow 6CO_2 + 6H_2O \tag{3.1}$$

This process yields $2870\,kj\,mol^{-1}$ glucose. Since respiration occurs in the presence of oxygen, this process is also called aerobic respiration of organic compounds.

Biochemically, the overall processes of aerobic respiration are carried out through glycolysis, the pentose phosphate pathway, the TCA cycle, and the electron transport pathway (Fig. 3.2). The oxidative pentose phosphate

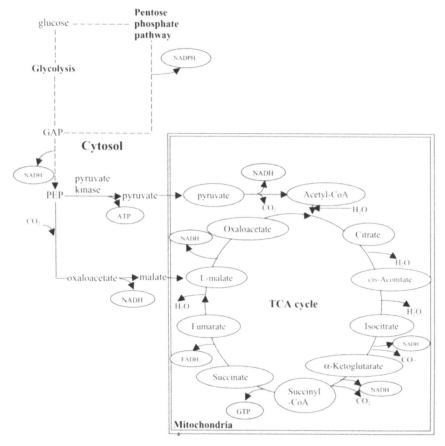

FIGURE 3.2 The respiratory pathways in living tissues include glycolysis, the pentose phosphate pathways, and the TCA cycle.

pathway is located in the plastids, and its primary function is to produce intermediates (e.g., amino acids and nucleotides) and nicotinamide adenine dinucleotide phosphate (NADPH) for the biosynthesis of tissue. The electron transport pathways are in the inner mitochondrial membrane associated with electron transfer and oxidative phosphorylation. CO_2 and adenosine triphosphate (ATP) production occur mainly in the glycolysis pathway and the TCA cycle. Glycolysis occurs in both the cytosol and plastids that convert glucose, via phosphoenolpyruvate (PEP), into pyruvate and malate. Pyruvate is the primary product of glycolysis in animals and microbes, whereas plant cells convert PEP mostly to malate (Lambers et al. 1998).

Oxidation of one glucose molecule in glycolysis generates two molecules of pyruvate or malate. Glycolysis produces two molecules of ATP when pyruvate is the product, and it has no net production of ATP when malate is the end-product. The production of malate in plant cells through glycolysis also incorporates one molecule of carbon dioxide. The malate and pyruvate formed in the cytosol are imported into the mitochondria, where the TCA cycle occurs to oxidize pyruvate and malate. Complete oxidation of one molecule of pyruvate results in three molecules of CO_2, four molecules of nicotinamide adenine dinonucleotide (NADH), one molecule of flavine adenine dinonucleotide (FADH$_2$), and one molecule of ATP. Complete oxidation of one malate molecule yields one additional molecule of CO_2 and NADH, which fully compensates the need of CO_2 during the synthesis of oxaloacetate and the need of NADH in the reduction of oxaloacetate in glycolysis (Fig. 3.2). Overall, the oxidation of one molecule of glucose during the glycolysis and TCA cycle produces the same amount of CO_2, regardless of whether pyruvate or malate is the intermediate product.

The malate that is imported into the mitochondria is oxidized partly via malic enzyme and partly via malate dehydrogenase. The reaction with malic enzyme produces pyruvate and CO_2. Pyruvate is then oxidated in the TCA cycle, so that malate is regenerated. The reaction with malate dehydrogenase generates oxaloacetate, a substrate of the TCA cycle. The energy and intermediates produced by respiratory processes are used to sustain plant growth, while the by-product, CO_2, is transported through the mesophyll and intercellular spaces before being released at the root or microbial surface.

The rate of respiration at the biochemical level is regulated by a combination of energy demand, substrate availability, temperature, and oxygen supply. In general, respiration positively responds to energy demand to meet energy requirements for the growth, maintenance, and transport processes. When tissues grow fast, take up ions rapidly, and/or have a fast turnover of proteins, they generally have a high rate of respiration. When substrate supply is low, however, the respiratory pathways become substrate-limited. In the long run, the respiratory capacity is adjusted through the gene transcription for respiratory enzymes to balance the demand for respiratory energy with the supply of respiratory substrate. Respiratory processes of roots respond strongly to short-term changes in temperature and generally acclimate to long-term changes in temperature (Atkin and Tjoelker 2003).

OTHER CO_2 PRODUCTION AND CONSUMPTION PROCESSES IN SOIL

When oxygen concentration is low, aerobic respiration is inhibited and anaerobic respiration takes place. The anaerobic respiratory processes occur during fermentation, which converts glucose (or other sugar compounds) to organic products. Fermentation uses internally produced organic electron donors and acceptors and is inefficient in energy production. Fermentation has multiple pathways, some of which produce CO_2 as a product (Table 3.1); many others do not produce CO_2. For example, the pathway of fermentation of glucose to ethanol produces two molecules of CO_2. The chemical reaction can be described by:

TABLE 3.1 Biochemical processes in roots and microorganisms that result in CO_2 production

Reductant	Oxidant	Products	Organism
Sugars	O_2	CO_2, H_2O	Roots, protozoa, fungi, many bacteria
Sugar and related compounds	Organic compounds	Lactic acid, ethyl alcohol, CO_2	Lactic acid bacteria
Sugars	Organic compounds	Ethyl alcohol, CO_2	Yeasts
Sugars	Organic compounds	Acetic, succinic and lactic acids, formic acid or H_2, and CO_2 ethyl alcohol	*Escherichia*
Sugars	Organic compounds	Butanediol, lactic acids, formic acid or H_2, and CO_2 ethyl alcohol	*Enterobacter*
Sugars, organic acids	Organic compounds	Propionic, succinic and acetic acids, CO_2	*Propionibacterium, Veillonella*
Sugars, starch, pectin	Organic compounds	Butyric and acetic acids, CO_2, H_2	*Clostridium*
Amino acids	Organic compounds	Acetic acids, NH_3, CO_2	*Clostridium*

Modified with permission from Pearson Education Ltd.: Richards (1987).

$$C_6H_{12}O_6 \rightarrow 2C_2H_5OH + 2CO_2 \qquad (3.2)$$

Methanotrophs generate a trace amount of CO_2 by oxidizing methane (CH_4) in aerobic environments (Lidstrom 1992):

$$CH_4 + 2O_2 \rightarrow CO_2 + 2H_2O \qquad (3.3)$$

This reaction occurs in the surface layers of wetland soils, unsaturated upland soils, and other aerobic conditions. Methanogens can use acetate as substrate during fermentation in anaerobic conditions to generate CO_2:

$$CH_3COOH \rightarrow CH_4 + CO_2 \qquad (3.4)$$

However, methanogens can also use CO_2 as an electron acceptor to produce methane:

$$CO_2 + H_2 \rightarrow CH_4 \qquad (3.5)$$

Methanogens are a group of anaerobic Archaea (Whitman *et al.* 1992). They are obligate anaerobic microorganisms, requiring redox potentials less than $-100\,mV$ in flooded soils. Since both acetate and hydrogen are by-products of fermentation, methanogenesis takes place in a complex food web and is strongly regulated by the organic material supply.

In addition, a trace amount of CO_2 may evolve from the carbonation reaction during rock weathering as:

$$H_2O + CO_2 \longleftrightarrow H^+ + HCO_3^- \longleftrightarrow H_2CO_3 \qquad (3.6)$$

This process is driven by the formation of carbonic acid, H_2CO_3, in the soil solution. In general, production and consumption of CO_2 by anaerobic metabolism and weathering are relatively trivial in comparison with that by aerobic respiration. Most of the studies on soil respiration do not consider the anaerobic metabolism and weathering.

RESPIRATORY QUOTIENT

Soil respiration refers to the metabolic processes of living organisms that produce CO_2 and consume O_2. A ratio of CO_2 produced to O_2 consumed in a respiring system can be used to define the respiratory quotient (RQ) as:

$$RQ = \frac{CO_2 \ produced}{O_2 \ consumed} \qquad (3.7)$$

The respiratory quotient is an index of potential changes in the source of substrate used for respiration and/or the pathway of respiration (Lipp and Anderson 2003). In nonphotosynthetic tissues, RQ is expected to be 1.0 if sucrose is the only substrate for respiration and is fully oxidized to CO_2 and H_2O. Measured RQ values often differ from 1.0 (Table 3.2) because respiratory substrates are compounds other than sucrose and/or because the respiratory intermediates are used for biosynthesis. When respiration is completely anaerobic, RQ can theoretically rise to infinity because no O_2 is consumed, while CO_2 may be produced.

If organic acids are used as the substrate of respiration, RQ is usually greater than 1.0, because organic acids are more oxidized than sucrose, producing more CO_2 per unit of O_2. If lipids, proteins, and other compounds that are more reduced than sucrose are the major substrate, the RQ is less than 1.0. Root respiration usually uses photosynthate as the primary substrate. Thus, root RQ is often found to be close to 1.0 (Table 3.2). During starvation

TABLE 3.2   The RQ of root respiration of several species based on Lambers *et al.* (2002) wiit modification

Species	RQ	Special Remarks	Reference
Acer saccharum	0.8	Field measurements	Burton *et al.* (1996)
Allium cepa	1.0	Root tips	Berry (1949)
	1.3	Basal parts	Berry (1949)
Sugar-maple	0.92	Seedling	Carpenter and Mitchell (1980)
Mixed deciduous forest	0.75	—	Edwards and Harris (1977)
Dactylis glomerata	1.2	NO₃ fed	Scheurwater *et al.* (1998)
Festuca ovina	1.0	NO₃ fed	Scheurwater *et al.* (1998)
Galingsoga parviflora	1.6	NO₃ fed	I. Scheurwater, unpublished
Helianthus annuus	1.5	NO₃ fed	I. Scheurwater, unpublished
Holcus lanatus	1.3	NO₃ fed	I. Scheurwater, unpublished
Hordeum distichum	1.0	NO₃ fed	Williams and Fawar (1990)
Lupinus albus	1.4	NO₃ fed	Lambers *et al.* (1980)
	1.6	N₂-fixing	Lambers *et al.* (1980)
Oryza sativa	1.0	NH₄⁺-fed	Brambilla *et al.* (1986)
	1.1	NO₃ fed	Brambilla *et al.* (1986)
Pinus ponderosa	0.84	Attached roots	Andersen and Scagel (1997)
	0.85	Detached roots	Lipp and Andersen (2003)
	0.39–1.02	Varied with ozone	
Pisum sativum	0.8	NH₄⁺-fed	De Visser (1985)
	1.0	NO₃ fed	De Visser (1985)
	1.4	N₂-fixing	De Visser (1985)
Zea mays	1.0	Fresh tips	Saglioimd Pradet (1980)
	0.8	Starved tips	Saglioimd Pradet (1980)

Note: All plants were grown in nutrient solution, with nitrate as the nitrogen source, unless stated otherwise. The *Pisum sativum* (pea) plants were grown with a limiting supply of combined N, so that their growth matched that of the symbiotically grown plants.

of excised root tips or under low-light environments, roots do not use simple carbohydrates from photosynthesis as respiratory substrates, and RQ is likely less than 1.0. Lipp and Anderson (2003) found that RQ ranged from 0.80 to 0.95, regardless of root excision and changes in shoot light environment. Organic acids (malate) produced during the reduction of nitrate in leaves can be transported and decarboxylated in the roots, resulting in the release of CO_2 and an increase in RQ (Ben Zioni et al. 1971). Values of RQ are lower in plants that use NH_4^+ as a nitrogen source than in plants grown with NO_3^- or symbiotically, with N_2 (Table 3.2).

The RQ for soil CO_2 production and O_2 consumption is much less well studied than root respiration. Soil RQ represents relative activities of aerobic and anaerobic microbial metabolism. It increases with the level of anaerobic respiration, since it can produce CO_2 without consuming O_2. Linn and Doran (1984) measured CO_2 production and O_2 consumption from residue-amended Crete-Butler soil from Nebraska. Both CO_2 production and O_2 consumption increase with water-filled porosity (WFP) until it reaches 60%, beyond which the CO_2 production and O_2 consumption decrease. Soil RQ is about 1.0 when WFP is less than 80% and increases up to 1.7 as WFP reaches 97% (Fig. 3.3). Measured CO_2 production and O_2 consumption in cropped and fallow soils in southern England yield soil RQ values ranging from 0.99 to 1.22 (Currie 1970).

3.2. ROOT RESPIRATION

Root respiration usually accounts for approximately half of the total soil respiration but varies from 10 to 90% among different studies (Hanson et al. 2000). Root respiration consumes approximately 10 to 50% of the total carbon assimilated each day in photosynthesis (Lamber et al. 1996). As a consequence, measured soil respiration is well correlated with fine-root density along a gradient from an open area to lichen or vaccinium areas in a central Siberian Scots pine forest in Russia (Fig. 3.4) (Shibistova et al. 2002) and in loblolly pine plantations in North Carolina, with and without irrigation and/ or fertilization (Maier and Kress 2000).

The amount of CO_2 produced through root respiration is determined by the root biomass and specific root respiration rates. Root biomass in an ecosystem depends on ecosystem production and allocation patterns of plant species, and it varies with growth environments and seasons. Forests and sclerophyllous shrublands have a root biomass of $5\,kg\,m^{-2}$, whereas croplands, deserts, tundra, and grasslands have a lower root biomass, usually less than $1.5\,kg\,m^{-2}$ (Jackson et al. 1996). Cold deserts have three times the root biomass of warm deserts. The greatest root biomass that have been documented in

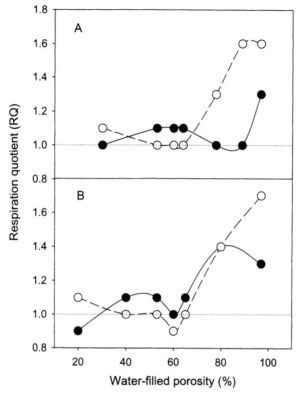

FIGURE 3.3   RQs of a residue-amended silty clay loam as related to percent water-filled porosity and soil bulk density of 1.40 Mg m^{-3} (panel A) and 1.14 Mg m^{-3} (panel B) at 3 (solid lines)
and 13 days (dashed lines) of incubation time after water treatments (Plotted with permission
from data by soil science society of America Journal: Linn and Doran 1984).

the literature were found in a Venezuelan caatinga rainforest (Klinge and
Herrera 1978) and in the California chaparral (Kummerow and Mangan
1981).

   At the individual plant level, carbohydrate allocation to root growth varies
with plant species, age, and growth environments. Usually, root to shoot
(root-shoot) ratio decreases with age due to ontogenic change during organ
development. In general, root-shoot ratio is high under low levels of nutrient
supply, low water availability in soil, and high levels of light. Effects of growth
temperature and CO_2 concentration on root-shoot ratio are circumstantial,
and no clear patterns have been generalized across various studies (Rogers
et al. 1996, Luo et al. 2006). On the ecosystem scale, root allocation is usually
higher in cold than in hot deserts and higher in grasslands than in forests.

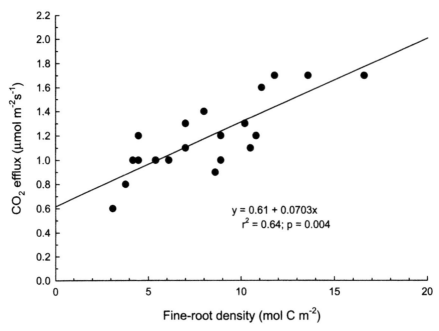

FIGURE 3.4 The relationship between root density in the top 20 cm of soil and the CO_2 efflux rate in early spring 1999 (Redrawn with permission from Tellus B: Shibistova *et al.* 2002).

Specific root respiration rate is the respiration rate per unit of root biomass, which varies greatly among species and with environmental factors. Measured respiration rates of excised roots from *Atriplex confertifolia* in northwestern Utah range from 0.2 to 4.3 $\mu mol\,kg^{-1}\,s^{-1}$ (Holthausen and Caldwell 1980). Root respiration is approximately 0.2 $\mu mol\ CO_2\ g^{-1}$ roots min^{-1} for loblolly pine seedlings at 20°C and decreases by 12% when plants are exposed to ozone (Edwards 1991). Bryla *et al.* (1997) measured root respiration of *Citrus volkameriana*, which varied from 2 to 3.5 $\mu mol\,m^{-2}\,s^{-1}$ during the study period of 110 days. They did not find that root respiration increases after prolonged exposure to drought and increased soil temperature.

Specific root respiration rates reflect the need for energy from many processes, including (1) biosynthesis of new structural biomass, (2) translocation of photosynthate, (3) uptake of ions from soil, (4) assimilation of nitrogen and sulfur into organic compounds, (5) protein turnover, and (6) cellular ion-gradient maintenance (Thornley 1970, Amthor 2000). Thus, root respiration is regulated by a number of biotic and abiotic factors that are related to the status, life history, and environment of the plants (Amthor 1991, Wang and Curtis

2002). For example, root respiration linearly increases with root nitrogen concentration for sugar-maple roots of various diameter classes collected at different soil depths in two forests in northern Michigan in late August (Pregitzer *et al.* 1998). Roots of smaller diameter in shallower depths have higher nitrogen concentration and higher respiration rates. Similarly, root respiration is linearly correlated with nitrogen concentration for seedlings of nine boreal species grown at either 5% or 25% of full sunlight (Reich *et al.* 1998).

Slow-growing plants usually have lower specific root respiration rates but consume a much higher percentage of the photosynthetic product than fast-growing plants. This happens regardless of whether the growth rates are inherently low or are limited by nutrient supply (Van der Werf *et al.* 1992). However, light-induced changes in growth rates do not affect root respiration very much. Specific root respiration rates generally decrease with root longevity (Eissenstat *et al.* 2000).

Respiration increases with temperature, resulting from the temperature sensitivity of enzymatically catalyzed reactions involved in respiration and the sensitivity of the increased ATP requirements as metabolic rates increase. The temperature stimulation of respiration also reflects the increased demand for energy necessary to support the increased rates of biosynthesis, transport, and protein turnover that occur at high temperatures. The rate of respiration at any given measurement temperature also depends on the growth temperature to which a plant is acclimated. Temperature acclimation results in homeostasis of respiration. The flexibility of root-respiratory acclimation to temperature is species-dependent.

Other environmental factors that influence respiratory processes include flooding, salinity, water stress, nutrient supply, irradiance, pH values, and partial pressure of CO_2 (Lambers *et al.* 1998). Flooding inhibits root respiration except in the case of wetland plants, which have evolved mechanisms of aeration. Sudden exposure of plants to salinity or water stress often enhances their respiration due to an increased demand for respiratory energy. Long-term exposure of sensitive plants to salinity or drought gradually decreases respiration, as a result of the general decline in carbon assimilation associated with slow growth under these conditions.

When plants are grown at a low supply of nutrients, their rate of root respiration is lower than that of plants that are well supplied with mineral nutrients, due to reduced growth rates and ion uptake. Root respiration rates were lower in dry soil than in wet soil during the 110 days of study (Bryla *et al.* 1997). Bouma *et al.* (1997) found that root respiration of citrus is not affected by a soil CO_2 concentration within the range of 400 to 25,000 ppm, in contrast to earlier findings for the Douglas fir (Qi *et al.* 1994).

Respiration is often conceptually separated into two components: growth respiration and maintenance respiration. Growth respiration yields the energy and building blocks (i.e., metabolic intermediates) for the biosynthesis of structural compounds. The maintenance respiration produces the energy required by the normal activities of living cells. McCree (1970) proposed the concept of growth and maintenance respiration, which have been examined by many studies in the context of basic plant biology and plant/ecosystem modeling.

3.3. RHIZOSPHERE RESPIRATION WITH LABILE CARBON SUPPLY

The respiration of microorganisms is greatly stimulated by an abundance of carbonaceous materials (mucilage, sloughed-off cells, and exudes) in the rhizosphere. The rhizosphere is a zone immediately next to the root surface with its neighboring soil, where a close plant-microbe interaction occurs (Fig. 3.5). The concept of the rhizosphere was first introduced by L. Hiltner in 1904 (Richards 1987) and describes the thin zone about 10 to 20 μm thick, surrounded by the mucilaginous layer. The chemical compounds in the rhizosphere vary from relatively simple oligosaccharides to a complex pectic acid polymer permeated by loose cellulose microfibrilis. The space between the root cell walls and mineral soil particles is filled with a gelatinous material known as mucigel (Greaves and Darbyshire 1972). The rhizosphere offers a highly favorable habitat for microorganisms. And the microbial community in this zone is usually quite distinct from that in the general soil. Interactions between plants and microorganisms in the rhizosphere play a critical role in regulating microbial activity, nutrient availability, decomposition of litter, and dynamics of SOM (Fig. 3.5).

Roots continuously release various substances to soil. According to the mode of release, there are three groups of rhizodeposition: (1) water-soluble exudates (sugars, amino acids, hormones, and vitamins), which leak from the root without involvement of metabolic energy; (2) secretions (polymeric carbohydrates and enzymes), which depend on metabolic processes for their release; and (3) lysates, released when cells autolyse (Lynch and Whipps 1990). The root exudates of maize, for example, were mainly water soluble (79%). Among the water-soluble exudates, carbohydrates account for about 64%, amino acids/amides for 22%, and organic acids for 14% (Hutsch et al. 2002).

Estimated amounts of carbon lost as exudates and secretions vary considerably with plant species, experimental facilities and sites, and measurement methods. Annual crops that grow in controlled facilities have been found to transfer 30 to 60% of their net fixed carbon to roots (Lynch and Whipps 1990). Carbon transfers to root as exudates, as indicated by respiration,

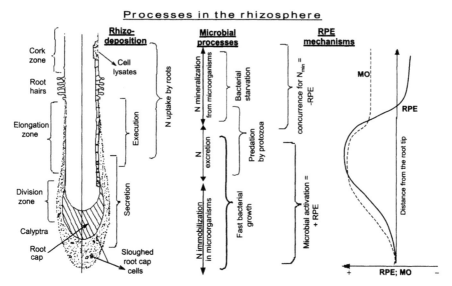

FIGURE 3.5 Schematic diagram of rhizosphere processes, including rhizodeposits, microbial growth, C and N turnover processes, and rhizosphere priming effects (RPE) along the growing root. RPE and change of microorganisms (MO) amount compared to the fallow soil is shown on the right (Redrawn with permission from *Journal of Plant Nutrition and Soil Science:* Kuzyakov 2002).

accounts for 10 to 70% of total carbon assimilation in 10 of the 11 studies (Lynch and Whipps 1990). In general, the fraction of net carbon transferred to root is higher for perennial plants than for annual plants (Grayston *et al.* 1996). The total root-derived carbon increases with the age of tree seedlings, ranging from 5% of net carbon uptake at 3 months to 21% at 19 months for chestnut trees (Rouhier *et al.* 1994). Hutsch *et al.* (2002) demonstrated with different plant species that up to 20% of photosynthetically fixed carbons are released into the soil during the vegetation period.

Most studies of root deposition were conducted in hydroponic and pot environments (Bekku *et al.* 1997a, DeLucia *et al.* 1997, Groleau-Renaud *et al.* 1998). It is still not feasible to measure the amount of rhizodeposits in natural ecosystems despite their importance in regulating plant and ecosystem carbon balance. Based on the kinetics of the ecosystem carbon processes, Luo *et al.* (2001b) quantified root exudation through a deconvolution analysis of soil respiration in response to a step increase in carbon influx in an elevated CO_2 experiment in the Duke Forest, North Carolina. Dynamics of the observed soil respiration in the first three years of the CO_2 fumigation suggests that root rhizodeposition is of minor importance in the loblolly pine forest.

However, root exudation may be an important pathway of carbon transfer to the rhizosphere in other ecosystems. For example, measured soil surface respiration gradually increases up to 35% by the end of a 58-day exposure of sunflower plants to elevated CO_2 compared with those in ambient CO_2 (Hui et al. 2001), implying substantial carbon transfer by root exudation.

The substances delivered from roots to the rhizosphere are decomposed primarily by bacteria. The small size and large surface-to-volume ratio of bacteria enable them to absorb soluble substrates rapidly. Thus, bacteria can grow and divide quickly in substrate-rich, rhizosphere zones. Bacteria also play an important role in the breakdown of live and dead bacterial and fungal cells. The major functional limitation results from its low mobility. Individual bacteria depend largely on the substrates that move to each one. The substrate at a particular location in the soil is supplied in one of the three major forms: diffusion, mass flow through water movement, and carry-over via root elongation. As roots grow, the rhizosphere moves, leading to successional change in the microbial community (Fig. 3.5).

In general, the microbial community structure in the rhizosphere is distinct from that in bulk soil. Three genera—*Pseudomonas, Achromobacter,* and *Agrobacterium*—are common bacteria in the rhizosphere. Anaerobic bacteria are also present in the rhizosphere more frequently, probably due to greater oxygen consumption by root and microbial respiration than in the bulk soil. Bacteria growth in the rhizosphere is stimulated more by simple substrate compounds, particularly by amino acids, than by complex organic compounds. For example, Vance and Chapin (2001) showed that microbial respiration responded more strongly to sucrose than to cellulose addition. In contrast, the rhizosphere does not influence fungi community as strongly as it influences the bacterial community. *Fusarium* and *Cylindrocarpon* are among the prominent inhabitants in the rhizosphere, but other genera, such as the zygomycetes *Mucor* and *Rhizopus,* are also represented.

Root-infecting fungi—mycorrhizae—are the widespread microorganisms that are associated with roots of nearly all families of flowering plants (Smith and Read 1997). They play a critical role in carbon and nutrient cycling in terrestrial ecosystems. According to the review by Allen (1991), mycorrhizal fungi consume 10 to 20% of net photosynthesis with a range from 5 to 85% among ecosystems. Mycorrhizae usually have short life spans (Friese and Allen 1991) and high nitrogen concentrations (Wallander et al. 1999), favoring decomposition of fungi tissues. Thus, carbon cycling through mycorrhizae is relatively fast. Nonetheless, mycorrhizae generate compounds such as chitin and glomalin, which are not readily decomposed and may form recalcitrant SOM (Rillig 2004).

While a large percentage (64 to 86%) of these root-borne substances are rapidly respired by microorganisms, about 2 to 5% of the net carbon assimilation remains in soil (Hutsch *et al.* 2002). Under nonsterile conditions, the exuded compounds are rapidly stabilized in water-insoluble forms and preferably bound to the soil clay fraction. The binding of root exudates to soil particles also improves soil structure by increasing aggregate stability. The release of organic materials from roots, even though it represents a small proportion of the total rhizodeposition, plays a critical role in the formation and decomposition of SOM through a rhizosphere-priming effect. Living plants can either increase by three- to fivefold or decrease by 10 to 30% the rate of SOM decomposition (Kuzyakov 2002). Such short-term rate changes in SOM decomposition are due to the priming effect in the direct vicinity of the living roots (Cheng and Coleman 1990, Liljeroth *et al.* 1994). Root growth dynamics and photosynthesis intensity are the most important plant-mediated factors affecting the priming effect (Kuzyakov and Cheng 2001). Environmental factors, the amount of decomposable carbon in soil, and mineral nitrogen content also influence microbial activation, preferential substrate utilization, and the rhizosphere-priming effect.

3.4. LITTER DECOMPOSITION AND SOIL ORGANISMS

Litter decomposition contributes to a significant amount of CO_2 production at the soil surface and in the soil (Jenny *et al.* 1949, Olson 1963). Removal of soil surface litter reduces annual soil respiration by 15% in an undisturbed grassland in central California and by 27% in a lemon orchard in the adjacent disturbed site (Wang *et al.* 1999). To understand CO_2 production during litter decomposition, it is necessary to describe litter production, litter pool sizes, and the decomposition processes.

Litter production is the amount of biomass that transfers from live plant parts to litter pools per unit of time. Litter production is positively correlated with net ecosystem productivity. Except for a fraction of NPP that is lost to herbivory and fire, all the plant biomass eventually becomes litter that is delivered to the soil as dead organic matter. Measured aboveground litterfall amounts to 550 to $1200\,g\,m^{-2}\,yr^{-1}$ in tropical forests (Vitousek and Sanford 1986), 300 to $650\,g\,m^{-2}\,yr^{-1}$ in a temperate forest (Johnson and Lindberg 1992, Finzi *et al.* 2001, Ehman *et al.* 2002), and 140 to $400\,g\,m^{-2}\,yr^{-1}$ in boreal forests (Buchmann 2000, Longdoz *et al.* 2001). In the Sonoran Desert, the annual litterfall varied from $60\,g\,m^{-2}\,yr^{-1}$ in the open desert and $157\,g\,m^{-2}\,yr^{-1}$ in the thornscrub to $357\,g\,m^{-2}\,yr^{-1}$ in the most productive sites (Martinez-Yrizar *et al.* 1999). On average, in a nine-year study in montane forests, leaf litter accounts for 65.1%, twig litter for 18.6%, and the follower/fruit litter for 14.4% (Liu *et al.* 2002b). The production of woody litter tends to increase with forest age. In grassland ecosystems where the aboveground biomass production is mostly not in perennial tissues, the annual litterfall is approximately equal to annual net primary production.

Estimated global litter production ranges from 38 to $68\,Pg\,C\,yr^{-1}$ with different extrapolation methods (Matthews 1997). Estimates of the major input to litter production according to net primary production are highly consistent with the estimates from dominant short-term disposition. Following the approach of modeling net primary production, Meentemeyer *et al.* (1982) used actual evapotranspiration to predict global patterns of plant litterfall and estimated $54.8\,Pg\,C\,yr^{-1}$ as the annual production of aboveground litterfall worldwide. Global patterns in the deposition of plant litterfall are similar to global patterns in net primary production (Esser *et al.* 1982).

Turnover of fine roots contributes a large amount of detritus to the soil in many ecosystems. The turnover quantifies the amount of deceased roots relative to the stock of live fine roots. Root turnover rates increased exponentially with mean annual temperature for fine roots in grasslands and forests, and for total root biomass in shrublands (Gill and Jackson 2000). On the broad scale, there is no correlative relationship between precipitation and root turnover. The average root turnover rates are slowest for entire tree root systems (10% annually), 34% for shrubland total roots, 53% for grassland fine roots, 55% for wetland fine roots, and 56% for forest fine roots. Root turnover rates decreased from tropical to high-latitude ecosystems for all plant function groups. The longevity of individual roots also correlates positively with mycorrhizal colonization and negatively with nitrogen concentration, root maintenance respiration, and specific root length (Eissenstat *et al.* 2000).

The balance between litter production and decomposition is the pool size of litter in an ecosystem. Litter production in tropical rainforests, for example, is among the highest (Schlesinger 1997). However, a high rate of litter decomposition in tropical regions results in a low accumulation of litter at the forest floor. In contrast, boreal forests have a relatively low litter production but accumulate much more litter biomass at the forest floor than in the tropical forests, due to the low decomposition rate in the cold regions. Estimates of the global litter pool vary greatly, ranging from 50 to $75\,Pg\,C$ at its low end (Schlesinger 1977, Hudson *et al.* 1994, Friedlingstein *et al.* 1995) to 150 to $200\,Pg\,C$ at its high end (Esser *et al.* 1982, Potter *et al.* 1993, Foley 1994). The lowest estimate of the total litter pool is $42\,Pg\,C$ (Bonan 1995), and the highest is $382\,Pg\,C$ (Esser *et al.* 1982). Estimation of the global litter pool generally does not include coarse wood debris, which can be substantial (Harmon *et al.* 1986).

Litter materials have various compositions, including soluble components, hemicellulose, cellulose, and lignin. For example, aboveground maize residues are composed of 29.3% soluble compounds, 26.8% hemicellolose, 28.4% cellulose, 5.6% lignin, and the rest ash (Broder and Wagner 1988). Woody litter from the Scots pine is composed of ethanol-soluble compounds (300 mg g^{-1}), lignin (383 mg g^{-1}), cellulose (111 mg g^{-1}), and lignin (65 mg g^{-1}) (Eriksson *et al.* 1990). Different components of litter each have distinct decomposition rates. Therefore, it is important to analyze litter compositions, because litter does not decompose as whole units. Rather, individual soil microbes produce a distinct set of degradative enzymes such that a suite of soil microbes would be able to decompose various groups of organic compounds in litter.

Litter decomposition is usually measured as the mass remaining of original litter after a period of incubation either in the laboratory or in the field. The mass remaining usually decreases rapidly at the beginning of the incubation and then more slowly as the incubation time goes on (Fig. 3.6). The time course of litter decomposition results from the fact that litter decomposition involves three processes: the leaching, fragmentation, and chemical alteration of dead organic matter to produce CO_2, mineral nutrients, and remnant complex organic compounds that are incorporated into SOM (Fig. 3.7). Leaching by water transfers soluble materials away from decomposing organic matter into the soil matrix. The soluble materials include free amino acids, organic acids, and sugars. These soluble compounds are readily decomposable by the vast majority of soil microbes, particularly by bacteria and "sugar fungi" (Zygomycetes such as *Mucor* spp. and *Rhizopus* spp.). Rapidly growing gram-negative bacteria specialize in labile substrates secreted by roots. Those microorganisms

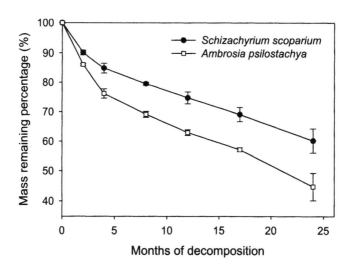

FIGURE 3.6   Mass remaining percentage of litter of C_3 forb *Ambrosia psilostachya* and C_4 grass *Schizachyrium scoparium* from the southern U.S. Great Plains. The difference in decomposition processes between the two species results from different litter quality (Modified from Su 2005).

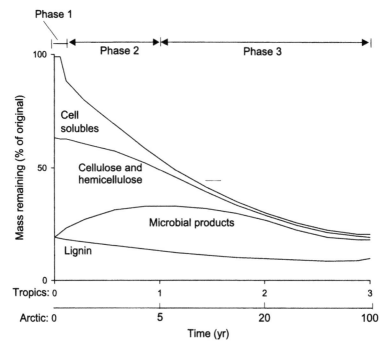

FIGURE 3.7 Typical time courses of litter decomposition to show that mass remaining declines over time and the decay rates vary with litter quality and temperature (Redrawn with permission from Springer-Verlag: Chapin *et al.* 2002).

can rapidly take up those compounds for catabolic and anabolic activities. The water-soluble compounds that are not used by microbes can pass to soil to react with the minerals or are lost from the system in solution.

Fragmentation is a process in which soil animals break down large pieces of litter. Soil animals influence decomposition by fragmenting and transforming litter, grazing populations of bacteria and fungi, and altering soil structure (Fig. 3.8). The microfauna are made up of the smallest animals (less than 0.1 mm). They include nematodes; protozoans, such as ciliates and amoebae; and some mites. Protozoans are single-cell organisms that ingest their prey primarily by phagocytosis, that is, by enclosing them in a membrane-bound structure that enters the cell. Protozoans are particularly important predators in the rhizosphere and other soil microsites that have a rapid bacterial growth rate (Coleman 1994). Nematodes are an abundant and trophically diverse group. Each of the nematode species specializes in bacteria, fungi, roots, or other soil animals. Bacterium-feeding nematodes in the forest can consume about $80\,g\,m^{-2}\,yr^{-1}$ of bacteria. The mesofauna, a taxonomically diverse group of soil animals 0.1 to

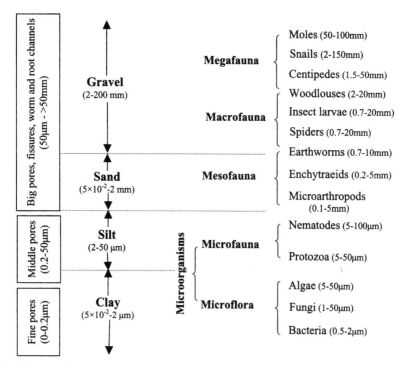

FIGURE 3.8 Classification of soil micro- and macroorganisms in relation to the size of pores and particles in soils (Modified with permission from John Wiley & Sons Ltd: Baldock 2002).

0.2 mm in length, have the greatest effect on decomposition. Those animals fragment and ingest litter coated with microbial biomass, producing large amounts of fecal material that has greater surface area and moisture-holding capacity than the original litter. Macrofauna include earthworms and termites that can alter resource availability by modifying the physical properties of soils and litter. Meanwhile, soil animals foraging for food sources fragment the litter and create fresh surfaces for microbial colonization.

The chemical alternation of litter is primarily a consequence of the activity of bacteria and fungi. Those microorganisms metabolically function as chemoorganotrophs. They are generally heterotrophic and obtain carbon and energy while degrading organic compounds added to soil, including plant residues and dead soil organisms. Those microorganisms secrete exoenzymes (extracellular enzymes) into their environment to initiate the breakdown of litter, which consists of compounds that are too large and insoluble to pass through microbial membranes. These exoenzymes convert macromolecules into soluble products that can be absorbed and metabolized by microbes.

Microbes also secrete products of metabolism, such as CO$_2$ and inorganic nitrogen, and produce polysaccharides that enable them to attach to soil particles. When microbes die, their bodies become part of the organic substrate available for decomposition. Actinomycetes are slow-growing, gram-positive bacteria that have a filamentous structure similar to that of fungal hyphae. Like fungi, actinomycetes produce lignin-degrading enzymes and can break down relatively recalcitrant substrates. They often produce fungicides to reduce competition.

Fungi are a diverse group of multicellular organisms with an incredible array of vegetative and reproductive morphologies with different life cycles. They are more abundant, on a mass basis, in soils than any other group of microorganisms. Their biomass ranges from 50 to 500 g wet mass m^{-2} (Metting 1993). Fungi can inhabit almost any niches containing organic substrates and are thus active participants in ecosystems as degraders of organic matter, agents of disease, beneficial symbionts, agents of soil aggregation, and an important food source for humans and many other organisms. Fungi are the main initial decomposers of terrestrial dead plant material. Fungi have a network of hyphae (i.e., filaments) that enable them to grow into new substrates and transport materials through the soil over distances of centimeters to meters. Hyphal networks enable fungi to acquire their carbon in one place and their nitrogen in another, much as plants gain CO$_2$ from the air and water and nutrients from the soil. Fungi that decompose litter on the forest floor, for example, may acquire carbon from litter and nitrogen from the mineral soil. Fungi are the principal decomposers of fresh plant litter, because they secrete enzymes that enable them to penetrate the cuticle of dead leaves or the suberized exterior of roots to gain access to the interior of a dead plant organ.

The amount of CO$_2$ produced during litter decomposition in an ecosystem is determined by the litter pool sizes (X) and specific decomposition rates (k). The relationship is expressed by:

$$\frac{dX}{dt} = -kX \qquad (3.8)$$

Equation 3.8 states that the litter decomposition rate is proportional to the mass. The specific decomposition rate, k, is the amount of litter mass lost per unit of time per unit of litter mass. The change of litter mass can be expressed by its integral equation as:

$$X = X_0 e^{-kt} \qquad (3.9)$$

where X_0 is the initial litter mass. The mass remaining of litter decreases exponentially with time (Fig. 3.6). Equations 3.8 and 3.9 can well describe experimental data from litter decomposition studies for periods from several months to a few years.

Litter decomposition is regulated by many factors, including (1) climatic factors such as annual mean temperature, annual mean precipitation, and annual actual evapotranspiration (Fogel and Cromack 1977); (2) litter quality, such as N content (Yavitt and Fahey 1986), C:N ratio (Berg and Ekbohm 1991), lignin content (Gholz et al. 1985), and lignin:N ratio (Melillo et al. 1982); and (3) vegetation and litter types (Gholz et al. 2000, Prescott et al. 2000). Among all the climatic variables, temperature and precipitation are the most important factors in influencing litter decomposition. Although the relative importance of temperature versus moisture in affecting litter decomposition is a matter of dispute (Taylor and Parkinson 1988, Pillers and Stuart 1993), temperature and moisture are usually interdependent and interactively determine litter decomposition (Witkamp 1966, Reiners 1968, Wildung et al. 1975). Litter decomposition also varies with vegetation types, mainly resulting from differences in their associated litter quality and microclimates (Prescott et al. 2000). Consistent differences in decomposition rates exist between litters of different species, regardless of climatic conditions. These suggest that substrate quality is one prime determinant of decay rates (Swift et al. 1979).

Silver and Miya (2001) compiled data on decomposition rates of root litter and estimated 175 k values (see Equation 3.8), which range from 0.03 to more than $7.0 \, g g^{-1} yr^{-1}$. Estimated specific decomposition rates from 70 studies compiled by D. Zhang, D. Hui, and Y. Luo (unpublished data) yield a total of 293 k values, ranging from 0.006 to $4.993 \, g g^{-1} yr^{-1}$ with a mean of $0.581 \, g g^{-1} yr^{-1}$. In general, k values are highest at the Equator and decrease with latitude (Fig. 3.9a). The average k values of litter decomposition vary with vegetation types, ranging from $1.3 \, g g^{-1} yr^{-1}$ in rainforests to $0.18 \, g g^{-1} yr^{-1}$ in tundra (Fig. 3.9b). The estimated k values also vary with litter types and decrease in the following order: grass litter > moss litter > broadleaf litter > root litter > litter from coniferous forests > barks > branch litter > coarse woody litter (Fig. 3.9c). Temperature, moisture, and initial litter quality are additional factors determining k values.

3.5. OXIDATION OF SOIL ORGANIC MATTER (SOM)

SOM is the organic fraction of the soil and usually does not include plant roots and undecayed macroanimal and plant residues in soil. SOM supplies nutrients for plant growth, contributes to cation exchange capacity so as to maintain soil fertility, and improves soil structure. Recently, extensive research on SOM has been conducted to explain the potential of soil to sequester carbon in a form of organic matter.

The estimated size of the global soil organic carbon (SOC) pool ranges from $700 \, Pg C$ (Bolin 1970) to $3150 \, Pg C$ (Sabine et al. 2003). Although the

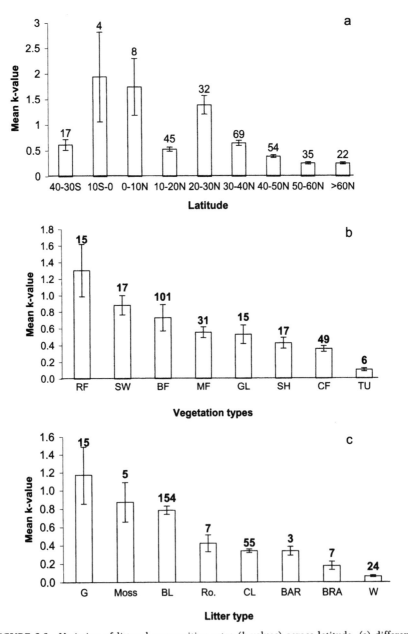

FIGURE 3.9 Variation of litter decomposition rates (k values) across latitude, (a) different vegetation types (b) and litter types (c) in global scale. Data at top of column are numbers of k values. Overall, k values decrease with latitude with high variability in low latitudinal regions. Vegetation types in panel b include rainforest (RF), swamp (SW), broadleaf forest (BF), mixed forest (MF), grassland (GL), hardwood forest (HD), shrub (SH), coniferous forest (CF), tundra (TU); litter types in panel c include grass leaf (G), moss, broadleaf litter (BL), roots (RO.), conifer needles (CL), bark (BAR), branch (BRA) and woody litter (W). Data are from D. Zhang, D. Hui, and Y. Luo (Unpublished).

generally accepted value of global SOC pool in the literature is around 1500 Pg C, a recent revision that includes organic carbon in permanently frozen soils and in deeper soil layers comes up with an estimate of 3150 Pg C. This value is approximately five times the total plant carbon pool. SOC in a particular ecosystem varies with several factors. Over a broad spatial scale, climatic factors such as temperature and precipitation play a major role in influencing SOC by regulating both inputs from live biomass and respiration back to the atmosphere.

SOM consists of humic and nonhumic substances. The nonhumic materials are unrecognizable organic residues of plants, animals, and microbes. They usually account for up to 20% of SOM. The remaining 80% or more of SOM are humic substances (i.e., humus), which are formed by secondary synthesis reactions. As litter undergoes biochemical alterations, microorganisms synthesize additional compounds, some of which polymerize or condense through either chemical or enzymatic reactions. A key mechanism of humus formation appears to be through enzymatic or autooxidative polymerization reactions involving phenolic compounds.

Humus is a complex mixture of chemical compounds with a highly irregular structure containing aromatic rings in abundance. Thus, SOM typically has a netlike, three-dimensional structure that coats mineral particles and can be electrochemically bound to clay and metal oxides in the soil. SOM and clay minerals can undergo nonenzymeatic chemical reactions to form more complex compounds, which become more difficult to break down. The carbon content of humus is approximately 58%, and nitrogen content varies from 3 to 6%, giving a C:N ratio of 10–20.

SOM can be separated into a few cohorts according to formation age and chemical compositions. A portion of SOM is easily decomposable, though most are stabilized by some physical, chemical, and/or biochemical protection from decomposition (Fig. 3.10) (Jastrow and Miller 1997, Six et al. 2002). Physical protection is rendered by soil aggregation, which reduces contacts between chemical compounds of SOM with microorganisms, enzymes, or oxygen. Chemical protection occurs when organic materials are associated with minerals either directly or indirectly through cation-bridging. Biochemical protection results from condensation and polymerization reactions, forming organic macromolecules. The macromolecules resist decomposition, because organisms are unable to make efficient use of them or lack the enzymes to degrade them. Thus, humus tends to accumulate in soil when exoenzymes cannot easily degrade its irregular structure (Oads 1989).

Breakdown of organic matter involves complex processes, including chemical alterations of organic matter, physical fragmentation, and releases of mineral nutrients. A variety of soil organisms—such as microorganisms, earthworms, microarthropods, ants, and beetles—are involved in this process

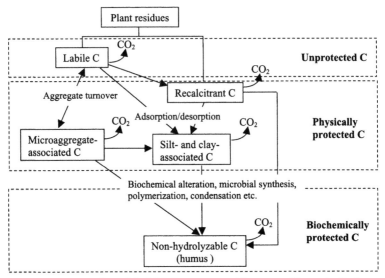

FIGURE 3.10   Conceptual model of SOM processes with measurable pools (Redrawn with permission from plant and soil: Six *et al.* 2002).

to perform chemical and physical changes at different stages. Organic matter breakdown is regulated by many factors, including soil moisture, thermal regimes, soil texture, bedrock type, nutrient status (cation exchange capacity), water capacity, illuviation and bioturbation rates, root penetration resistance, and the availability of oxygen to support aerobic microbial respiration. These variables tend to be coupled in such a way that soil texture becomes a useful proxy for most of them, with SOC levels negatively correlating with the particle sizes of the soil substrate. Disturbances such as deforestation, logging, agricultural and grazing practices, and biomass burning usually reduce SOC by either lessening carbon input or increasing carbon release. For example, plowing usually damages soil structure and accelerates the decomposition of SOM. Deforestation and biomass burning decrease carbon input into SOC pools. SOM consists of stable materials with a decomposition rate of 5% or less per year, depending on climatic conditions. An increase in soil temperature usually favors decomposition of humus materials. Increases in soil aeration favor oxidative decomposition. Adequate nitrogen supply usually increases the rate of decomposition of SOM. Mechanical disturbance by cultivation also favors decomposition. Under the anaerobic environment in wetlands, swamps, or marshes, litter decomposition is greatly reduced and organic residue accumulates, eventually forming histosol, an organic soil.

Histosols are usually called peats or bogs to indicate slow decomposition of plant litter. When water is drained, decomposition of SOM rapidly occurs, releasing large amounts of CO_2.

Wadman and de Haan (1997) measured the organic matter contents of 36 soils annually for 20 years in a pot experiment. The 36 soils were collected mainly from arable lands and varied in initial organic matter content from 1.31% in sandy soil to 51% in reclaimed peat soil. Despite the wide range of soil types studied, degradation of all SOM follows a similar pattern. Decomposition of SOM decreases with time and can be well described by:

$$Y(t) = b + cr^t \qquad (3.10)$$

where $Y(t)$ is the organic matter content in soil at time t, b is the size of the stable pool, c is the size of the decomposable pool of SOM with the first-order decay over time, and r is the relative decomposition rate. Estimated r values from the 36 soils vary from 0.649 to 0.995, with a mean of 0.885 and a standard deviation of 0.081.

CHAPTER 4

Processes of CO₂ Transport from Soil to the Atmosphere

Carbon dioxide produced in soil by roots and micro- and macroorganisms transfers through soil profiles to the soil surface. At the soil surface, CO_2 is released into the air by both diffusion and air turbulence. The released CO_2 is then mixed in plant canopy, partly absorbed by photosynthesis during daytime, and mostly released to the atmosphere through a planetary boundary layer (PBL). This chapter describes CO_2 transport from the site of production in soil to the bulk atmosphere along the four segments of the soil-atmosphere continuum. The four segments are the soil, soil surface, plant canopy, and PBL (Fig. 4.1). Although none of the transport processes may alter the total amount of CO_2 produced in soil, they are the fundamental mechanisms upon which most of the measurement methods for soil respiration are based (see Chapter 8). Thus, understanding the transport processes is critical for developing and evaluating measurement methodology. Transport processes are also sources of short-term fluctuation in soil surface CO_2 efflux which may bias measured soil respiration values.

4.1. CO₂ TRANSPORT WITHIN SOIL

The soil is a heterogeneous medium of solid, liquid, and gaseous phases, varying in its properties both across the landscape and in depth. Transport of gaseous CO_2 in the heterogeneous soil is driven largely by a concentration gradient along a profile from deep layers to soil surface.

CO₂ concentration has distinct vertical profiles, high in deep soil layers and low in the surface soil layers. For example, the CO_2 concentration is from

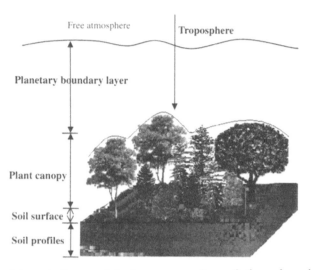

FIGURE 4.1 Schematic diagram of the four segments (i.e., soil, the surface, plant canopy, canopy boundary layer, and planetary boundary layer) in processes of CO$_2$ transport.

320 to 1000 µmol mol^{-1} in the surface and 17,500 to 32,000 µmol mol^{-1} in the deep soil at two sites in California (Lewicki *et al.* 2003). The CO$_2$ concentration in the deep soil layers could be 100 times the concentration at the soil surface, reaching 6 to 8% (Buyanovasky and Wagner 1983). The steep vertical CO$_2$ concentration gradient is formed primarily from the slow upward movement of CO$_2$ from sources of production. Due to the vertical distributions of roots and SOM, CO$_2$ is produced more in the surface layer than in the deep layers by roots and soil micro- and macroorganisms along a soil profile (Fig. 4.2). The majority of the CO$_2$ thus produced is released to the atmosphere with a small fraction that leaches into groundwater as dissolved inorganic carbonate. The upward movement of CO$_2$ from deep soil layers to the soil surface via diffusion and mass flow requires a gradient. Air movement in soil is a very slow process, leading to a buildup of steep CO$_2$ gradients in spite of the fact that the profile of CO$_2$ production sources is the opposite of the CO$_2$ concentration gradients (Fig. 4.2). Another factor in the development of CO$_2$ concentration profile is CO$_2$ molecular weight that is heavier than air molecules. Naturally, CO$_2$ has the tendency to sink down along the soil profile.

The soil CO$_2$ concentration profile and its gradient vary with several factors: (1) soil texture and porosity, (2) precipitation and/or water infiltration, and (3) CO$_2$ production rate versus movement rate. If soil porosity is

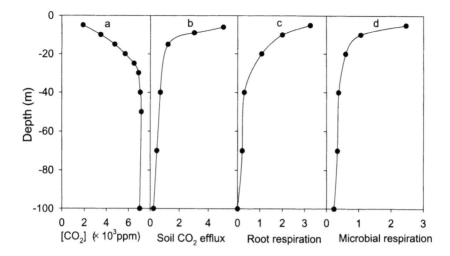

FIGURE 4.2 Gradients of soil CO_2 concentration (a), soil CO_2 efflux (b), root respiration (c), and microbial respiration (d) along soil profiles. The unit of soil CO_2 efflux, root respiration, and microbial respiration is $\mu mol\ m^{-2}\ s^{-1}$ (Developed by the reference to Hui and Luo 2004).

low, CO_2 concentration gradient is usually high. During the precipitation and infiltration, soil CO_2 is either forced out (degassing) or washed the vertical away, resulting in low CO_2 concentration along the profile. If CO_2 production is high, it requires a high CO_2 gradient to diffuse CO_2 to the soil surface.

The soil CO_2 profiles display a distinct seasonality. For example, the CO_2 concentration at a depth of 50 cm increases by about 4500 ppm from early June to late July in a young jack pine forest in Canada (Fig. 4.3) (Striegl and Wickland 2001). It decreases to the values similar to those measured at the beginning of the growing season by mid-August. The jack pine forest has an extensive lateral root system, largely in the upper 45 cm of soil (Carroll and Bliss 1982, Rudolph and Laidly 1990). The strong fluctuation in the CO_2 concentration over season is driven largely by changes in soil CO_2 production.

CO_2 movement in soil occurs through a continuous network of air-filled pores that connect the surface to the deeper layers of the soil, except in excessively wet or compacted conditions (Hillel 1998). Gaseous movement within the soil takes place primarily by mass flow and diffusion. The mass flow occurs when a gradient of total gas pressure exists between zones. The entire mass of air streams from the zone of the higher pressure to that of the lower pressure. Diffusion, on the other hand, is driven by a gradient of partial pressure (or concentration) of CO_2 molecules in the air. It causes unevenly

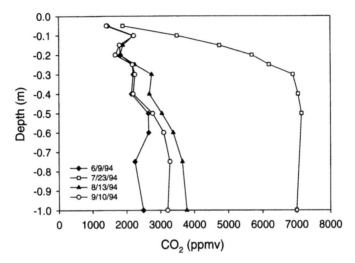

FIGURE 4.3   Soil CO_2 concentration versus depth and date to show seasonal shift in a young jack pine forest in Canada (Redrawn with permission from *Canadian Journal of Forest Research:* Striegl and Wickland 2001)

distributed CO_2 molecules to migrate from a zone of the higher concentration to a zone of the lower concentration, even though the gas as a whole may remain stationary.

Mass flow in the soil can occur through several mechanisms (Rolston 1986, Payne and Gregory 1988). Changes in temperature and atmospheric pressure cause soil air either to expand or contract. Rainwater entering the soil pushes out "old air" with a high CO_2 concentration. Plant water uptake creates a pressure deficit that draws air with a low CO_2 concentration into the soil. Wind gusts that blow over the surface may also pump air into, or suck air out of, the soil surface. The fluctuation of a shallow water table may push air upward or draw air downward. Tillage or compaction by machinery in agricultural practices can change soil air pressure, too.

The mass transport of CO_2 molecules in the unsaturated zone can occur in both the liquid and gas phases (Šimůnek and Suarez 1993). They can be described respectively by:

$$F_{ca} = -q_a c_a \qquad\qquad (4.1)$$

$$F_{cw} = -q_w c_w \qquad\qquad (4.2)$$

where F_{ca} and F_{cw} are the CO_2 fluxes caused by convection in the gas and the dissolved phases respectively (cm day^{-1}), q_a is the soil air flux (cm day^{-1}), q_w is the soil water flux (cm day^{-1}), and c_a and c_w are the volumetric concentration of CO_2 in the gas phase and dissolved phase respectively (cm^3 cm^{-3}).

The diffusive transport of gases in the soil occurs partly in the gaseous phase and partly in the liquid phase. Diffusion through the air-filled pores can be a major mechanism of CO$_2$ transport from the deep soil to the surface, driven by a steep CO$_2$ gradient along soil profile. Diffusion through water films of various thicknesses is a means of supplying oxygen to and disposing of CO$_2$ from live tissues, which are typically hydrated. For both portions of the pathway, the diffusion process can be described by the Fick's law:

$$F_{da} = -\theta_a D_a \frac{\partial c_a}{\partial z} \tag{4.3}$$

$$F_{dw} = -\theta_w D_w \frac{\partial c_w}{\partial z} \tag{4.4}$$

where F_{da} and F_{dw} describe the CO$_2$ fluxes caused by diffusion in the gas and the dissolved phases respectively (cm day^{-1}), D_a is the effective soil matrix diffusion coefficient of CO$_2$ in the gas phase (cm^2 day^{-1}), D_w is the effective soil matric dispersion coefficient of CO$_2$ in the dissolved phase (cm^2 day^{-1}), θ_a is the volumetric air content in the soil (cm^3 cm^{-3}), and θ_w is the volumetric water content in the soil (cm^3 cm^{-3}). The effective diffusion coefficient of CO$_2$ in the gas phase (D_a) is related to soil porosity and relative water content as:

$$D_a = D_{as} \frac{\theta_a^{7/3}}{\phi^2} \tag{4.5}$$

where D_{as} is the diffusion coefficient of CO$_2$ in free air (cm^2 day^{-1}) and ø is the total soil porosity (cm^3 cm^{-3}). Moldrup et al. (2000a and b) examined gas diffusion coefficients of various soil types and found that the D_a is highly predictable if we know the air-filled soil porosity when soil water potential is -100 cm H$_2$O.

The effective dispersion coefficient in the dissolved phase, D_w, varies with both hydrodynamic dispersion and diffusion as:

$$D_w = \lambda_w \left| \frac{q_w}{\theta_w} \right| + D_{ws} \frac{\theta_w^{7/3}}{\phi^2} \tag{4.6}$$

where λ_w is the dispersivity or dispersion length in the water phase (cm), which typically ranges from about 0.5 cm or less at the laboratory scale to about 10 cm or more for field-scale experiments (Nielsen et al. 1986). D_{ws} is the diffusion coefficient of CO$_2$ in free solution (cm^2 day^{-1}). The diffusion coefficient of CO$_2$ in the dissolved phase, D_{ws}, is about 10,000 times lower than that in the gas phase (D_{as}), and they vary with temperature (Fig. 4.4). In a standard condition of temperature (25°C) and pressure (normal atmospheric pressure), Table 4.1 shows diffusion coefficients for several gases in air and water. Thus, CO$_2$ diffusion in the liquid phase is usually negligible.

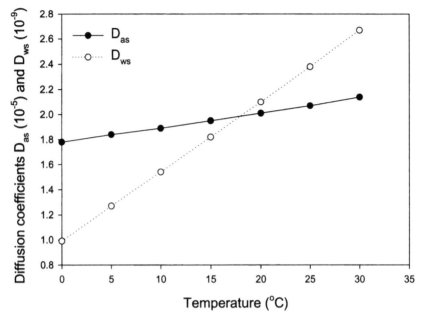

FIGURE 4.4 Changes of diffusion coefficients of CO$_2$ with temperature in air (D_{as}) and water (D_{ws}).

TABLE 4.1 Diffusion coefficients at standard temperature and pressure

Species	Media	Diffusion Coefficient (m^2 s^{-1})
CO$_2$	air	1.64×10^{-5}
O$_2$	air	1.98×10^{-5}
H$_2$O vapor	air	2.56×10^{-5}
CO$_2$	water	1.6×10^{-9}
O$_2$	water	1.9×10^{-9}
N$_2$	water	2.3×10^{-9}
NaCl	water	1.3×10^{-9}

Adapted with permission from Academic Press: Hillel (1998).

Each of these diffusion and mass transport processes can be formulated in terms of a linear rate law that the flux is proportional to the moving force. Thus, we can integrate the CO$_2$ transports in the unsaturated zone by mass flow and diffusive transport in both gas and aqueous phases, and by CO$_2$ production and/or removal (Patwardhan *et al.* 1988). We can get the one-dimensional CO_2 transport described by the following mass balance equation:

$$\frac{\partial c_T}{\partial t} = -\frac{\partial(F_{da} + F_{dw} + F_{ca} + F_{cw})}{\partial z} - Qc_w + S \tag{4.7}$$

where c_T is the total volumetric concentration of CO_2 (cm^3 cm^{-3}) and S is the CO_2 production/sink term. The term Qc_w represents the dissolved CO_2 removed from the soil by root water uptake, assuming that plants take up water together with the dissolved CO_2. The total CO_2 concentration, c_T, is the sum of CO_2 in the gas and dissolved phases:

$$c_T = c_a\theta_a + c_w\theta_w \tag{4.8}$$

Thus, a change in total concentration of CO_2 in a soil layer is determined by CO_2 fluxes into or out of the layer plus CO_2 production and/or minus CO_2 removal.

Among the transport processes, molecular diffusion can account for most exchanges of soil gases, particularly in deep soil layers. The gradient of CO_2 concentration adjusts to accommodate variable production/consumption rates and variable diffusion coefficients. Thus, unequal production and consumption of gases, changes in liquid water content, and temperature effects usually play a minor role in deep-layer CO_2 transport. At the surface layer of soil, CO_2 transport is regulated by a different set of forces.

4.2. CO₂ RELEASE AT THE SOIL SURFACE

While CO_2 transport along the soil profile is determined primarily by diffusivity of soil matrix and the steepness of the CO_2 gradient, CO_2 releases at the soil surface are strongly influenced by gusts and turbulence. It has long been documented that water loss at the soil surface via evaporation is strongly regulated by wind. For example, Hanks and Woodruff (1958) demonstrated that evaporation through soil, gravel, and straw mulches increases with wind velocity in a wind tunnel experiment. Benoit and Kirkham (1963) and Acharya and Prihar (1969) observed that the evaporation rate increases when air movement increases over soil columns covered by a layer of mulch.

Both barometric pressure fluctuations and pressure fluctuations caused by wind or air turbulence can alter soil gas exchange. According to the estimate by Kimball (1983), barometric pressure fluctuations can cause up to a 60% variation in the diffusion rate of gases in deep soils. Wind or air turbulence can increase gas fluxes to various degrees, according to soil surface texture. In an experiment with a specially designed vapor exchange meter, Kimball and Lemon (1971) demonstrated that pressure fluctuation caused by wind or air turbulence can increase gas exchange several times compared with diffusion through straw mulches and coarse gravels (Fig. 4.5). In the silt loam soils with a low porosity, pressure fluctuation can increase gas fluxes by at least 25%. Effects of air turbulence on surface CO_2 probably occur through very shallow depths of soils. The transport coefficient for soil gas exchange typically ranges from 0.01 to 0.1 cm² s⁻¹ (Kimball 1983). The lower limit of the transport coefficient is the molecular diffusion coefficient. Above and within plant canopies where turbulent mixing of air is the primary mechanism for gas exchange, the transport coefficient typically ranges from 100 to 10,000 cm² s⁻¹. Any turbulence at the soil surface that penetrates into soil layers will increase the effective value of the transport coefficient above this lower limit of molecular diffusion.

Measured CO_2 efflux by chambers placed over the soils results mainly from CO_2 release at the soil surface. The effects of pressure inside the chamber caused by flow restrictions were first demonstrated by Kanemasu et al. (1974) and carefully studied by Fang and Moncrieff (1996, 1998), Lund et al. (1999), and Longdoz et al. (2000). Underpressurization or overpressurization of the

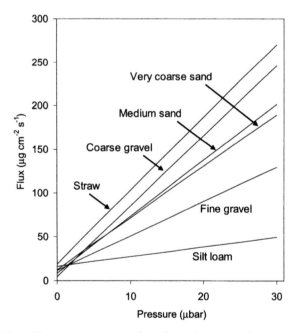

FIGURE 4.5   Flux of heptane evaporation from beneath 2-cm surface coverings of various porous media against root mean square pressure fluctuation for 1-min. periods (Redrawn with permission from *Soil Science Society of American Proceeding*: Kimball and Lemon 1971).

chambers can cause large bias in measured CO$_2$ fluxes at the soil surface (Davidson *et al.* 2002b). Wind outside the chamber also causes fluctuation in measured CO$_2$ fluxes (Lund *et al.* 1999). Using data from eddy-covariance measurements, Baldocchi and Meyers (1991) demonstrated that CO$_2$ efflux rates at the soil surface increase markedly with increasing levels in the standard deviation in static pressure (σ_p), suggesting a role for pressure fluctuations in regulating forest CO$_2$ exchange (Fig. 4.6). Fluctuations in σ_p are related to convective air movements in the PBL due to sensible heat flux from a warming surface (Stull 1997). Static pressure fluctuations promote diffusion of gas through coarse soils and loose litter through pumping action (Kimball 1983, Kimball and Lemon 1971) and enhance effluxes of both water vapor and CO$_2$ from litter layers.

Synchronous changes in soil surface temperature and velocity fluctuations over the diurnal time course may strongly regulate the diurnal cycle of soil CO$_2$ efflux. At night cooler temperatures decrease CO$_2$ production and reduce turbulence, which results from the stable thermal stratification of the atmospheric surface layer. Turbulence and temperature increase during the day due to surface heating. The buildup of the convective PBL generates turbulence, while surface heating increases respiratory activity. The two modes of action promote the transfer of CO$_2$ effectively between the soil surface and the atmosphere during the daytime.

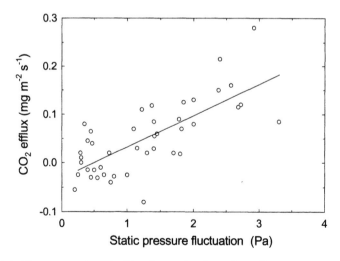

FIGURE 4.6   The response in CO$_2$ efflux from a deciduous forest floor to static pressure fluctuations (Redrawn with permission from Journal of Geophysical Research: Baldocchi and Meyers 1991).

Litter layers increase resistance of CO_2 diffusion from soil to the atmosphere. Measured soil CO_2 concentration at 15 cm of mineral soil is 950 ± 200 $\mu mol\, mol^{-1}$ in the unfertilized plots with thin litter layers and 1250 ± 220 $\mu mol\, mol^{-1}$ in the fertilized plots with thick litter layers in a loblolly pine forest in North Carolina (Maier and Kress 2000). Litter removal increases efflux soil CO_2 due to reduced resistance of CO_2 diffusion at the soil surface. The increments themselves are linearly correlated with the litter amount at the soil surface.

4.3. CO₂ TRANSFER IN PLANT CANOPY

CO_2 released from the soil surface is mixed within the canopy. Since canopies have multiple sources and sinks along the profile, part of the respiratory-released CO_2 from the soil may be absorbed by photosynthesis during the daytime. Most of it will be mixed with the aboveground plant respiratory CO_2 before being transported to the canopy above.

The transfer of CO_2 molecules within the canopy depends on profiles of CO_2 concentration and wind speed. At night wind speed is low, air is calm, and no photosynthesis occurs. CO_2 concentration is highest at the surface and declines along the profile within an idealized uniform canopy (Fig. 4.7). Along the profile, the density of CO_2 sources in any horizontal plane, $S(z)$, is related to the change in the CO_2 flux (F) across that plane (Monteith and Unsworth 1990):

$$S(z) = \frac{\partial F}{\partial z} \qquad (4.9)$$

The total flux across the whole canopy at height z as given by an integral from the ground to height z is:

$$F(z) = F(0) + \int_0^z S(z)dz \qquad (4.10)$$

where F(0) is the flux from the ground at z = 0, which is CO_2 efflux from soil. With a well-developed monotonic profile, fluxes F within the canopy can be related to the so-called K-theory by:

$$F(z) = -K(z)\frac{\partial c}{\partial z} \qquad (4.11)$$

where c is CO_2 concentration and $K(z)$ is transfer coefficient of CO_2, which varies with source distribution at different heights. The source density $S(z)$ can be estimated by substituting Equation 4.11 into Equation 4.9:

$$S(z) = -\left(\frac{\partial K(z)}{\partial z}\frac{\partial c}{\partial z} + K(z)\frac{\partial^2 c}{\partial z^2} \right) \qquad (4.12)$$

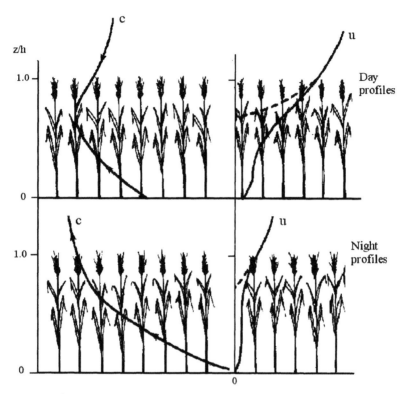

FIGURE 4.7   Idealized profiles of CO_2 concentration (c) and wind speed (u) in a field crop growing to a height h plotted as a function of z/h (Modified with permission from Edword Arnold: Monteith and Unsworth 1990).

Equations 4.10 to 4.12 can be used to estimate CO_2 fluxes and source strengths at the soil surface and different heights within a canopy, given well-measured profiles of CO_2 concentration and wind speed at night.

The one-dimensional gradient-diffusion model (i.e., the K theory) is unlikely to apply to the daytime profiles of CO_2, when turbulence is usually strong. With strong solar radiation input into the canopy in daytime, sources and sinks of CO_2, water vapor, heat, and momentum are variable along the vertical profile (Fig. 4.8). The transfer process is dominated by turbulent wind flow. Gusts are the strongest turbulent events in a wide range of canopy types (Raupach 1989a). The gusts are the energetic, downward incursions of air into the canopy space from the fast-moving air above. These intermittent gusts are responsible for more than 50% of energy transferred in events occupying less than 5% of the time. As a result, countergradient fluxes are

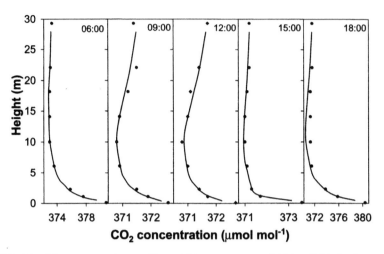

FIGURE 4.8 CO₂ concentration profiles within a canopy on 25 May 2000 (Redrawn with permission from Tellus B: Styles *et al.* 2002).

very common for heat, water vapor, and CO₂ fluxes within the canopy (Denmead and Bradley 1987). As shown in Figure 4.9, wind speed in the forest stand with little understory vegetation reaches a secondary maximum near ground, resulting in countergradient fluxes.

Although the K theory cannot well approximate the countergradient transport within canopies, the transport processes in the turbulent canopy are nevertheless constrained by the mass conservation (Wyngaard 1990). Considering an infinitely small box within a canopy, a change in the CO₂ concentration in the box over time is related to exchanges of CO₂ through convective and diffusive transfers into and out of the box. And sources or sinks of CO₂ result from plant photosynthesis and respiration. The mass conservation equation is:

$$\frac{\partial c}{\partial t} + u_j \frac{\partial c}{\partial x_j} = v \frac{\partial^2 c}{\partial x_j^2} + S \tag{4.13}$$

where t is time, u_j is the wind speed in any of the three directions j of the coordinate system, x_j is the distance in direction j, v the molecular diffusion coefficient of CO₂, and S the local source or sink term of CO₂. The four terms in Equation 4.13 represent (1) change in CO₂ concentration over time, (2) mass transfer of CO₂ via advection, (3) CO₂ transfer through diffusion due to gradient in CO₂ partial pressure, and (4) sources/sinks within the box respectively. The variation in source density with height, $S(z)$, depends on physical

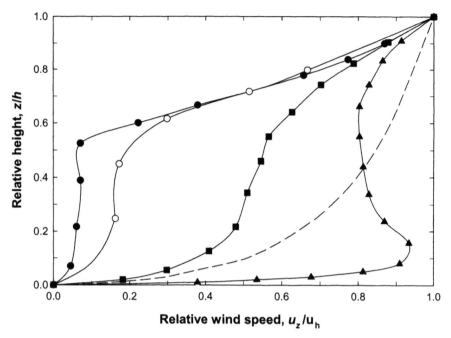

FIGURE 4.9 Normalized vertical profiles of wind speeds within canopy for a dense stand of cotton (●), dense hardwood jungle with understory (■), isolated conifer stand with no understory (▲), and a corn crop (○). Dashed line is for a logarithmic profile (Redrawn with permission from Cambridge University Press: Jones 1992).

and physiological processes of aboveground biomass, while the concentration profile $c_i(z)$ depends on the turbulence that distributes CO_2 molecules.

Raupach (1987, 1989a,b) developed a so-called localized near-field theory to distinguish between two regimes of dispersion: the near field and the far field. Dispersion in the near field is dominated by turbulent eddies. In this region, particles tend to maintain their initial speed and direction. In contrast, dispersion in the far field behaves as a random walk that is well described by gradient diffusion theory. The localized near-field theory (Raupach 1989a,b) is a semi-Lagrangian theory and provides an approximate means of the concentration profile $c(z)$ from a given source density profile $S_c(z)$, given the large-scale, coherent nature of turbulent eddies in vegetation canopies. All individual elements in a canopy are considered independent point sources, from which material (e.g., CO_2) is released into small parcels of air as they pass. The semi-Lagrangian approach estimates a statistical probability of independent parcels released into the air stream from all these sources reaching a specific point at a particular time. Thus, transport depends on the turbulence structure of the airflow.

Transfer of CO_2 and other mass above the plant canopy occurs in the canopy boundary layer. The canopy boundary layer is the zone above the canopy surface, where the mean velocity of wind is reduced substantially below that of the free stream due to the sheering stress. The wind speed within the boundary layer increases with height above the canopy. An idealized relationship between wind speed and height follows a natural logarithmic equation (Monteith and Unsworth 1990). Assume that there are no CO_2 sources or sinks within the boundary layer and no advection, transfer of CO_2 and other mass above the canopy can be described by the standard gradient-diffusion equation (i.e., Equation 4.11).

Fluxes may be estimated if the concentration gradient and K_i at any height are known. The coefficients of the turbulence transfer in the air are the same for momentum, heat, water vapor, and gases in neutral stability and proportion to friction velocity.

4.4. CO_2 TRANSPORT IN THE PLANETARY BOUNDARY LAYER

The PBL is the layer above soil and/or vegetation where vertical transports by turbulence play a dominant role in transfers of momentum, heat, moisture, CO_2, and other gases. The height of the PBL ranges from 100 to 3000 m and varies with time, location, and weather conditions. Vegetation roughness, solar heating, and evapotranspiration are major factors influencing turbulent strength over the earth's surface. Because turbulent flows are several orders of magnitude more effective at transporting gases than is molecular diffusion, they result in rapid CO_2 transport in the PBL. This CO_2 transport in PBL is also affected by a covariance between the biospheric flux (i.e., photosynthesis and respiration) and turbulent transport (Denning et al. 1996). Both photosynthesis and thermal convection are driven by solar radiation on the diurnal and seasonal scales. During the growing season, photosynthetic uptake of CO_2 is associated with a deep PBL with strong thermal convection. The rapid transport and plant uptake together result in relatively low and uniform distributions of CO_2 in PBL. In winter, PBL is shallow with weak thermal convection. Ecosystem respiration becomes the dominant component of biosphere CO_2 fluxes. Thus, CO_2 transport is slow with a steep gradient of CO_2 concentration within the shallow PBL in winter.

During the daytime, when surface heating generates buoyant convection over land, PBL is referred to as a convective boundary layer (CBL). The turbulence efficiently mixes the bulk of CBL, producing a uniform average CO_2

concentration in a thin surface layer. Thus, CO_2 efflux can be estimated by the boundary layer budget method (Denmead *et al.* 1996, Levy *et al.* 1999) according to the mass conservation Equation 4.13. This method causes minimal disturbance to the ecosystem environment over several km² and provides "area-integrated" fluxes.

At night, when the surface is cooler than the air over land, PBL can become stably stratified. It is often known as nocturnal boundary layer (NBL), which extends to heights of only tens of meters and is bounded by a low-level, radiative inversion. The inversion restricts vertical mixing, so that emissions of CO_2 from plant and soil are contained in a shallow NBL whose concentration changes considerably (Fig. 4.10). Based on the mass conservation principle, CO_2 fluxes might be estimated from the rate of concentration change below the inversion, when turbulent flux can be neglected (Demead *et al.* 1996, Eugster and Siegrist 2000). Thus, the surface flux can be calculated from:

$$-F = \int_{0}^{z} \frac{ds}{dt} dz \qquad (4.14)$$

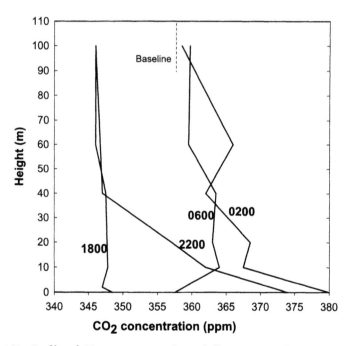

FIGURE 4.10   Profiles of CO_2 concentration during balloon ascents when using NBL to estimate CO_2 production by pasture (Redrawn with permission from Global Change Biology: Denmead *et al.* 1996).

where F is the gas flux $(\mu\mathrm{mol\,m^{-2}\,s^{-1}})$, $\dfrac{\partial s}{\partial t}$ is the rate of change in CO_2 concentration with time $(\mu\mathrm{mol\,mol^{-1}\,s^{-1}})$, z is the height of the air layer (usually NBL) whose concentration is affected by the emission, and s is the atmospheric CO_2 concentration.

In an experiment at the Wagga site, Australia, in October 2004, Denmead *et al.* (1996) used a helium-filled balloon to carry an airline aloft in a series of vertical traverses up to a height of 100 m for measurements of atmospheric CO_2 concentration (Fig. 4.10). Due to an inversion developed early in the evening, most of the emitted CO_2 from soil and vegetation between 1800 and 2200 hours was trapped between the surface and a height of 40 m. From the CO_2 enrichment in that layer, an average surface emission rate was estimated to be 0.05 mg m^{-2} s^{-1}, which is consistent with eddy correlation measurements of the nocturnal CO_2 flux at the site. Due to increases in wind speeds, turbulence mixes air from greater heights. The vertical profiles of CO_2 concentration measured at 0200 and 0600 hours could not be used to estimate respiratory CO_2 fluxes from the surface. Similarly, vertical profiles of CO_2 concentrations along a height up to 2000 m above the land surface showed the diurnal change, and the PBL height changes with the formation and disappearance of NBL (Eugster and Siegrist 2000).

PART **III**

Regulation

Controlling Factors

Respiration is fundamentally a biochemical process and occurs in cells for all living organisms—plants, animals, and microorganisms alike. Yet ecologists measure soil respiration on the scales of plot and ecosystem and are ultimately interested in its role in carbon cycling on regional and global scales. At each of the hierarchical orders from cell to globe, respiration involves different sets of chemical, physical, and biological processes. The latter processes are, in turn, influenced by an array of biotic and abiotic factors. Among the factors are substrate supply, temperature, moisture, oxygen, nitrogen (C:N ratio), soil texture, and soil pH value. This chapter accordingly identifies major factors at various hierarchical levels and evaluates their relative importance in determining soil respiration.

5.1. SUBSTRATE SUPPLY AND ECOSYSTEM PRODUCTIVITY

Respiratory release of CO_2 results from the breakdown of carbon-based organic substrates. At the biochemical level, therefore, CO_2 production by respiration has a 1:1 molar relationship with substrate consumption in terms of carbon atoms. At the ecosystem level, soil respiration is a composite of multiple processes, consuming substrates from various sources (see Chapter 3). Root respiration uses intercellular and intracellular sugars, proteins, lipid, and other substrates. Soil microorganisms consume all kinds of substrates, ranging from simple sugars contained in fresh residues and root exudates to complex humic acids in SOM. Although respiratory CO_2 release is linearly proportional to substrate availability, the rate at which the substrates are converted to CO_2 varies with substrate types (Berg *et al.* 1982). Simple sugars can be readily converted to CO_2 by roots and microbes with short residence times. It can be very difficult for humic acids to be decomposed and converted to CO_2 with residence times of hundreds or thousands of years. Substrates with intermediate residence times include celluloses, hemicelluloses, lignins, and phenols. The heterogeneity in substrate quality and multiple sources of supply make it extremely difficult to derive simple relationships between substrate supply and respiratory CO_2 production, which can be potentially incorporated into models.

Evidence from recent experiments demonstrates that substrate supply directly from canopy photosynthesis exerts a strong control on soil respiration. A tree-girdling experiment that severed carbon supply from aboveground photosynthesis to roots in a Scots pine forest in northern Sweden demonstrated a rapid decline in soil respiration by approximately 50% within one to two months (Högberg *et al.* 2001). Clipping and shading experiments in grasslands in the U.S. Great Plains decreased soil respiration by nearly 70% within one week (Craine *et al.* 1999, Wan and Luo 2003, Fig. 5.1), indicating a direct and dynamic link between soil respiration and substrate supply from the aboveground photosynthesis.

The direct control of soil respiration by the aboveground photosynthesis is also demonstrated by a mesocosm experiment of a model grassland ecosystem at constant temperatures and soil moisture content (Verburg *et al.* 2004). The experiment spanned over two growing seasons in 1999 and 2000.

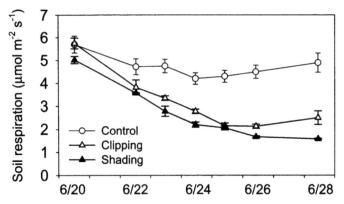

FIGURE 5.1 The responses of soil respiration to clipping or shading in tallgrass prairie in 2001 (Redrawn with permission from Global Biogeochemical Cycles: Wan and Luo 2003).

The day and nighttime temperatures were controlled at 28°C and 22°C respectively, and soil water content was maintained at a relatively constant level of 70% field capacity. Measured soil respiration rates increased from near zero without plants to $4\,\mu mol\,m^{-2}\,s^{-1}$ without N fertilization at the peak growing season in 1999 and to $7\,\mu mol\,m^{-2}\,s^{-1}$ with N fertilization in 2000 (Fig. 5.2). Given that the temperature and water content regimes were controlled at constants, the strong seasonal variation in soil respiration can result only from changes in substrate supply from the aboveground parts of plants.

The tight connections of soil respiration to aboveground photosynthesis have also been demonstrated by other studies. Root and soil respirations, for example, respond to aboveground herbivory (Ruess *et al.* 1998), availability of nutrients (Nadelhoffer 2000, Burton *et al.* 2000), light (Craine *et al.* 1999), and other factors that govern plant carbon gain. On the other hand, the belowground environment strongly influences root growth and carbohydrate demand from the aboveground photosynthesis. Root respiration increases exponentially with increases in soil temperature (Burton *et al.* 1996), resulting in peak fine-root elongation (Ruess *et al.* 1998, Tryon and Chapin 1983) and root respiration (Högberg *et al.* 2001) in boreal regions in mid- to late summer, when soil temperatures are warmest. The interaction between the demand for carbohydrates, as regulated by the soil environment, and the aboveground capacity to supply carbohydrates, as determined by photosynthesis, together govern the belowground carbon flux and therefore root and soil respiration.

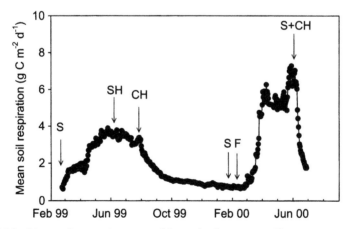

FIGURE 5.2   Mean soil respiration in a model grassland ecosystem. The arrows point to times of seeding (S), shoot harvest (SH), root crown harvest (CH), fertilizer application (F), and shoot and root crown harvest (S + CH) (Redrawn with permission from Global Change Biology: Verburg *et al.* 2004).

Despite the fact that ample experimental evidence demonstrates the intimate connections of soil respiration with aboveground photosynthesis, it is difficult to develop a quantitative relationship that directly links them. Indirect indices have been used to link soil respiration with aboveground substrate supply. For example, Reichstein *et al.* (2003) used leaf area index (LAI) as a surrogate of aboveground vegetation productivity and found strong correlations between normalized soil respiration (18°C, without water limitation) and peak LAI (Fig. 5.3).

In addition to the direct control of soil respiration by the aboveground photosynthesis, litter provides substantial amounts of carbon substrate to microbial respiration. As a consequence, soil respiration usually increases with the amount of litter. For example, Maier and Kress (2000) manipulated the aboveground litterfall at the soil surface in a loblolly pine forest and found a linear relationship between the increase in soil respiration and the amount of litter added to the soil surface (Fig. 5.4). Similar relationships between the litter amount and soil respiration have been found in other ecosystems (Boone *et al.* 1998, Bowden *et al.* 1993, Sulzman *et al.* 2005).

Soil respiration is also strongly regulated by carbon substrate in SOM, as demonstrated by many laboratory incubation studies. For example, when Franzluebbers *et al.* (2001) collected soil samples from four climate regions in North America for an incubation study, they found that basal soil respira-

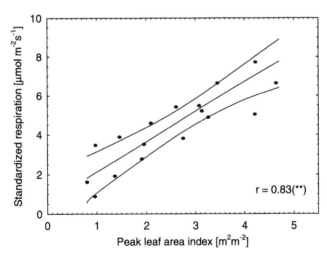

FIGURE 5.3 The relationship between normalized soil respiration (corrected for temperature and moisture effects) and peak leaf area index (LAI, Redrawn with permission from Global Biogeochemical Cycles: Reichstein *et al.* 2003).

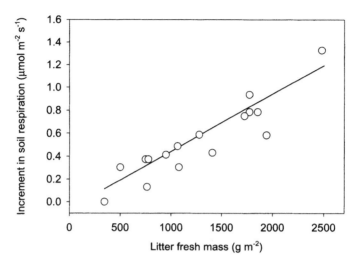

FIGURE 5.4   The relationship between the increment of soil respiration and litter mass added to the soil surface (Redrawn with permission from Canadian Journal of Forest Research: Maier and Kress 2000).

tion linearly correlated with the content of SOC (Fig. 5.5). Regression coefficients that indicate how fast carbon in SOC is released via microbial respiration during the incubation period are much higher for soil from warm (i.e., Georgia and Texas) than cold regions (i.e., Alberta and Maine) and slightly higher for soil from dry (Texas and Alberta) than wet regions (i.e., Georgia and Maine). The differences in the regression coefficients are determined by fractions of biological active soil carbon. In the cold regions, suboptimal temperatures limit biologically activity for a large portion of the year, resulting in the accumulation of partially decomposed organic carbon. The partially decomposed materials may undergo chemical transformations to recalcitrant SOC. It is also possible that species and functional composition of microbial communities are significantly different between the warm and the cold regions, leading to the different responses of microbial respiration to substrate supply.

Even if soil samples are from the same location, substrate availability may vary with physical environments, such as drying and freezing, and thus affect soil respiration. Rewetting air-dried soils, for example, results in a large respiratory flush directly related to the amount of amino acids and other nitrogenous material released by the drying process (Stevenson 1956, Birch 1958 Borken et al. 1999, McInerney and Bolger 2000, Fierer and Schimel 2003) Freezing causes a marked increase in the total amount of free extractable

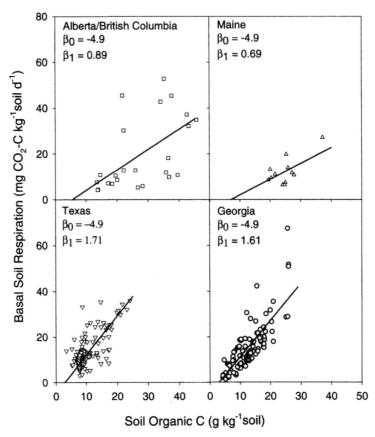

FIGURE 5.5   Relationship of basal soil respiration with soil organic carbon in surface soils from cold-dry (Alberta/British Columbia), cold-wet (Maine), warm-dry (Texas), and warm-wet (Georgia) climates (Redrawn with permission from *Soil Biology and Biochemistry*: Franzluebbers *et al.* 2001).

amino acids and sugars and a considerable increase in soil respiration (Ivarson and Sowden 1970, Morley *et al.* 1983, Schimel and Clein 1996).

On regional scales, soil respiration correlates with ecosystem productivity. In a comparison of 18 European forests, Janssens *et al.* (2001) demonstrated that annual GPP is the primary factor influencing soil respiration over years and across sites (Fig. 5.6). Reichsten *et al.* (2003) suggested that measures of vegetation productivity are necessary to reliably model large-scale patterns of soil respiration. In general, root respiration is coupled to shoot photosynthetic activity via allometric relationships (Heilmeier *et al.* 1997). Also, the largest fraction of heterotrophic respiration originates from decomposition of young organic matter (dead leaves and fine roots). Thus, both the root

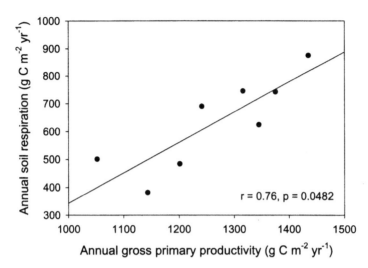

FIGURE 5.6 Annual soil respiration vs. annual gross primary productivity across less disturbed European forests (Redrawn with permission from Global Change Biology: Janssens *et al.* 2001).

respiration and heterotrophic respiration are dependent on primary productivity over broad spatial scales. However, the types of relationships that usually emerge on a large scale (across sites, regional, and global) between soil respiration and primary productivity may not be applicable to a specific site across years (Davidson *et al.* 2002a).

On a global scale, mean rates of soil respiration correlate positively with NPP across different vegetation biomes (Raich and Schlesinger 1992). Furthermore, annual soil respiration rates correlate positively with aboveground net primary productivity (ANPP) in northern peatlands (Moore 1986) and with aboveground litter production in forest ecosystems (Schlesinger 1977, Raich and Nadelhoffer 1989). These studies indicate a tight linkage between plant productivity and soil respiration, due to the fact that primary production provides the organic fuel that drives soil metabolic activity.

5.2. TEMPERATURE

Temperature affects almost all aspects of respiration processes. At the biochemical level, a respiratory system involves numerous enzymes that drive glycolysis, the TCA cycle, and the electron transport train (see Chapter 3). Biochemical and physiological studies usually demonstrate a general temperature-response curve that respiration increases exponentially with temperature in its low range, reaches its maximum at a temperature of 45 to 50°C, and then declines. In the low-temperature range, the maximum activity (V_{max}) of respiratory enzymes is probably the most limiting factor. Low temperatures can limit the capacity of both soluble and membrane-bound enzymes, although the transition from a gel-like state to a fluid state in membranes may be particularly important (Atkin and Tjoelker 2003). In the high-temperature range, adenylates (adenosine monophosphate [AMP], adenosine diphosphate [ADP], and adenosine triphosphate [ATP]) and substrate supply play a greater role in regulating respiratory flux (Svensson et al. 2002, Douce and Neuburger 1989, Atkin et al. 2002, Atkin and Tjoelker 2003). In extreme high temperatures, enzymes may degrade and respiratory activity become depressed.

The relationship between temperature and biochemical processes of respiration is usually described by an exponential equation or an Arrhenius equation. Van't Hoff (1885) proposed a simple empirical exponential model to describe chemical reactions in response to change in temperature as:

$$R = \alpha e^{\beta T} \tag{5.1}$$

where R is respiration, α is the respiration rate at 0°C, β is a temperature-response coefficient, and T is temperature. Arrhenius (1898) modified van't Hoff's empirical equation with an activation energy (i.e., the minimum energy needed to create a chemical reaction) parameter:

$$R = d e^{\frac{-E}{\Re T}} \tag{5.2}$$

where d is a constant, E is activation energy, $\Re$ is the gas constant, and T is temperature (degrees in Kelvin). Both equations describe an exponential increase in respiration with increasing temperature. Van't Hoff's equation is commonly accepted for biological systems over a limited temperature range. The Arrhenius equation can represent the behavior of many chemical systems and even some rather complex biological processes (Laidler 1972).

Root respiration also increases exponentially with temperature in its low range when the respiration rate is limited mainly by biochemical reactions (Berry 1949, Atkin et al. 2000). At high temperatures, the transport of substrates and products of the metabolism (e.g., sugar, oxygen, CO_2), mainly via diffusion processes, becomes a limiting factor. At temperatures above 35°C, the protoplasm system may start to break down. Limitation of respiration through the physical processes of diffusive transport may also occur at lower temperatures if the oxygen concentration is low. Responses of root respiration are more sensitive to temperature for young roots than old roots (Fig. 5.7). Temperature also indirectly influences root respiration via its effects on root growth. Roots grow faster at higher temperatures in annual crop plants

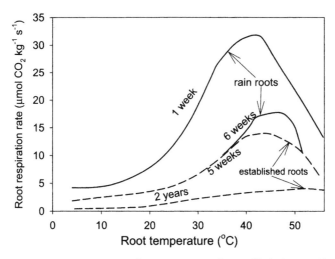

FIGURE 5.7 Temperature responses of root respiration for established roots (dashed lines) that were two years old or five weeks old and for rain roots (solid lines) that were six weeks old or one week old (Redrawn with permission from Journal of Experimental Botany: Palta and Nobel 1989).

(Kasper and Bland 1992) and perennials (Lieffers and Rothwell 1986, McMichael and Burke 1998, King *et al.* 1999, Weltzin *et al.* 2000, Kutsch *et al.* 2001). Controlled experiments also demonstrate optimal temperatures for root-length extension, with growth rates accelerating up to an optimum temperature and then declining at supraoptimal temperatures (Barney 1951, Merritt 1968, McMichael and Burke 1998). Optimal temperatures for root growth vary widely among different taxa, partly due to temperature regimes to which plants have adapted (Larson 1970, Tryon and Chapin 1983, McMichael and Burke 1998). Root growth in natural plant communities often correlates with photosynthetically active radiation (PAR) rather than soil temperature (Aguirrezabal *et al.* 1994), as demonstrated in studies either along altitudinal gradients (Fitter *et al.* 1998) or with soil warming (Fitter *et al.* 1999, Edwards *et al.* 2004). Root respiration may become less sensitive to soil temperature over seasons of a year, resulting from its rapid thermal acclimation (Edwards *et al.* 2004).

According to their temperature requirements, microorganisms are divided into three groups—cryophiles, mesophiles, and thermophiles—with their respective optimum temperatures being <20, 20 to 40, and >40°C. In natural conditions, soil contains many cohorts of microorganisms, and soil respiration usually responds to temperature exponentially within a very broad range. Rates of soil microbial respiration measured from frozen organic soil of three moist upland tundra at a temperature range of −10 to 0°C and thawed soil at temperature from 0 to +14°C can be well described by a simple, first-order exponential equation (Mikan *et al.* 2002). Similarly, microbes at different soil depths respond to temperature changes exponentially (Fierer *et al.* 2003) in a broad range. Dehérain and Demoussy (1896) found that a maximum of CO_2 efflux occurs at 65°C. However, Flanagan and Weum (1974) found the maximal rate of soil microbial respiration at a temperature of 23°C (Fig. 5.8).

At the level of soil aggregate, temperature may influence soil respiration indirectly via its effects on substrate and/or O_2 transport. Diffusion of both gases and solutes across soil water films is determined by both soil diffusivity and the volumetric water content (Equation 4.5). On the one hand, soil diffusivity increases with temperature at a given soil water content (Nobel 2005). On the other hand, an increase in temperature over a period of time likely reduces soil water content and the thickness of soil water films. The soil water content influences diffusion in the high order of power. From a dynamic view, therefore, the net, indirect effects of temperature via changes in soil water content are usually negative on soil respiration in uplands. In wetlands, a temperature-induced decrease in soil moisture has a larger effect on O_2 concentration and redox conditions than on solute diffusion. Since oxygen, rather than organic solutes, is usually the limiting substrate for respiration in wetlands, soil drying due to increased temperature (e.g., global warming) could stimulate soil respiration.

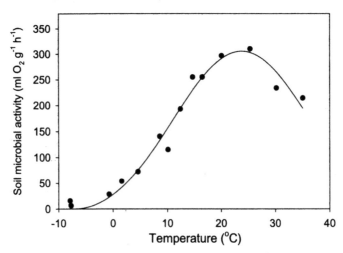

FIGURE 5.8   The relationship between soil microbial respiration and temperature (Flanagan and Veum 1974).

On the ecosystem scale, temperature, in concert with light and other co-varying factors, influences the seasonality of substrate supply to the below-ground system and then partially determines soil respiration. Although radiation is one main driving variable for seasonal changes in photosynthesis, temperature plays a distinctive role in the seasonality of substrate supply by its effects on the phenology of shoot and root growth (Fitter et al. 1995, Schwartz 1998, Dunne et al. 2003). Changes in temperature by one or two degrees in spring trigger a large, sometimes abrupt change in leaf area index, photosynthetic activities, and soil respiration during the leafing-out period in deciduous forests (Curiel Yuste et al. 2004). Root biomass, rhizosphere activities, and litter carbon input to soil also display strong seasonality (Ekblad et al. 2005). For example, monthly root biomass is highest in June and lowest in February in Tanzania's Serengeti grasslands averaged across 11 sites (McNaughton et al. 1998). The seasonality is often more pronounced for root growth of deciduous than coniferous trees (Steele et al. 1997, Coleman et al. 2000). Seasonal variation in root growth affects respiration of roots, mycorrhizae, and rhizosphere microorganisms, likely leading to distinct rhizosphere phonological patterns (Lyr and Hoffmann 1967). The indirect effects of temperature on soil respiration via plant phenology are often species-specific, depending on developmental stages of plants (Fu et al. 2002). For example, respired CO_2 from soybean or sorghum roots increases significantly from vegetative to flowering stages and declines thereafter. Respiration of amaranthus roots is highest at the vegetative stage and declines with the plant stage. Root respiration of sunflowers does not vary significantly with plant developmental stages. Phenological variation in shoot and root activities can significantly contribute to the seasonality of soil respiration (Curiel Yuste et al. 2004).

The sensitivity of respiratory processes to temperature is often described by Q_{10}—a quotient of change in respiration caused by change in temperature by 10°C, as defined by:

$$Q_{10} = \frac{R_{T_0+10}}{R_{T_0}} \tag{5.3}$$

where R_{T_0} and R_{T_0+10} are the respiration rates at reference temperature T_0 and temperature $T_0 + 10$°C, respectively. When the relationship between temperature and soil respiration is fitted by an exponential function, Q_{10} can be estimated from coefficient b in equation 5.1 as:

$$Q_{10} = {}^e10b \tag{5.4}$$

At the biochemical level, measured Q_{10} is usually around 2. That is, the respiration rate doubles for every 10°C increase in temperature. Since it is very difficult to measure the temperature sensitivity of each respiratory process individually, Q_{10} values for soil respiration are often derived from its seasonal temperature variation. Thus, the estimated Q_{10} values are the product of multiple processes in response to changes in temperature.

The estimated values of Q_{10} for soil respiration vary widely from little more than 1 (low sensitive) to more than 10 (high sensitive), depending on the geographic locations and ecosystem types (Peterjohn et al. 1993, 1994; Lloyd and Taylor 1994; Kirschbaum 1995; Simmons et al. 1996; Chen et al. 2000). Based on data compiled nearly 15 years ago, the global median value of Q_{10} is 2.4, with a range of 1.3 to 3.3 (Raich and Schlesinger 1992). Q_{10} values range from 2.0 to 6.3 for European and North American forest ecosystems (Davidson et al. 1998, Janssens et al. 2003). Reanalysis of data by Lloyd and Taylor (1994) suggested that variation in Q_{10} values reported in Raich and Schlesinger (1992) results largely from differences between studies in effective mass of carbon per unit area. The corrected respiration from different studies follows a similar temperature-respiration response function (Fig. 5.9).

High Q_{10} values result largely from the confounding effects of temperature on multiple processes with covarying variables such as light and moisture (Davidson et al. 1998, 2006). Seasonal variation in air temperature is highly coincident with the seasonal patterns of solar radiaton. The latter is the primary environmental driver of seasonal variation in substrate supply. The

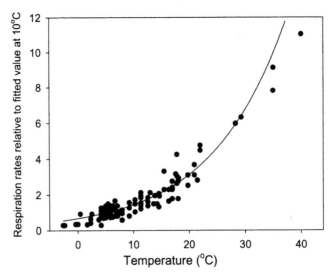

FIGURE 5.9   The relationship between soil respiration rate and temperature (Redrawn with permission from Functional Ecology: Lloyd and Taylor 1994).

estimated Q_{10} is confounded with the effects of radiation, and reaches its annual minimum in midsummer and the annual maximum in winter (Xu and Qi 2001b). Moreover, root growth also affects respiratory sensitivity to temperature. For example, Hanson *et al.* (2003) reported a Q_{10} value of 2.5 for soil respiration in an oak forest in Tennessee, when data points associated with root growth observed in minirhizotrons were excluded. The apparent Q_{10} would be inflated if data from springtime root-growing periods were included.

Temperature sensitivity of soil respiration is affected by moisture conditions. Dörr and Münnich (1987) found that the Q_{10} values range from 1.4 to 3.1, with the low values in the wet years and the high values in the dry years in a multiyear study of a grassland and a beech-spruce forest in Germany. However, other results showed that the Q_{10} values are lower in the well-drained sites than the wetter sites (Davidson *et al.* 2000, Xu and Qi 2001b, Reichstein *et al.* 2003). In addition, Silvola *et al.* (1996) found that the average Q_{10} value is 2.9 with water tables of 0 to 20 cm and 2.0 with water tables below 20 cm.

Temperature sensitivity of soil respiration varies among different components. Boone *et al.* (1998) showed that root and rhizosphere respiration in a mixed temperate forest is more sensitive to changes in temperature than the respiration of bulk soil (Table 5.1). Several studies corroborate this conclusion (Atkin *et al.* 2000, Pregitzer *et al.* 2000, Maier and Kress 2000, and Pregitzer 2003). Furthermore, Liski *et al.* (1999) suggested that temperature sensitivity

TABLE 5.1   The responses of Q_{10} values to litter manipulation at the Harvest Forest

Treatment	Q_{10}	R^2
Control	3.5 (0.4)	0.91
Double litter	3.4 (0.4)	0.90
No litter	3.1 (0.3)	0.91
No roots	2.5 (0.4)	0.73
No inputs	2.3 (0.2)	0.89
OA-less	2.6 (0.3)	0.82
"Roots"	4.6 (0.5)	0.95

Note: Control = normal litter input, no litter = aboveground litter excluded from plots annually; double litter = aboveground litter doubled annually; no roots = roots excluded from plots by fibergrass-lined trenches; no input = no aboveground litter and no roots; and OA-less = organic (O) horizons and upper mineral soil (A) horizon (to 20 cm depth) removed and replaced with subsoil (Modified with permission from Nature: Boone *et al.* 1998).

of decomposition is lower for old SOM than for litter based on soil carbon storage data along temperature gradients of high- and low-productivity forests. This conclusion that old SOM is less sensitive to temperature changes has been very controversial (Giadina and Ryan 2000, Fang *et al.* 2005, Knörr *et al.* 2005), largely due to the lack of long-term data from controlled experiments to isolate different components of the temperature sensitivity.

5.3. SOIL MOISTURE

Soil moisture is another important factor influencing soil respiration. The common conceptual relationship states that soil CO_2 efflux is low under dry conditions, reaches the maximal rate in intermediate soil moisture levels, and decreases at high soil moisture content when anaerobic conditions prevail to depress aerobic microbial activity (Fig. 5.10). The optimum water content is usually somewhere near field capacity, where the macropore spaces are mostly air-filled, thus facilitating O_2 diffusion, and the micropore spaces are mostly

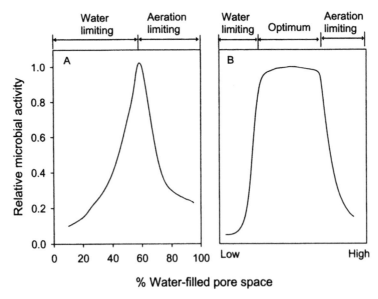

FIGURE 5.10 The idealized relationship between water-filled pore space and relative amount of microbial respiration. The idealized relationship in panel A assumes that there is one optimal soil moisture content based primarily on Papendick and Campbell (1981). The idealized relationship in panel B assumes that there is one plateau of optimal soil moisture content primarily based on Liu *et al.* (2002a) and Xu *et al.* (2004).

water-filled, thus facilitating diffusion of soluble substrates. The maximal rate of soil CO_2 efflux, for example, occurs at -15 kPa (50% of the water-holding capacity) in humid acrisols and a boreal mor layer (Ilstedt et al. 2000). In the high soil moisture conditions, effects of soil water on respiration are regulated primarily by oxygen concentration. Although laboratory studies suggest the maximal rate of soil respiration at optimal soil water content (Fig. 5.10a), many of the field observations suggest that soil moisture limits soil CO_2 efflux only at the lowest and highest levels (Bowden 1993, Bowden et al. 1998, Liu et al. 2002a, Xu et al. 2004). There may be a plateau of responses of soil respiration to a broad range of soil moisture, with steep decreases at either very low or very high soil moisture content (Fig. 5.10b).

Soil moisture influences soil respiration directly though physiological processes of roots and microorganisms, and indirectly via diffusion of substrates and O_2. Soil microorganisms as a community have a great flexibility to adapt a wide spectrum of soil water environments. Although some microorganisms lack the physiological mechanisms to adjust internal osmotic potential in response to water stress, many microorganisms possess osmoregulatory strategies for growth and survival under soil water stress (Harris 1981). The osmoregulatory microorganisms usually have cell wall-membrane complex and hence are capable of constitutive production of compatible solutes and/or induce additional compatible solutes. Thus, those organisms can withstand extreme downshock (plasmolytic) and upshock (plasmoptic) water stress and can sustain growth under low soil water conditions.

Effects of water stress on microbial growth vary with rates of biosynthesis, energy generation, and substrate uptake, as well as the nature and mode of water perturbation. Extreme dry conditions induce dormancy or spore formation in soil microorganisms (Griffin 1981, Harris 1981, Schjønning et al. 2003) and/or cell dehydration (Stark and Firestone 1995). Soil fungi are active at a water potential as low as -15 MPa through bridging air-filled pores by hyphae extension, whereas bacteria are inactive below $-1.0 \sim -1.5$ MPa (Swift et al. 1979). At low moisture content, bacteria maintain only a basic metabolism as in dormancy. Dormancy can result in substantial reductions in respiration per unit of biomass or reductions in total respiratory biomass.

In nonextreme dry or logging conditions, soil moisture regulates respiration primarily through substrate and O_2 diffusion (Linn and Doran 1984). The substrate supply is the main rate-limiting process for aerobic microbial activity in dry soil, whereas O_2 diffusion controls the activity in wet soil. The physical configuration of water in dry soil may influence the motility of microorganisms and diffusion of nutrients and exudates to sites of biological activity. The limitation to motility is particularly important for microorganisms that lack a hyphal system to bridge air spaces. The movement of microfauna and motile bacteria may be also limited if the water-filled pores or pore necks in the soil are too small to permit passage. In addition, the air-water interface itself can affect movement of the organisms. Water in soil pores at high water content affects exchanges of gaseous O_2 and CO_2 at sites of microbiological and root activities. The diffusion coefficients of O_2 and CO_2 are about 0.161 and 0.205 $cm^2 s^{-1}$ in air, respectively. In water, the diffusion rates of both gases decrease by 10,000 times that in air (Table 4.1). Therefore, the effective area for diffusive movement of either gas decreases in proportion to the pore space occupied by water. In a sandy soil, the decrease in gas diffusion coefficients is much less than in a clay soil at a given soil water potential.

Solutes move to and away from microorganisms by mass flow and diffusion. Mass flow is important in replenishing nutrients in the bulk soil solution during the water infiltration and redistribution. Diffusion is the main process that supplies substrates to microbes. The diffusion of soluble substrates to the surface of a soil microbial cell was given by Papendick and Campbell (1981):

$$J = \frac{(c_o - c_b)D_o k\theta^3}{s} \tag{5.5}$$

where J is flux, c_o is the solute concentration at a cell surface, c_b is the solute concentration in bulk soil, D_o is diffusivity, k is a constant, θ is the volumetric water content, and s is the diameter of a bacterial cell. Note that soil water content influences substrate diffusion in the third order of power. Papendick and Campell (1981) integrated the substrate diffusion processes with microbial metabolic rates and showed that nitrification rates of microorganisms drop rapidly at the soil water content of 10 to 20% when substrate concentrations are high. At low substrate concentrations, relative effects of soil waters on microbial processes become smaller.

Driven by stochastic events of rainfall, soil water content in the field is very dynamic and fluctuates over time. Right after rainfall, water infiltration recharges soil water content to a high level. In the subsequent period, water evaporation at the soil surface and transpiration from the foliage canopy gradually deplete soil water, causing a decline in soil water content. The stochastic events of rainfall and great fluctuation in soil moisture content usually result in strong variations in soil respiration in natural ecosystems, particularly in arid regions. When soil is dry before rainfall, soil respiration is usually very low. Rainfall, even with a very small amount of water added to dry soil surfaces, can result in bursts of CO_2 releases from the soil (Liu et al. 2002a, Xu et al. 2004). As soil water loses via evapotranspiration and soil becomes dry over time after rainfall, rates of soil CO_2 efflux decline (Fig. 5.11). Although the temporal pattern of soil CO_2 efflux is similar in response to different amounts of rainfall, the rate of CO_2 efflux varies greatly. High

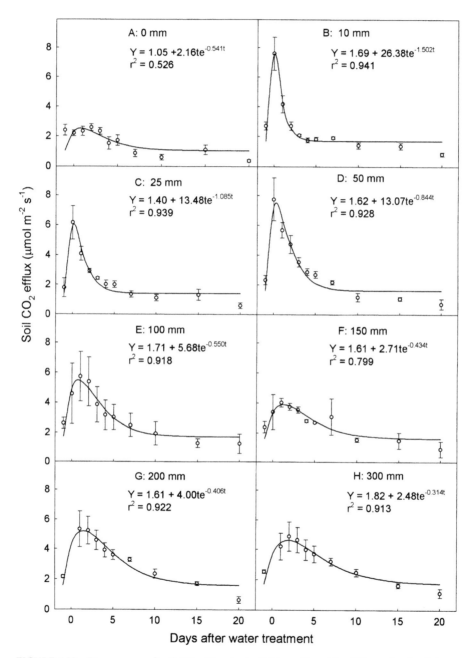

FIGURE 5.11 Time course of soil CO_2 efflux in the field as affected by different levels of the water treatment. While soil moisture content varies substantially during the experimental period, air temperature is relatively constant. Open circles are the measured data and shown as mean $\pm$ SE. Curves represent the equation $Y = Y_0 + ate^{-bt}$ to describe experimental data (Adapted with permission from Plant and Soil: Liu *et al.* 2002a).

rates of CO_2 efflux occur at low soil moisture contents, presumably resulting from degassing right after an amount of water added to the soil surface (Fig. 5.11b). When large amounts of water are added to soil, soil moisture contents are recharged to high levels, but rates of CO_2 efflux are not very high (Figs. 5.11g and h). The low rates of CO_2 efflux at the high soil water contents are probably attributable to inhibition of gaseous movement in water-saturated soil soon after precipitation. As a consequence, the relationship derived from data collected within one wetting-drying cycle with different amounts of water addition is widely scattered between soil CO_2 efflux and moisture (Liu et al. 2002a).

During a wetting-drying cycle, multiple mechanisms regulate soil CO_2 efflux. During the rainfall, water infiltration fills soil pores and replaces CO_2 highly concentrated air, resulting in degassing. Degassing is the fastest response to precipitation. It usually happens within minutes of precipitation and may last up to a few hours. In the strict sense, degassing is not soil respiration but rather releases the stored CO_2 in soil from past microbial and root respiration.

Several hours to a few days after rain falls onto dry soil, microbe activities are activated (Gliński and Stepniewski 1985), resulting in an increase of soil CO_2 efflux. Rewetting of extremely dry soil usually causes a strong increase in CO_2 emission, most likely because (1) a considerable proportion of soil microorganisms dies during drought (van Gestel et al. 1991), leading to quick decomposition of dead cells; (2) availability of organic substrates increases through desorption from the soil matrix (Seneviratne and Wild 1985); and (3) exposure of organic surfaces to microorganisms increases (Birch 1959). Fierer and Schimel (2003) used ^{14}C labeling to identify carbon sources of the pulse CO_2 release after rewetting. Their results suggest that the CO_2 pulse release is generated entirely by mineralization of microbial biomass carbon. Since they did not observe substantial microbial cell lysis on rewetting, microorganisms likely mineralize the large amount of intracellular compounds in response to the rapid increase in soil water potential. They also found that drying and rewetting release physically protected SOM, increasing the amount of extractable SOM-carbon by up to 200%.

Several days after addition of water to dry soil, specific root respiration and root growth increase. It takes seven days for desert plants to initiate new root growth after rewet (Huang and Nobel 1993). A couple of weeks after rainfall in arid lands, foliage becomes greener (Liu et al. 2002a) and more carbohydrates are supplied to roots and the rhizosphere. In wet regions or rainy seasons, rainfall events usually do not induce these mechanisms and do not trigger strong responses of soil respiration to a small amount of water addition.

Long-term effects of water availability on soil respiration are mediated largely by ecosystem production and soil formation. At the global scale, soil

respiration linearly increases with precipitation (Fig. 5.12). Along a hydrological gradient, soil respiration correlates strongly with the moisture content
of the litter in a Norway spruce stand in southern Sweden (Gårdenäs 2000).
The correlation reflects the effects of moisture content on both ecosystem
production to supply carbon substrate and decomposition to release CO_2.

The relationship between CO_2 efflux and soil water content is very complex,
involves numerous mechanisms, and varies with regions and time-scales. Our
understanding of the relationship and underlying mechanisms is extremely
limited. In practice, the relationship has been described by linear, quadratic,
parabolic, exponential, and hyperbolic equations (see Chapter 10, Table 10.2).
The soil water conditions have been expressed by matric potential, gravimet-
ric water content, volumetric water content, fractions of water-holding capacity, water-filled pore space, precipitation indices, and depth to water table. In
general, the relationship between soil respiration and moisture is scattered
(Liu *et al.* 2002a) and developed mostly from observations of seasonal variation (Luo *et al.* 1996, Mielnick and Dugas 2000) or along spatial gradients
(Davidson *et al.* 2000) in water content. Such a relationship is usually confounded by other environmental factors due to concomitant variations in soil
temperature and root and microbial activities over seasons or along the gradients. To understand how soil moisture affects soil CO_2 efflux, it is impera-
tive to conduct experiments that manipulate soil moisture alone, while other
factors, such as soil temperature and biological conditions, are controlled.

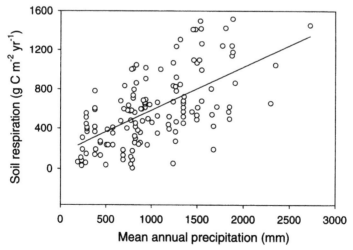

FIGURE 5.12 The relationship between soil respiration and mean annual precipitation
(Redrawn with permission from Tellus B: Raich and Schlesinger 1992).

5.4. SOIL OXYGEN

When soil water content exceeds optimal conditions, soil respiration is depressed due to limitation of oxygen (O_2). Soil O_2 environment becomes a main limiting factor of soil respiration in wetlands, flooding areas, and rainforests (Stolzy 1974, Gambrell and Patrick 1978, Crawford 1992). Silver *et al.* (1999) measured soil O_2 concentration in three subtropical wet forests in the Luquillo Mountains, Puerto Rico. The annual precipitation increases from 3500 mm in the low elevation forest to 5000 mm in high elevation forest. As a consequence, the O_2 concentration decreases from 21% in the low-elevation Tabonuco forest to 13% in the midelevation Colorado forest to 8% at the depths of 10 cm and 6% at 35 cm in the high-elevation Cloud forest (Fig. 5.13). Even in one forest, soil microsites experience low soil O_2 concentration (0 to 3%) for up to 25 consecutive weeks. Compaction and nontillage can result in poor aeration and anaerobic conditions, reducing root and microbial respiration (Linn and Doran 1984, Rice and Smith 1982).

Soil O_2 concentration greatly affects root and microbial respiration. When plants of *Senecio aquaticus* grow in anoxic conditions, root growth respiration is one-third of that in the aerated culture (Lambers and Steingrover 1978). The rate of root respiration is zero in the absence of O_2 and reaches its maximum value at about 5% O_2 for newly grown roots in response to rain

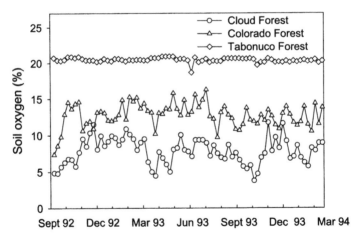

FIGURE 5.13   Mean soil O_2 concentration over time at 10 cm depth in the high-elevation Cloud forest, midelevation Colorado forest, and low-elevation Tabonuco forest in the Luquillo Experimental Forest, Puerto Rico (Redrawn with permission from Biogeochemistry: Silver *et al.* 1999).

and 16% for established roots of both *Ferocactus acanthodes* and *opuntia ficus-indica* (Nobel and Palta 1989).

Microorganisms are divided, according to their oxygen needs, into obligatory aerobes, facultative anaerobes, and obligatory anaerobes. For obligatory aerobes, a sharp decrease in respiratory CO_2 release occurs at O_2 concentrations below 0.01 to $0.02 \, m^3 \, m^{-3}$. Facultative anaerobes can use either oxygen or organic acids as electron receptors and thus can carry out respiration at low or null O_2 concentration. Respiration of obligatory anaerobes takes place only at an oxygen concentration close to zero (Fig. 5.14). In soils that normally contain all the groups of microbes, the relationship of respiration to O_2 concentration is similar to that of facultative anaerobes (Gliński and Stepniewski 1985).

5.5. NITROGEN

Nitrogen directly affects respiration in several ways. Respiration generates energy to support root nitrogen uptake and assimilation. Uptake of one unit

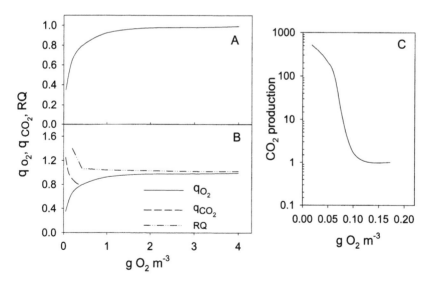

FIGURE 5.14 Idealized patterns of respiratory processes by microbes as a function of O_2 content at 20°C. A—for obligatory aerobes (q_{O_2} calculated from Michaelis-Menten equation assuming the highest K_m value given for bacteria by Longsmuir [1954]), B—for facultative anaerobes at the same Km value, C—for obligate anaerobes. In Panel B, q_{O_2} is for CO_2 production, q_{O_2} is for O_2 uptake, and RQ is for respiratory quotient (Redrawn with permission from CRC Press Inc.: Gliński and Stepniewski 1985).

of NO_3^- may cost at least 0.4 units of CO_2 (Bouma *et al.* 1996). Once NO_3^- is taken up by roots, it is reduced to NH_3 before the nitrogen can be assimilated into amino acids. Reduction of NO_3^- to NH_3 requires slightly more than 2 CO_2 per NO_3^- (Amthor 1994, 2000). Assimilation of NH_3 into amino acids bioenergetically does not cost much. Nitrogen fixation from N_2 to NH_3 is catalyzed by nitrogenase within symbionts. It costs at least 2.36 CO_2 per NH_3 (Pate and Layzell 1990). Nodule growth and maintenance have an additional cost for nitrogen fixation.

High nitrogen content in tissues is usually associated with high protein content (typically 90%), resulting in high maintenance respiration for protein repair and replacement (Penning de Vries 1975, Bouma *et al.* 1994). Maintenance respiration per unit nitrogen at 15°C ranges from 1.71 to 3.70 μmol $CO_2 s^{-1}$ for foliage (Ryan 1991, 1995; Ryan *et al.* 1996) and is about 2.6 μmol $CO_2 s^{-1}$ for fine roots (Ryan *et al.* 1996).

High nitrogen content is generally associated with high growth rates, leading to high growth respiration. Thus, respiration rates have been consistently observed to correlate with tissue nitrogen concentration in both the site comparison and ingrowth core experiments (Burton *et al.* 1996, 1998). For example, a site comparison in Michigan, where precipitation is similar but annual mean temperature varies, shows an nitrogen-respiration relationship (Burton *et al.* 1998) as:

$$R_{CO_2} = (0.058N + 0.622M)e^{0.098T} \qquad (5.6)$$

where R_{CO2} is the root respiration rate in μmol $CO_2 g^{-1}$ fine-root biomass s^{-1}, N is the root nitrogen concentration in $g kg^{-1}$, M is soil matric potential in Mpa, and T is the soil temperature at 15 cm. Thus, differences in nitrogen availability among sites and changes in nitrogen availability through nitrogen deposition (Aber *et al.* 1989) or global change (Pastor and Post 1988, Cohen and Pastor 1991) can alter root respiration rates.

Nitrogen affects litter decomposition and thus microbial respiration in a complex pattern (Magill and Aber 1998, Saiya-Cork *et al.* 2002). Litter decomposition is enhanced by high nitrogen availability—either through higher concentrations in litter or elevated mineral nitrogen concentrations in throughfall and soil solutions—in early stages and is repressed in later stages during which lignin is degraded (Fog 1988, Berg and Matzner 1997). The mechanisms underlying nitrogen effects on decomposition remain unclear (Sinsabaugh *et al.* 2002). Degradation of cellulose is an nitrogen-limited process and generally increases with nitrogen. The oxidative activities associated with recalcitrant litter or SOM are usually repressed by nitrogen, presumably because the microdecomposers of recalcitrant materials are generally adapted to low nitrogen conditions. High nitrogen availability might shift extracellular enzyme activity away from nitrogen limitation and toward phosphorus limitation (Sinsabaugh *et al.* 2002), randomize bond structures, reduce the efficiency of ligninolytic enzymes (Berg 1986), inhibit lignolytic activity in a number of fungi by NH_4^+ (Kaal *et al.* 1993), and suppress the production of the ligninolytic enzyme systems (Keyser *et al.* 1978, Tien and Meyer 1990) by white rot basidiomycetes (Carreiro *et al.* 2000). Saiya-Cork *et al.* (2002) found that nitrogen amendment decreases phenol oxidase activity by 40% in soil and increases it by 63% in litter. Condensation of nitrogen-rich compounds with phenolics can make SOM more recalcitrant, resulting in decreases in microbial respiration (Haider *et al.* 1975). Addition of NH_4^+ salts can also inhibit microbial activity (Gulledge *et al.* 1997).

Nitrogen also indirectly affects soil respiration through ecosystem production. Nitrogen additions stimulate plant primary production (Vitousek and Howarth, 1991), which supplies more substrate for soil respiration. In nitrogen-sufficient or -rich environments, nitrogen fertilization could exacerbate conditions of "nitrogen saturation," resulting in nitrogen leaching and runoff and causing little change in soil respiration.

5.6. SOIL TEXTURE

There are 12 soil texture types characterized on the basis of the percentages of sand, silt, and clay they contain (Eswaran 2003). Soil texture is related to porosity, which in turn determines soil water-holding capacity, water movement and gas diffusion in the soil, and ultimately its long-term fertility. Thus, soil texture influences soil respiration mainly through its effects on soil porosity, moisture, and fertility.

Soil moisture and respiration correlated significantly at sandy sites, but not at clayish sites in managed mixed pine forests in southeastern Georgia when soil water content was above the wilting point threshold (Dilustro *et al.* 2005). Soil respiration at the sandy sites is suppressed during the warm, dry periods, whereas finer soil texture at the clayish sites buffers soil moisture effects on soil respiration due to a slow release of soil moisture. In three different soil mixtures from a fine sandy soil in Lake Alfred, Florida, and a silt-clay loam in Centre County, Pennsylvania, respiration rates in the sandy soils after rewetting return to pre-watering levels nearly twice as fast as in the finer-textured soils, probably because lower soil water content in the sandy soils would allow CO_2 to diffuse more freely through air-filled pores (Bouma and Bryla 2000).

Soil texture also influences rooting systems and thus indirectly soil respiration. Generally, root growth is slower in soil of coarser texture (more sandy) than of finer texture (less sandy) due to lower fertility, lower unsaturated hydraulic conductivity, and lower water storage capacity. High root biomass and production result in high rates of root respiration and the associated microbial respiration in the rhizosphere (Högberg *et al.* 2002). In addition, root litter decomposition is sensitive to soil texture, with faster rates in the clay soil than in the sandy loam soil (Silver *et al.* 2005).

Total carbon and nitrogen pools correlated positively with clay content in the Great Plains of North America (Kaye *et al.* 2002) and correlated negatively with soil sand content from shortgrass steppes (Hook and Burke 2000). Labile constituents of organic matter are preferentially adsorbed onto fine clay particles and may be a significant source of energy for the soil microbes (Anderson and Coleman 1985). Thus, fine-textured soils tend to have higher labile carbon and nitrogen pools, and higher nitrogen mineralization, than coarse-textured soils.

Water infiltration and gas diffusion, which affect motility of microbial propagules and supply of air and moisture to microbial growth, vary greatly with soil texture and thus influence soil CO_2 production. In laboratory experiments, CO_2 production from the clayish soil is 20 to 40% less than from the silty loam at 10°C or 20°C and under soil 4- or 16-day wet-dry cycles (McInerney and Bolger 2000). CO_2 production is nearly 50% greater from clay loam soil than from sandy soil (Kowalenko *et al.* 1978). However, the proportion of total carbon respired and microbial biomass in total carbon is lower in soils with high silt and clay contents than in soils with low silt and clay content, in spite of the fact that soil texture has a strong relationship with total carbon (Fig. 5.15). In addition to microbial biomass, soil texture and SOC strongly regulate composition of microbial community, denitrification, and N mineralization rates along the Yenisei River in Siberia (Šantrůčková *et al.* 2003). The contents of fungi, actinomycetes, aerobic and nitrite-oxidizing bacteria, and cellulolitic microorganisms in the rhizosphere of sorghum plants are significantly different between pliocenic clay and alluvial sandy soils (Fig. 5.16).

5.7. SOIL PH

Soil pH regulates chemical reactions and a multiplicity of enzymes in microorganisms. A bacteria cell usually contains about 1000 enzymes; many of these are pH-dependent and associated with cell components, such as membranes. In the soil matrix, adsorption of enzymes to the soil humus shifts their pH optima to higher values. Most of the known bacterial species grow within the pH range of 4 to 9. The fungi are moderately acidophilic, with a pH range of 4 to 6. Thus, soil pH has a marked effect on the growth and proliferation of soil microbes as well as soil respiration. Plants can acidify their rhizosphere soil by as much as two pH units due to release of organic acids

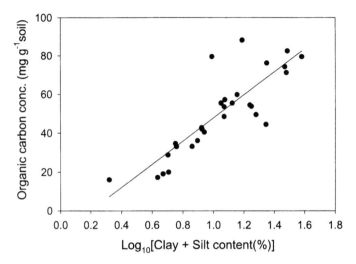

FIGURE 5.15   The relationship between concentration of total carbon in surface (0 to 10 cm) soil and logarithm of clay + silt content (%) (Redrawn with permission from Australian Journal of Soil Research: Mendham *et al.* 2002).

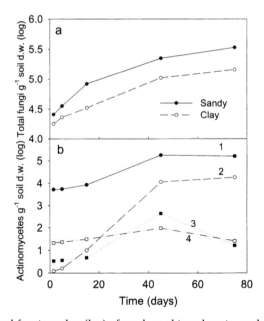

FIGURE 5.16   Total fungi number (log) of sorghum rhizosphere in sandy and clay soils (a) and actinomycetes number (log) in sandy (1) and clay (2) soils and nitrite-oxidizing bacteria number (log) in sandy (3) and clay (4) soils (Redrawn with permission from Plant and Soil: Pera *et al.* 1983).

in exudates and higher root uptake of cations than anions, leading to root excretion of H^+ ions (Glinski and Lipiec 1990).

Soils with pH 3.0 produce 2 to 12 times less CO_2 than the soils at pH 4.0 (Sitaula *et al.* 1995), due to the adverse effect of low pH on soil microbial activity. Production of CO_2 usually increases with pH when pH is less than 7 and decreases with pH at soil pH beyond 7 (Kowalenko and Ivarson 1978). Emission of CO_2 decreases by 18% at pH 8.7 and 83% at pH 10.0 compared with that at pH 7.0 (Rao and Pathak 1996). Xu and Qi (2001a) found that pH values in the top 10 cm correlated negatively with soil CO_2 efflux, accounting for 34% of variation in soil CO_2 efflux.

5.8. INTERACTIONS OF MULTIPLE FACTORS

Soil respiration is often interactively affected by multiple factors, although it is often difficult to separate their interactions. Soil respiration, like many other physiological processes of plants and microbes, usually responds to the most limiting factor. Soil respiration is not sensitive to moisture under low temperatures (below 5°C) but more responsive at high temperatures (10 to 20°C). Similarly, soil respiration is not sensitive to temperature under low moisture (below 7.5% volumetrically) but is more responsive to temperature under high moisture content (10 to 25%) (Carlyle and Bathan 1988). Similarly, soil respiration in a tallgrass prairie is more sensitive to temperature changes in relatively wet than dry soils (Harper *et al.* 2005). When both temperature and moisture are not at their extremes, the two factors interactively influence soil respiration and together can account for most of its variability observed in the field.

Other factors may interact with temperature and moisture to influence soil respiration. For example, Vanhala (2002) evaluated the effects of temperature, moisture, and pH on seasonal variations of soil respiration in coniferous forest soils. Soil respiration is regulated by moisture and pH when the soil respiration rate is measured at a constant temperature (14°C). When moisture content is kept constant at 60% of a water-holding capacity, soil respiration is controlled mainly by the amount of organic matter and pH. The respiration rate per unit of nitrogen concentration varies mainly with pH values.

Substrate supply also interacts with other factors to regulate soil respiration. Newly synthesized carbohydrate by canopy photosynthesis is mostly partitioned into labile pools before a small fraction of it is converted to recalcitrant carbon in soil. The temperature sensitivity of soil respiration varies with pools from which respired carbon comes, although which pool is more sensitive to changes in temperature is a matter of controversy (Boone *et al.* 1998, Giardina and Ryan 2000, Knörr *et al.* 2005). Assuming that labile carbon is more sensitive to temperature changes than is recalcitrant carbon, soil respiration would be more influenced by temperature when substrate supply from labile pools is ample. Thus, soil respiration varies more with temperature during active growing seasons than in dormant seasons or under elevated than ambient CO_2.

CHAPTER 6

Temporal and Spatial Variations in Soil Respiration

It has been well documented that soil respiration greatly varies with time and space. The spatial and temporal variations in soil respiration result from variations of environmental variables (see Chapter 5), biochemical processes of respiration (see Chapter 3), and transport processes of CO_2 gas (see Chapter 4). A high degree of spatial and temporal variability in soil respiration not only causes measurement errors (Parkin and Kaspar 2004) but also makes it very difficult to extrapolate point measurements to estimate regional and global carbon budgets (Law *et al.* 2001, Tang and Baldocchi 2005). This chapter aims to synthesize results reported in the literature in an attempt to search for temporal and spatial patterns of soil respiration.

6.1. TEMPORAL VARIATION

As shown in Figure 1.2, soil respiration rates display strong temporal variation over time. In general, the temporal variability can be characterized
on four time-scales: diurnal/weekly, seasonal, interannual, and decadal/
centennial.

DIURNAL AND WEEKLY VARIATION

Over one day, soil CO_2 efflux usually increases in the morning with an
increase of soil temperature, reaches a peak at noon to midafternoon as the
soil temperature keeps increasing, and then declines in the afternoon and
throughout the night as the temperature decreases (Fig. 6.1, Makarov 1958,
Bijracharya *et al.* 2000, Xu and Qi 2001a). In most situations the diurnal variation in soil respiration can be explained as a close function of soil temperature, because this is the variable that changes strongly on the diurnal scale

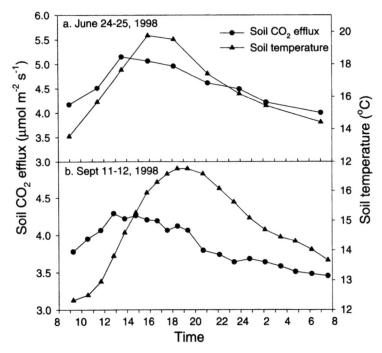

FIGURE 6.1   Diurnal trend of soil CO_2 efflux and soil temperature (10 cm) in mid- and post-
growing season of 1998 (Redrawn with permission from Global Change Biology: Xu and Qi
2001).

(Rayment 2000). Nevertheless, soil respiration is also correlated with photosynthesis with a time delay by 7 to 12 hours (Tang et al. 2005a, Tang and Baldocchi 2005). Thus, substrate supply can be another important factor that regulates diurnal variation of soil respiration. In addition, abrupt increases in soil CO_2 efflux can occur in response to rainfall events on a diurnal scale, especially after a long drought (Rochette et al. 1991, Jensen et al. 1996, Curtin et al. 2000). Fluctuation in atmospheric pressure and humidity may also affect the diurnal patterns of CO_2 emission from soils (Baldocchi et al. 2001).

Diurnal variation may not be apparent for soil respiration in heavily shaded areas in forests because of the lack of variation in soil temperature (Davidson et al. 2000; Jensen et al. 1996). Rates of soil respiration at night may be even higher than during the daytime in arid ecosystems, due to increased relative humidity at night (Medina and Zelwer 1972). High humidity favors activities of microorganisms.

The diurnal variation can be a source of errors if it is not accounted for appropriately when point measurements of soil respiration are used to estimate annual soil carbon efflux. In general, the midmorning effluxes closely approximate the 24-hour mean efflux (Larionova et al. 1989, Davidson et al. 1998). For example, Xu and Qi (2001a) found that the measurements taken between 0900 and 1100, which have a sampling error of 0.9 to 1.5%, better represent the daily mean soil respiration than do the entire daytime measurements, which tend to overestimate the daily mean rates by 4 to 6%. If measurements made at the warmest part of the day are used to estimate daily means, estimated daily or monthly rates of soil respiration can be substantially biased.

On a weekly time-scale, fluctuations in soil CO_2 efflux may be induced from synoptic weather changes associated with the passage of high and low pressure systems and fronts (Fig. 6.2, Subke et al. 2003). Synoptic weather events cause distinct periods of clear sky, overcast, and partly cloudy conditions, all of which alter the amount of available light to an ecosystem and cause changes in air temperature, humidity, and atmospheric pressure. The multidimensional changes in climatic variables associated with synoptic weather events can directly and interactively influence photosynthesis and respiration (Gu et al. 1999). Changes in photosynthetic assimilation in turn affect root and soil respiration with a time delay on a weekly scale. The temporal variation of the $\delta^{13}C$ signals of soil respiration is to a large extent accounted for by variations in weather conditions two to six days before sampling (Ekblad et al. 2005). The rates of root respiration depend largely on the availability of recently produced photosynthates during the previous 7 to 12 hours (Tang et al. 2005a), 1 to 6 days (Ekblad and Högberg 2001, Bhumpinderpal-Singh et al. 2003, Ekblad et al. 2005), or 5 to 10 days (Bowling et al. 2002).

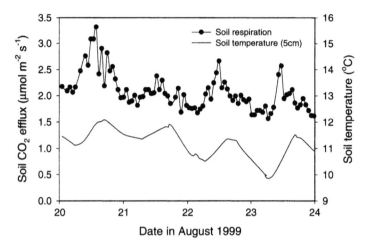

FIGURE 6.2 Soil CO_2 efflux and soil temperature at the depth of 5 cm measured over 4 days in August 1999 (Redrawn with permission from Soil Biology and Biochemistry: Subke *et al.* 2003).

SEASONAL VARIATION

Seasonal variation in soil CO_2 efflux has been observed in almost all ecosystems. Soil respiration rates are usually highest during summer and lowest in winter. The seasonal variation is driven largely by changes in temperature, moisture, photosynthate production, and/or their combinations. The main controlling factors in seasonal variation of soil respiration may depend on the type of ecosystems and climate. In a U.S. southern Great Plains grassland, for example, neither temperature nor moisture is limiting in spring, resulting in fast plant growth and high soil respiration (Fig. 1.2). In summer, moisture becomes limiting, whereas in winter the limiting factor is temperature. As a result, soil respiration declines in summer and is low in winter. In mesic ecosystems, such as tropical rainforests, temperate forests, and grasslands, soil respiration generally follows seasonal trends in soil temperature and/or radiation (Anderson 1973, Buyanovsky *et al.* 1985, Hanson *et al.* 1993, Billing *et al.* 1998, Epron *et al.* 2001, Borken *et al.* 2002).

In arid and semiarid ecosystems, soil moisture is the main factor limiting soil respiration. Thus, seasonal patterns of soil respiration closely follow dynamics of soil moisture (Fig. 6.3, Davidson *et al.* 2000). In the Amazon basin, where the seasonal variation in temperature is not large, while variation in soil water content is substantial, soil respiration in pastures and forests correlates significantly with water–filled pore space in soil (Salimon *et al.* 2004). In Mediterranean climate regimes with cold, wet winters and hot, dry summers, water usually constrains biological activity in summer. Seasonal

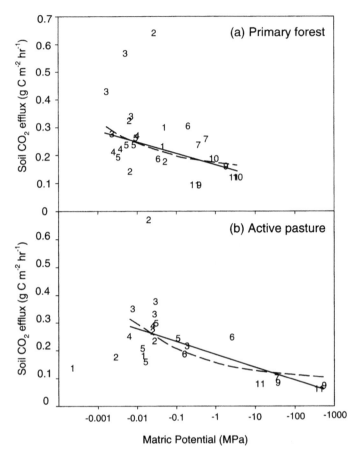

FIGURE 6.3 Correlation between seasonal variation in water content shown by the logarithm of soil matric potential and soil CO_2 flux in primary forest (a) and active pasture (b) in Brazil. Plotting symbols indicate the month of the year (Jan = 1 to Dec = 12) (Redrawn with permission from Biogeochemistry: Davidson *et al.* 2000).

patterns of soil respiration are largely determined by soil water availability. Soil respiration rates correlate positively with soil water content and negatively with soil temperature in sandstone and serpentine grasslands (Luo *et al.* 1996) and a young ponderosa pine plantation in northern California (Xu and Qi 2001a).

On a global scale, soil CO_2 efflux reaches the maximum during the summer season when plant growth in most active in both temperate zones and near-equatorial regions (Raich and Potter 1995, Raich *et al.* 2002). In general, the factors favoring plant growth usually favor soil metabolic activity. Plants also allocate considerable substrate to roots and microbes during active growing seasons, stimulating soil respiration.

Seasonality in soil respiration is also regulated by vegetation types (Grogan and Chapin 1999). Evergreen and deciduous species show distinct seasonal patterns in productivity, primarily due to differences in leaf longevity (Schulze 1982). As a consequence, plant phonology has an important influence on soil respiration, mainly through different timing of root growth, root turnover, and litterfall (Curiel Yuste *et al.* 2004). The amplitude of the seasonal changes in soil respiration correlates positively with the seasonal changes leaf area index, a measure of the deciduousness of the vegetation. Furthermore, seasonal increases in the CO_2 effluxes are closely related to the increase in root production and biomass (Thomas *et al.* 2000). The soil surface CO_2 effluxes increase approximately linearly with stem production, which continues throughout the year, with the lowest rates of increase over the winter in young *Pinus radiata* trees in Christchurch, New Zealand.

INTERANNUAL VARIABILITY

The significant year-to-year variability in soil respiration has been observed in a variety of ecosystems: grasslands (Fig. 1.2, Frank *et al.* 2002), a beech forest (Epron *et al.* 2004), mixed temperate forests (Fig. 6.4, Savage and

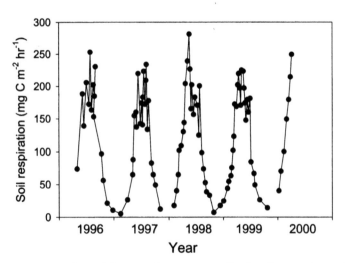

FIGURE 6.4 Interannual variability in soil respiration in Howland forest in Maine (Modified with permission from Global Boigeochemical Cycles: Savage and Davidson 2001).

Davidson 2001), ponderosa pine forests (Irvine and Law 2002), and forest plantations (Fig. 7.1, King *et al.* 2004). The interannual variability in soil respiration appears to be a ubiquitous phenomenon and results from (1) year-to-year changes in climatic variables (e.g., temperature, summer drought, winter snow depth, and the time of snowmelt) (Griffis *et al.* 2000, Scott-Denton *et al.* 2003, Epron *et al.* 2004); (2) changes in physiological and ecological processes (e.g., growing season length, stand structure, and timing of leaf emergence) in response to climatic variability and disturbance regime (Weber *et al.* 1990, Goulden *et al.* 1996, Hui *et al.* 2003); and (3) changes in nutrient availability (King *et al.* 2004). In most studies, interannual variability in soil respiration is attributed to climatic variations. Soil temperature and/or soil water content are commonly used to describe the interannual difference in soil respiration. Indeed, spring and summer climate conditions explain a great portion of interannual variations in soil respiration. On a global scale, annual soil CO_2 effluxes correlate with mean annual temperature with a slope of 3.3 Pg $C yr^{-1} {}^{\circ}C^{-1}$ (Raich *et al.* 2002). However, within seasonally dry biomes (savannas, shrublands, and deserts), interannual variability in soil CO_2 effluxes correlates significantly with interannual differences in precipitation.

Physiological changes in plants in response to interannual climatic variability and disturbance regimes also influence interannual differences in soil respiration (Hui *et al.* 2003). Braswell *et al.* (1997) showed that climate-induced physiological changes are greater than the direct effect of climatic variability on net ecosystem exchange (NEE). A study of soil respiration for five years in Harvard Forest and four years in Howard Forest in New England showed that the major sources of interannual variation in soil respiration are related to the occurrences of spring and summer droughts and the onset of springtime increases in respiration (Savage and Davidson 2001). Variations in the onset of spring from year to year contribute to 33 to 59% of the interannual variability in soil respiration. In addition, interannual variations in soil respiration in the Harvard Forest are as high as 0.23 kg C m^{-2} yr^{-1}, exceeding the interannual variation of 0.14 kg C m^{-2} yr^{-1} in NEE. Thus, interannual variation in soil respiration can be a major cause of the interannual variability of NEE. Developing forests are likely more responsive to variations in weather and resource availability (King *et al.* 1999), resulting in higher variability in soil respiration than is found in old forests (King *et al.* 2004).

DECADAL AND CENTENNIAL VARIATION

Successional changes explain much of the variation in soil respiration over time-scales of decades to centuries (Chapin *et al.* 2002). Soil respiration rates at the start of primary succession are near zero, because there is little or no SOM. As ecosystems develop, soil respiration increases slowly. In midsuccession, soil respiration increases substantially in response to increases in plant productivity and litter production. In late succession, soil respiration levels off when the ecosystem reaches a steady state (Fig. 6.5a). For example, the mean soil respiration rates in July and August 1995 increased with primary succession to 6.2, 44, and 63 mg CO_2 m^{-2} h^{-1} respectively among three sites with ages from 30 to 2000 years in a high Arctic glacier foreland in Ny-Ålesund, Svalbard (Bekku *et al.* 2004a). The microbial respiration rates measured in the laboratory also follow this trend (Bekku *et al.* 2004b).

During secondary succession, soil respiration rises sharply in early successional stages because disturbances that trigger secondary successional processes, such as forest clear-cutting, usually transfer large amounts of labile carbon to soils and create an environment that is favorable to decomposition. The burst of soil respiration generally lasts for one or a few years before it subsides to a lower level. In midsuccession, soil respiration is relatively low, because regenerating vegetation reduces soil temperature by shading the soil surface and may have moderate rates of primary production. Soil respiration increases again in late succession due to increased root respiration and litter production with high primary productivity (Fig. 6.5b). This general trend has been observed along a successional gradient of four ages (0 to 60 years old) in a jack pine forest, except that soil respiration decreases in the first year after clear-cut (Fig. 7.9, Striegl and Wickland 2001).

However, trends of soil respiration over successional sequences may not display clear patterns, due to diverse soil and environmental conditions in different-aged stands. The highest rates of soil respiration, for example, occur in forests 12 years old, followed by forests of 40, 4, and 75 years old across a chronosequence of four different-aged Scots pine forests in southern Finland

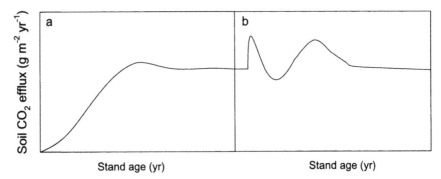

FIGURE 6.5 Idealized patterns of changes in soil CO_2 emissions in primary (a) and secondary (b) forest succession.

(Kolari *et al.* 2004). Heterotrophic soil respiration (R_h) changes slightly in boreal forests with increases of age classes (0 to 10, 11 to 30, 31 to 70, 71 to 120, and >120 years old) (Pregitzer and Euskirchen 2004). In temperate forests, soil respiration rates decline from $970\,g\,C\,m^{-1}\,yr^{-1}$ in the youngest age class (0 to 10 years) to $280\,g\,C\,m^{-1}\,yr^{-1}$ in the oldest forests (>120 years). Gulledge and Schimel (2000) observed an inverse trend of soil CO_2 efflux with the successional age of the sites, with the greatest fluxes in the early successional alder stand ($464\,g\,C\,m^{-2}\,yr^{-1}$), intermediate in the midsuccessional birch/aspen stand ($279\,g\,C\,m^{-2}\,yr^{-1}$), and lowest in the late successional white spruce stands (212 and $177\,g\,C\,m^{-2}\,yr^{-1}$). Over the successional series, the temperature sensitivity index, Q_{10}, of soil respiration under the condition of no moisture limitation was lowest for alder (1.9), moderate for birch/aspen (2.8), and highest for the white spruce site (3.4 to 12). Soil CO_2 efflux shows only a weak trend with increasing stand ages from 15 to 54 years, with the highest rates observed in the cultivated meadow (Thuille *et al.* 2000). To avoid confounding effects, soil and environmental conditions must be carefully considered when a chronosequence is selected to study successional changes in soil respiration.

6.2. SPATIAL PATTERNS

Spatial variability in soil respiration occurs on various scales, from a few square centimeters to several hectares (ha) up to the globe (Rochette *et al.* 1999, Rayment 2000). While variability in square centimeters can be dealt with by an appropriate chamber design, variability on the scales of square meters or larger must be tackled with appropriately designed sampling strategies (i.e., replicates, area covered, and locations of collars or chambers, etc.) and by using suitable upscaling techniques. The spatial variability of soil CO_2 efflux has to be understood to derive a representative estimate of regional carbon budget. To characterize the spatial variability in soil respiration, we have to recognize its patterns at various scales and identify underlying causes. Toward that goal, this section discusses the spatial variability in soil respiration on four spatial scales: stands, landscapes, regions, and biomes.

STAND LEVEL

A large spatial variability in soil CO_2 efflux occurs at a stand level, even in relatively homogeneous soils such as agricultural fields or mesocosms with homogenized soils. In a mesocosm experiment, soil respiration rates ranged from 4 to $25\,\mu mol\,m^2 s^{-1}$ from 150 measurements on an area of $3.6\,m^2$ over two days (Griffin et al. 1996a). A similar variability occurred in a box-lysimeter experiment with homogenized soil and no plants (Nay and Bormann 2000). Due to the large heterogeneity in the natural soil, spatial differences in soil respiration have been observed in various ecosystems with high coefficients of variation (CV), including the following: grasslands (CV = 35%, Pol-van Dasselaar et al. 1998); temperate forests (CV = 10 to 100%, Hanson et al. 1993, Jensen et al. 1996, Law et al. 1999); rainforests (CV = 15 to 70%, Schwendenmann et al. 2003); pine plantations (CV = 21 to 55%; Fang et al. 1998, Xu and Qi 2001a); agricultural fields (CV = 150%, Cambardella et al. 1994); and homogeneous patches of Scots pine (CV = 30 to 65%, Janssens and Ceulemans 1998). To represent the spatial variability of soil respiration over a whole stand, sound sampling strategies, such as random sampling and stratified sampling with adequate replicates, should be employed (Rayment 2000).

The high spatial variability in soil respiration results from large variations in soil physical properties (e.g., soil water content, thermal conditions, porosity, texture, and chemistry), biological conditions (e.g., fine-root biomass, tunneling soil animals, fungi, and bacteria), nutrient availability (e.g., deposit litter and nitrogen mineralization), and others (e.g., disturbed history and weathering). In a young ponderosa pine plantation in northern California, for example, most of the spatial variation (84%) in soil CO_2 efflux can be explained by fine-root biomass, microbial biomass, and soil physical and chemical properties (i.e., soil temperature and moisture, soil nitrogen and organic matter, magnesium, bulk density, and pH) (Xu and Qi 2001a). Gärdenäs (2000) studied the degree to which spatial variation in soil moisture affects soil respiration rates for three weeks along a hydrological gradient in a Norway spruce stand in Skogaby, Sweden. Variation in the moisture content of the litter layer accounts for most of the spatial variation in soil respiration. Phosphorus concentration partially accounts for spatial differences in soil respiration in an old-growth neotropical rainforest in La Selva, Costa Rica (Schwendenmann et al. 2003).

Spatial variability in soil respiration exhibits some patterns along changes in environmental and biological factors. Spatial variation in soil respiration in a black spruce (Picea mariana) forest ecosystem, for example, is well correlated with the thickness of the dead moss layer (Rayment and Jarvis 2000). Observed rates of soil respiration decrease along the distance from the trunk, especially in sparse forests or savanna ecosystems (Scott-Denton et al. 2003, Wieser 2004). This spatial variation between trees is attributable to parallel gradients in litter mass and fine-root density, given that soil carbon content does not change much along the gradient (Table 6.1).

TABLE 6.1 Changes of litter mass, fine root, soil carbon, baseline soil respiration at a temperature of 10°C (R_{10}), temperature sensitivity (Q_{10}) with the distance to trunk in a 95-year-old cembran pine stand of Innsbruck, Austria

Location	Distance to Trunk (m)	Litter (g m⁻¹)	Fine Root (mg cm⁻³)	Soil Carbon (mg g⁻¹)	R_{10} (μmol m⁻² s⁻¹)	Q_{10}
Close to stem	0.5	505 ± 371	8.5 ± 5.6	482 ± 14	0.347	4.26
Distance to stem	1.5	209 ± 23	4.2 ± 2.1	452 ± 47	0.159	4.10
Open gap	4.0	163 ± 22	3.0 ± 1.9	473 ± 21	0.074	3.67

Note: Equation, $R_s = R_{10} e^{Q_{10} \frac{(T-10)}{10}}$, is used to derive R_{10} and Q_{10} (Modified with permission from Tree Physidogy: Wieser 2004).

LANDSCAPE LEVEL

Because landscapes are spatially heterogeneous areas with elements of patches, corridors, and matrices on scales ranging from hectares to hundreds of square kilometers (Turner 1989), large variability naturally occurs for soil respiration on the scale. However, the spatial variability in soil respiration has been much less studied on the landscape scale than on the ecosystem and regional scales. The limited information is used here to identify factors controlling soil respiration on this scale.

The spatial variability in soil respiration on the landscape scale is caused largely by variations in climate, topography, soil characteristics, vegetation types, areas and edges of patches, and disturbance history. Various patches have different controlling factors on soil respiration, leading to diverse spatial patterns between patches. For example, soil respiration varies greatly among six dominant patch types (mature northern hardwoods, young northern hardwoods, clear-cuts, open-canopy jack pine barrens, mature jack pine, and mature red pine) within a managed northern Wisconsin landscape (Euskirchen et al. 2003). Litter depth is a better predictor of mean soil respiration among the patch types than soil temperature and moisture (Fig. 6.6a). Litter decomposition rates also differ substantially among patches, largely due to variations in canopy cover, litter composition, and litter quality (Saunders et al. 2002). Along a gradient from a river through buffer zones to crop fields within a riparian landscape in central Iowa, annual soil respiration rates correlate strongly with SOC content and fine-root biomass (Tufekcioglu et al. 2001). Soil respiration rates correlate positively with soil microbial biomass and also relate to soil physiochemical characteristics such as soil carbon content and water-holding capacity across nine landscape regions in the Serengeti National Park, Tanzania (Fig. 6.6b, Ruess and Seagle 1994). Soil respiration and

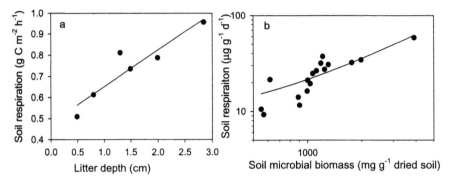

FIGURE 6.6   Soil respiration rate as a function of litter depth (a) and soil microbial biomass carbon (b) on a landscape scale (Redrawn with permission from Ecosystems: Euskirchen *et al.* 2003 and Ecology: Ruess and Seagle 1994 respectively).

decomposition rates both increase with mean annual precipitation across the Great Plains of North America (McCulley *et al.* 2005). Overall, substrate availability has been identified by several studies (Janssens *et al.* 2001, Reichstein *et al.* 2003, Campbell *et al.* 2004) as the main factor in controlling soil respiration at landscape levels.

Both disturbance regimes (e.g., land use changes) and climatic change over time affect soil respiration at the landscape level. From 1972 to 2001 in a managed forest landscape of northern Wisconsin, for example, the mature forest covers declined by about 12%, while the nonforested and young, regrowth forest covers increased by 22 to 34%. Changes in land use composition during this period result in increases of 2.8 to 3.1% in landscape-level soil respiration, while a 2°C warming in the growing season's mean air temperature increases the soil respiration rates by 6.7 to 7%. Their combined effects on the soil respiration rates vary from 3.8 to 10% (Zheng *et al.* 2005). Landscape mean soil respiration is more sensitive to an increase in minimum temperature than an increase in mean or maximum temperature across this landscape.

REGIONAL SCALE

Regional- and continental-scale carbon effluxes from soils are the product of diverse ecosystems in response to interactive effects of climatic and edaphic conditions, biotic factors (e.g., canopy height, LAI, and productivity of different biomes), landscape patterns, natural disturbances, and land use management. Thus, the large spatial variability in soil respiration is considered inevitable on the regional scale. The regional patterns of soil respiration have been examined by synthesis of data from eddy-covariance flux networks (Valentini *et al.* 2000, Janssens *et al.* 2001) and by transect studies in grasslands (Murphy *et al.* 2002, McCulley *et al.* 2005), hardwood forests (Simmons *et al.* 1996), and Arctic tundra (McFadden *et al.* 2003). Generally, warmer and wetter regions exhibit greater rates of soil respiration and decomposition of organic matter than colder and drier regions do when other variables do not significantly vary over the regions. Among climatic factors, precipitation is often important to predict the regional variability in soil respiration.

In the U.S. Great Plains from eastern Colorado to eastern Kansas, for example, mean annual precipitation accounts for most of the regional variability in soil respiration (56%) and litter decomposition (89%) (McCulley *et al.* 2005). Both soil respiration and litter decomposition increase from semiarid shortgrass steppes to subhumid tallgrass prairies (Fig. 6.7). Other factors (e.g., soil temperature, landscape setting, and soil texture) also contribute to regional variations in soil respiration. Similarly, precipitation contributes more than either temperature or soil texture to spatial patterns of litter decomposition rates across the Great Plains (Epstein *et al.* 2002). It alone

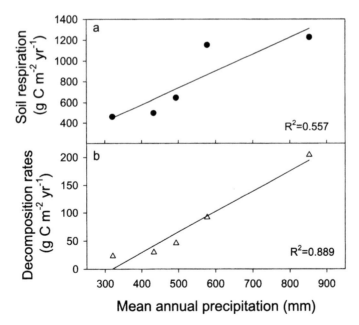

FIGURE 6.7   Regional regressions of soil respiration (a) and decomposition rates (b) versus the mean annual precipitation (Redrawn with permission from Ecosystems: McCulley *et al.* 2005).

explains more than 30% of the spatial variability in litter decomposition. In northern hardwood ecosystems along a regional climate gradient from northern to southern and coastal zones in Maine, leaf litter mass and CO_2 effluxes from leaf litter decomposition both positively correlate with mean annual precipitation (Simmons *et al.* 1996). However, soil respiration positively correlates with temperature, with the regression slopes increasing with latitude, indicating increased temperature sensitivity of soil respiration from warm to cold environments.

In some studies, regional variability of soil respiration cannot be explained by climatic variables but is modulated by gradients in biological activity and edaphic conditions. Basal soil respiration, a measure of overall soil microbial activity (Gray 1990), displays an increasing trend from south to north along a transect in the northeastern German lowland (Wirth 2001). Soil microbial biomass and edaphic conditions—for example, total nitrogen, organic carbon, cation exchange capacity (CEC), and pH—largely explain the spatial variability. Furthermore, soil moisture and vegetation type are more important in controlling soil CO_2 efflux than fire regime (i.e., disturbance) in savanna areas of central Brazil (Pinto *et al.* 2002).

As cross-site comparisons become available in the regional and global eddy flux networks, there is a growing appreciation of spatial variability in soil respiration. Forest productivity, for example, has been found to be much more important than temperature in regulating soil respiration across 18 European forests (Janssens *et al.* 2001). Similarly, soil CO_2 efflux correlates strongly with aboveground, belowground, and microbial biomass in lodgepole pine forests of Yellowstone Nation Park in Wyoming (Litton *et al.* 2003). Therefore, measures of vegetation productivity have to be incorporated into models for predicting large-scale patterns of soil respiration (Reichstein *et al.* 2003).

BIOMES: FORESTS, GRASSLANDS, TUNDRA, SAVANNAS/ WOODLANDS, DESERTS, CROP FIELDS, AND WETLANDS

Soil respiration varies greatly with different ecosystem types, reflecting intrinsic characteristics of those ecosystems in prevailing environments and biological activities. Mean rates of annual soil respiration differ twentyfold among major vegetation biomes (Table 6.2). Soil respiration is lowest in the cold tundra and northern bogs and highest in tropical moist forests, where both temperature and moisture availability are high year-round (Raich and Potter 1995). On a global scale, mean rates of annual soil respiration correlate positively with mean plant productivity among different biomes (Fig. 5.6). Primary production supplies organic substrate that drives root and microbial activities. A recent synthesis of 31 AmeriFlux and CarboEurope sites in

TABLE 6.2   Mean rates of soil respiration ($gCm^{-2}yr^{-1}$, mean ± SE) in different vegetation types

Vegetation type	Soil Respiration Rate	n	Significance
Tundra	60 ± 6	11	e
Boreal forests and woodlands	322 ± 31	16	cde
Temperate grasslands	442 ± 78	9	bcd
Temperate coniferous forests	681 ± 95	23	b
Temperate deciduous forests*	647 ± 51	29	b
Mediterranean woodlands and heath	713 ± 88	13	b
Croplands, field, etc.	544 ± 80	26	bc
Desert scrub	224 ± 38	3	de
Tropical savannas and grasslands	629 ± 53	9	bc
Tropical dry forests	673 ± 134	4	b
Tropical moist forests	1260 ± 57	10	a
Northern bogs and mires	94 ± 16	12	e
Marshes	413 ± 76	6	bcd

*Including mixed broadleaf and needleleaf forests (Raich and Schlesinger 1992).

temperate ecosystems in the northern hemisphere by Hibbard *et al.* (2005) demonstrated that soil respiration averaged over the growing season is lowest in grasslands and woodland/savanna, intermediate in deciduous broadleaf forests, and highest in evergreen needleleaf forests (Fig. 6.8).

The global and regional syntheses compile results from different sites with many confounding factors of climate, soil, and biology in influencing soil respiration. To isolate the effects of vegetation type alone on soil respiration, Raich and Tufekcioglu (2000) conducted a pairwise comparison of soil respiration by selecting published data measured by the same authors with the same methods from the same soil parent material and in similar topographic positions. Under comparable conditions, soil respiration rates are consistently approximately 20% greater in grasslands than in forests. Grasslands usually allocate more photosynthates to belowground than do forests. Forests allocate more carbon to wood production. Among forests, soil respiration rates in coniferous forests are 10% lower on average than those in broadleaf forests located on the same soil types. The two forest biomes have different carbon allocation patterns, litter production rates, litter quality, and relative contributions of root respiration to soil respiration (Weber 1985, 1990). Crop fields have rates of soil respiration approximately 20% higher than those of the adjacent fallow fields. Grasslands have soil respiration rates about 25% higher than those of the adjacent crop fields. A similar trend occurs in three adjacent crop fields, forests, and grasslands, with cumulative CO_2 production from soil

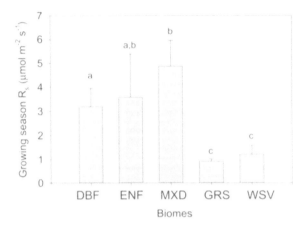

FIGURE 6.8 Average and standard deviation of growing season soil respiration for five biomes (DBF-deciduous broadleaf, ENF-evergreen needleleaf, MX-mixed deciduous/evergreen, GRS-grassland, and WSV-woodland/savanna). Different letters denote significant differences (p < 0.05) between biomes (Redrawn with permission from Biogeochemistry: Hibbard *et al.* 2005).

incubation to be 390 ± 18.9, 1300 ± 62.3, and $1800 \pm 84.9\,\mathrm{mg\,kg^{-1}}$ respectively (Saviozzi *et al.* 2001).

Forest biomes include boreal, temperate, and tropical forests. Forests cover about 4.1 billion hectares of the earth's land surface and have a total carbon pool of about 1150 Pg, of which 49% is stored in the boreal forests, 14% in temperate forests, and 37% in tropical forests (Dixon 1994). Generally, rates of annual soil respiration are low in boreal forests, intermediate in temperate forests, and high in tropical forests. For example, annual soil respiration rates are 592, 753, and $1650\,\mathrm{g\,C\,m^{-2}\,yr^{-1}}$ respectively for a Canadian boreal forest, a North American deciduous temperate forest, and an Amazonian tropical rainforest (Malhi *et al.* 1999). Heterotrophic soil respiration releases 157, 290, and $456\,\mathrm{g\,C\,m^{-2}\,yr^{-1}}$ respectively for mature boreal, temperate, and tropical forests (Pregitzer and Euskirchen 2004).

Boreal forests cover about 11% of the earth's land area (Bonan and Shugart 1989) and are located in a circumpolar belt of high northern latitudes. In boreal forests, soil moisture and temperature conditions vary greatly during a growing season, causing a great seasonality in soil respiration (Singh and Gupta 1977, Howard and Howard 1993). In general, annual soil CO_2 efflux in the boreal forests ranges from 150 to $600\,\mathrm{g\,C\,m^{-2}\,yr^{-1}}$. For example, soil respiration releases 464, 212, 279, and $177\,\mathrm{g\,C\,m^{-2}\,yr^{-1}}$ respectively from flood-plain alder, floodplain spruce, upland birch/aspen, and upland spruce stands (Gulledge and Schimel 2000). In eastern Canada, annual soil respiration rates are 200 to $350\,g\,C\,m^{-2}\,yr^{-1}$ in the mixed hardwood stand, a spruce wood stand, and their adjacent fields (Risk *et al.* 2002). In the Alaskan interior, soil respiration rates are 267, 227, and $144\,g\,C\,m^{-2}\,yr^{-1}$ respectively on three bryophytes of lichen, feather moss, and sphagnum moss on a black spruce forest floor in 2002 (Kushida *et al.* 2004). Soil CO_2 efflux during the winter of 1994–1995 ranged from 40 to $55\,g\,C\,m^{-2}$ in a boreal forest near Thompson, Manitoba (Winston *et al.* 1997). However, estimates of annual soil respiration were 905 and $870\,g\,C\,m^{-2}\,yr^{-1}$ respectively in 1994 and 1995 in a boreal aspen (*Populus tremuloides*) forest (Russell and Voroney 1998), which is greater than other estimates for boreal forest ecosystems.

Temperate forests are generally found at the middle latitudes (between 20° and 50° in both the southern and northern hemispheres), where precipitation is adequate to support tree growth. Deciduous tree species normally dominate in mild temperate climates, while coniferous tree species dominate temperate forests in cold regions or with cold winters. In deciduous forests, substrate supply from litterfall may play an important role in causing larger temporal fluctuation of soil respiration than it does in evergreen forests. The range of annual soil CO_2 efflux in temperate forests compiled by Raich and Schlensinger (1992) is from 400 to $1000\,g\,C\,m^{-2}\,yr^{-1}$, with averages of 681 and $647\,g\,C\,m^{-2}\,yr^{-1}$ for coniferous and deciduous forests respectively. Annual carbon efflux from the soil, for instance, is 840, 970, 910, and $750\,g\,C\,m^{-2}\,yr^{-1}$ for the pedunculate oak forest without understory, oak forests with understory species of *Prunus serotina*, *Rhododendron ponticum*, and *Fagus sylyatica* plus *Sorbus aucuparia* respectively (Curiel Yuste *et al.* 2005). Soil CO_2 efflux is $509\,g\,C\,m^{-2}\,yr^{-1}$ at a productive black cherry–sugar maple forest in northwest Pennsylvania (Bowden *et al.* 2000). Annual soil CO_2 release from loblolly pine forests in North Carolina is much higher than the above estimates, being 1263, 1489, 1293, and $1576\,g\,C\,m^{-2}\,yr^{-1}$ in control, irrigated, fertilized, and fertilized and irrigated plots respectively (Maier and Kress 2000). Annual soil CO_2 release from 9- and 29-year-old slash pine plantations in Florida is 820 and $1300\,g\,C\,m^{-2}\,yr^{-1}$ respectively (Ewel *et al.* 1987).

Tropical forests cover approximately 17% of the terrestrial ecosystems across the earth's warm, moist equatorial regions (Lieth and Werger 1989). Nutrient availability may be the main factor in controlling soil respiration, since high temperature and abundant precipitation occur in tropical forests. The tropical forests account for an estimated 43% of global NPP and 27% of the carbon storage in soils (Brown and Lugo 1982, Melillo *et al.* 1993). The high NPP and considerable carbon storage in soils and vegetations lead to high rates of CO_2 efflux from soil (Silver 1998). Tropical moist forests have the highest rates of carbon efflux from soil, with a range of 890 to $1520\,g\,C\,m^{-2}\,yr^{-1}$ and an average of $1260\,g\,C\,m^{-2}\,yr^{-1}$ (Table 6.2). Soil respiration in tropical dry forests is lower, with a range of 350 to $1000\,g\,C\,m^{-2}\,yr^{-1}$ and an average of

$670 \, g \, C \, m^{-2} yr^{-1}$. Annual soil respiration rates are 980 and $690 \, g \, C \, m^{-2} yr^{-1}$ in pine plantations at 800 and 1050 m elevation respectively in Indonesia (Gunadi 1994). Much higher rates of soil respiration ($1400 \, g \, C \, m^{-2} yr^{-1}$) were observed in two Australian rainforests at 800 m elevation (Maggs and Hewett 1990). However, observed soil respiration in three Hawaiian rainforests is low and ranges from 650 to $890 \, g \, C \, m^{-2} yr^{-1}$ (Raich 1998).

Grasslands account for more than 20% of the terrestrial lands and 10% of the carbon storage on the global scale (Schimel 1995, Schlesinger 1997). Annual soil carbon effluxes estimated by Raich and Schlesinger (1992) range from 400 to $500 \, g \, C \, m^{-2} yr^{-1}$ for grasslands. Recent studies report much higher rates of annual soil CO_2 efflux, probably due to improved measurements with more intensive, year-round sampling. Annual soil respiration rates are 1131 and $877 \, g \, C \, m^{-2} yr^{-1}$ in a tallgrass prairie of Oklahoma in 2002 and 2003 respectively, due to the difference in precipitation (Zhou et al. 2006) and 1350, 1100 and $1120 \, g \, C \, m^{-2} yr^{-1}$ respectively in unclipped, early-season clipped, and full-season clipped plots on Konza Prairie from June 1996 to June 1997 (Knapp et al. 1998, Bremer et al. 1998). Annual soil CO_2 efflux increases with precipitation in a Texas grassland and is 1600, 1300, 1200, 1000, 2100, and $1500 \, g \, C \, m^{-2} yr^{-1}$ respectively from 1993 to 1998 (Mielnick and Dugas 2000). However, in low production grasslands in California, annual soil respiration rates are 340 to $480 \, kg \, C \, m^{-2} yr^{-1}$ (Luo et al. 1996).

Tundra contains 14% of the global soil carbon pool (Post et al. 1982), but the carbon flux from soil is low due to low temperature. In the Eurasian and Greenland Arctic tundra, soil CO_2 efflux is significantly affected by temperature and depth of water table but little affected by thaw depth, soil nitrogen, and organic matter concentrations (Christensen et al. 1998). In addition, winter CO_2 release in the Arctic region can be substantial and reaches 111 to $189 \, g \, C \, m^{-2}$ in the Alaskan tundra (Grogan and Chapin 1999). In comparison, the rate of winter soil CO_2 release from a boreal forest in northern Russia was $89 \, g \, C \, m^{-2}$ (Zimov et al. 1996) and $69 \, g \, C \, m^{-2}$ for tussock tundra at Toolik Lake in the winter of 1993–1994 (Oechel et al. 1997). The alpine tundra in Colorado releases $153 \, g \, C \, m^{-2}$ from Julian day 168 to 218 in 1993 and $233 \, g \, C \, m^{-2}$ from day 175 to 235 in 1994 (Saleska et al. 2002). Climate has strong effects on soil CO_2 release in both summer and winter, whereas vegetation type has little impact on CO_2 efflux in winter but is the principal control in summer (Grogan and Chapin 1999).

Savannas/woodlands cover an area of $17 \times 10^6 \, km^2$ of the earth's surface, a greater area than that occupied by temperate forests, and are second only to tropical forests in their contribution to the earth's terrestrial primary production (Atjay et al. 1987). Much less attention, however, has been paid to the carbon balance of savanna and woodlands than other ecosystems, particularly soil CO_2 effluxes. Indeed, savannas and woodlands are potentially a significant carbon sink, because savannas and seasonally dry tropical forest ecosystems contribute 15% of the annual global carbon sink (Taylor and Lloyd 1992). Soil moisture and fire regimes have overriding influences on soil respiration in savannas and woodlands, particularly during the dry and warm seasons. During the wet seasons, temperature plays a significant role in regulating soil respiration. Soil respiration rates were 0.4 and $0.5\,g\,C\,m^{-2}\,d^{-1}$ in open savanna plots and in woody savanna plots respectively during a period of extreme drought in a semiarid savanna of the Kruger National Park, South Africa (Zepp et al. 1996). Annual soil respiration rates in wooded communities are lower $(533.6\,g\,C\,m^{-2}\,yr^{-1})$ than in grasslands $(858.4\,g\,C\,m^{-2}\,yr^{-1})$ in a paired juniper woodland and a C_4-dominated grassland in eastern Kansas (Smith and Johnson 2004). However, McCulley et al. (2004) observed the opposite results and found that the wooded communities have higher annual soil respiration than the remnant grasslands (745 vs. $611\,g\,C\,m^{-2}\,yr^{-1}$ respectively), probably due to gradients in precipitation and SOC content.

Deserts cover about one-fifth of the earth's surface and occur where rainfall is less than $50\,cm\,yr^{-1}$. The extreme environments limit plant production and then soil respiration. Among all the biomes, deserts have the lowest rates of soil respiration and fewest studies, probably due to the lesser importance of the deserts in regulating global carbon cycling. Soil moisture has an overriding influence on soil respiration. Annual soil respiration rates estimated from published measurements in deserts range from 184 to $300\,g\,C\,m^{-2}\,yr^{-1}$, with an average of $224\,g\,C\,m^{-2}\,yr^{-1}$ (Raich and Schlesinger 1992). However, rates estimated from a modeling study by Raich and Potter (1995) average $406\,g\,C\,m^{-2}\,yr^{-1}$. In the Antarctic dry valley of southern Victoria lands, soil CO_2 efflux ranges from -0.1 to $0.15\,\mu mol\,m^{-2}\,s^{-1}$ (Parsons et al. 2004). The negative flux is associated with a drop in soil temperature.

Crop fields occupy 1.7 billion hectares globally, with a soil carbon stock of about $170\,Pg$, slightly more than 10% of the total carbon inventory in the top $100\,cm$ of soil in upland ecosystems (Paustian et al. 1997). Compared with natural ecosystems such as grasslands and forests, crop fields release a relatively large amount of CO_2 from soils due to fertilization and intensive cultivation. Although many factors affect soil respiration, temperature is likely to be a dominant factor in a given region, because water and nutrients are often supplemented to the optimal levels for crop growth. In continuous maize (*Zea mays* L.) crop fields in the University of Nebraska-Lincoln east campus, for example, soil respiration releases $1155\,g\,C\,m^{-2}\,yr^{-1}$ (Amos et al. 2005). Soil CO_2 emissions reach $1160\,g\,C\,m^{-2}\,yr^{-1}$ in the double-crop wheat-soybean rotation on a typical soil of the rolling pampa in Argentina (Alvarez et al. 1995). However, Beyer (1991) observed relatively low rates of soil respiration that are 412 and $624\,g\,C\,m^{-2}\,yr^{-1}$ in two loamy Orthic Luvisols and 657 and $555\,g\,C\,m^{-2}\,yr^{-1}$ in two sandy Haplic Podzols of Schleswig-Holstein. Annual soil CO_2 emission is

$639\,g\,C\,m^{-2}\,yr^{-1}$ in wheat land of Missouri (Buyanovsky *et al.* 1987). The mean CO_2 efflux rate during an irrigation cycle is low in fallow field ($0.63\,\mu mol$ $m^{-2}\,s^{-1}$), intermediate in wheat field ($1.05\,\mu mol\,m^{-2}\,s^{-1}$), and high in alfalfa field ($2.26\,\mu mol\,m^{-2}\,s^{-1}$) in the desertic Sultanate of Oman (Wichern *et al.* 2004), due to differences in productivity and rooting systems.

Wetlands inhabit a transitional zone between terrestrial and aquatic habitats. The wetlands cover only about 3% of the land area (Roehm 2005) but store nearly 37% of the global terrestrial carbon (Bolin and Sukamar 2000), and are estimated to sequester 0.1 to $0.7\,Pg\,C\,yr^{-1}$ (Ovenden 1990, Gorham 1995, Wojick 1999). Wetlands are among the most productive ecosystems (Schlesinger 1997). NPP reaches a range of 1600 to $3220\,g\,C\,m^{-2}\,yr^{-1}$ for swamps, $1350\,g\,C\,m^{-2}\,yr^{-1}$ for rice, 1170 to $1990\,g\,C\,m^{-2}\,yr^{-1}$ for floodplains, 620 to $1400\,g\,C\,m^{-2}\,yr^{-1}$ for bogs, 430 to $970\,g\,C\,m^{-2}\,yr^{-1}$ for fens, 290 to $740\,g\,C\,m^{-2}\,yr^{-1}$ for marshes, and 50 to $100\,g\,C\,m^{-2}\,yr^{-1}$ for lakes (Aselmann and Crutzen 1990). Sources of carbon into wetlands are largely from plant photosynthesis and partially from sediments transported via river stream flows. The latter pathway provides both inorganic carbon and organic carbon to wetland ecosystems.

As a consequence of anoxic conditions, the rate of organic matter decomposition is slow, and carbon tends to accumulate in wetland soils (Gorham 1995). Organic soil carbon pool ranges from 35 to $90\,kg\,C\,m^{-2}$ in boreal wetlands, 35 to $80\,kg\,C\,m^{-2}$ in temperate wetlands, 5 to $10\,kg\,C\,m^{-2}$ in hardwood wetlands, 10 to $15\,kg\,C\,m^{-2}$ in conifer swamp (Trettin and Jurgensen 2003), 1.5 to $12\,kg\,C\,m^{-2}$ in the top 30 cm soil layer of a *Spartina alterniflora* marsh (Craft *et al.* 1999, 2002), and 8 to $26\,kg\,C\,m^{-2}$ in a coastal wetlands (Choi and Wang 2004). However, Armentano and Menges (1986) estimated higher soil carbon pool in wetlands of temperate zones, with a range of 60 to $144.7\,kg\,C$ m^{-2}. Accumulation of carbon in wetland soil is a significant component of the terrestrial soil carbon pool. Wetland soil carbon storage is sensitive to climatic changes, water table fluctuations, and human disturbances. Those perturbations easily result in a shift from CO_2 sink to source by altering the anoxic conditions. The magnitude of carbon sink or source in wetlands is driven to some degree by latitudinal gradients. For instance, cold ecosystems of the northern latitudes, namely peat lands, store great amounts of carbon in the peat due to slow decomposition (Roehm 2005).

Although CO_2 efflux from wetlands is potentially very important in regulating the global carbon cycle, it is poorly understood and usually excluded from global estimates and modeling studies (Raich and Potter 1995, Trettin *et al.* 2001). According to Roehm's review in 2005, the mean rates of CO_2 emissions of carbon from freshwater wetlands to the atmosphere range between 1.2 and $7.2\,g\,C\,m^{-2}\,d^{-1}$, with a global total of $11.59\,Pg\,C\,yr^{-1}$. However, carbon fluxes vary widely in different wetlands (Table 6.3) and are estimated

TABLE 6.3   Estimated CO_2 efflux from freshwater wetlands (Roehm 2005)

Type	Boreal Area		Temperate Area		Tropical Area	
	Area $(10^{12}m^2)$	Efflux $(gCm^{-2}d^{-1})$	Area $(10^{12}m^2)$	Efflux $(gCm^{-2}d^{-1})$	Area $(10^{12}m^2)$	Efflux $(gCm^{-2}d^{-1})$
Peat	3.1 (2.6–3.6)	4.8 (0.2–31.2)	0.17	7.2 (0.1–14.4)	3.4 (1.7–5.1)	2.9 (1.6–18.5)
Marsh and swamp	1.1 (0.6–1.5)	2.5 (0.5–6.5)	0.004	2.5 (0.5–6.5)	2.4 (2.1–2.8)	1.3 (0.2–10.4)
Total $(TgC\,yr^{-1})$		6.4 (0.4–44.4)		0.4 (0.01–0.9)		4.7 (1.2–44.7)

Note: Total global flux is $11.6\,PgC\,yr^{-1}$. Modified with permission from Oxford University Press: Roehm 2005.

to range from 0.13 to $9.12\,gCm^{-2}d^{-1}$ in studies of several estuaries (Abril and Borges 2005). This estimate of CO_2 efflux from freshwater wetlands is quite large relative to the value of 0.6 to $1.2\,PgC\,yr^{-1}$ estimated by Raich and Potter (1995).

Soil moisture controls the rate of oxygen diffusion into the soil and then strongly affects CO_2 efflux from wetlands. Hence, flooding or prolonged saturation tends to increase the reduction capacity of the soil and decrease decomposition of organic matter and CO_2 release rates. Cumulative rates of soil carbon mineralization in 16 northern Minnesota wetlands are estimated to be 1 to $8\,mg\,cm^{-3}\,59\,wk^{-1}$ and 0.25 to $1.8\,mg\,cm^{-3}\,59\,wk^{-1}$ under aerobic and anaerobic conditions respectively (Bridgham et al. 1998). Estimated annual rates of organic matter mineralization range from 96 to $4068\,gCm^{-2}yr^{-1}$ in the Westerschelde Estuary of the Netherlands (Middelburg et al. 1996).

Peat lands cover about 75% of the wetlands by area and are particularly important for the storage of soil carbon (Armentano and Menges 1986, Andriesse 1988). Peat lands are especially vulnerable to climatic warming, resulting from the changes in oxygen conditions or water table. Although peat lands occupy less than 3% of the earth's land area and total ecosystem productivity rates are low, they store up to $525\,PgC$ in the soils (Harden et al. 1992, Maltby and Immirzi 1993). Peat quality, temperature, and hydrological conditions are the primary factors that control carbon release (Jauhiainen et al. 2005). Annual CO_2 effluxes from peat lands have a great range, from $60\,gCm^{-2}$ year^{-1} at ombrotrophic sites dominated by Sphagnum fuscum to $340\,gCm^{-2}$ year^{-1} at sites with abundant understory vegetation in the boreal peat lands of Finland (Silvola et al. 1996). A tropical peat swamp forest releases CO_2 to the atmosphere by $953\,gCm^{-2}yr^{-1}$ (Jauhiainen et al. 2005). However, in the tropical peat lands of Sarawak, Malaysia, annual soil CO_2

efflux reaches $2100\,g\,C\,m^{-2}\,yr^{-1}$ in the forest, $1500\,g\,C\,m^{-2}\,yr^{-1}$ in oil palm, and $1100\,g\,C\,m^{-2}\,yr^{-1}$ in sago (Melling *et al.* 2005). The main factor that controls soil respiration is the relative humidity for forest peat land, soil temperature for sago, and water-filled pore space for oil palm peat lands.

6.3. VARIATION ALONG GRADIENTS

Natural gradients (e.g., latitudes, altitudes, topography, and successional ages), which vary systematically in climate or other variables, are very useful in understanding mechanisms of abiotic and biotic controls on the spatial variability of soil respiration and other ecosystem processes (Jenny 1980, Vitousek and Matson 1991). A number of gradient studies have been carried out to examine variations in soil respiration (Simmons *et al.* 1996, Conant *et al.* 1998, Austin and Sala 2002, Rodeghiero and Cescatti 2005), although results are often confounded by many covarying factors. An ideal study is to identify a gradient along which the primary factor in question varies, while all the other variables remain constant. Since such an ideal gradient may not exist, the challenge is interpreting results from gradient studies. While the gradient study along successional ages of forests is discussed in section 6.1 on decadal and centennial variation, this section examines gradient studies on soil respiration along latitudes, altitudes, and topographic forms.

LATITUDES

Latitude is not a driving variable per se that directly influences respiratory processes in ecosystems. However, it is a good proxy for joint actions of multiple factors, such as radiation, length of growing seasons, temperature, precipitation, and vegetation cover. Vegetation and climate, which are among the most critical factors in regulating the spatial variability of soil respiration, change vastly with latitudes. Overall, temperature, precipitation, length of growing season, and vegetation productivity decrease from the Equator to the North or the South Pole. Therefore, CO_2 effluxes from soils generally follow the trend and decrease with latitude. Annual soil CO_2 effluxes in forests, for example, linearly decrease from approximately $2000\,g\,C\,m^{-2}\,yr^{-1}$ in tropical regions to nearly zero in the polar region (Fig. 6.9). The similar trend is revealed in a study with a climate-driven regression model to estimate global soil CO_2 efflux (Raich *et al.* 2002). Soil microbial respiration rates also decrease with latitude along a latitudinal transect in Siberia, while soil texture and SOC exert dominant effects (Šantrůčková *et al.* 2003). However, Valentini *et al.* (2000) found that annual ecosystem respiration (i.e., soil and aboveground plant respiration) increases with latitude, while gross primary production

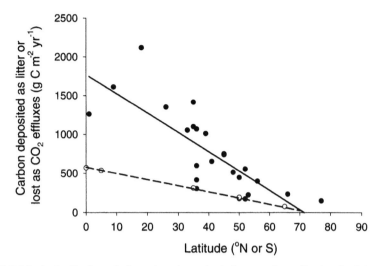

FIGURE 6.9   Latitudinal trends for carbon dynamics in forest and woodland soils of the world. The dashed line shows the mean annual carbon input to soils in litterfall; the solid line shows the pattern of carbon loss as CO_2 effluxes from soils. The linear regression for CO_2 effluxes is $CO_2 = -24.2$ (LAT) $+ 1721.5$; $R^2 = 0.60$, F $= 30.05$ (Redrawn with permission from the Annual Review of Ecology and Systematics Vol. 8 © 1977 by Annual Reviews www.annualreviews.org: Schlesinger 1977).

tends to be constant in 15 European forests, despite a general decrease in mean annual air temperature.

ALTITUDES

Vegetation and climate also change with an altitudinal gradient in a region. The cooler conditions occur in the higher elevation along altitude. Like latitude, altitudinal gradients are often used to examine environmental regulations of soil respiration. Along an elevation gradient in Japan, both litterfall and soil respiration are lowest at the highest site (Nakane 1975). Along an altitudinal gradient from 480 to 1450 m, both the length of the growing season and annual soil CO_2 efflux decrease in Olympic National Park, Washington (Fig. 6.10) (Kane *et al.* 2003). Along an altitudinal gradient from coniferous forests and mixed deciduous forests to meadows, grasses, and sedges in the Colorado Rockies, winter CO_2 effluxes from the soils are positively related to carbon availability (Brooks *et al.* 2005).

The primary factors that regulate soil respiration along an altitudinal gradient may vary at different elevations. For example, the mean soil CO_2 efflux decreases with altitude from 200 to 1050 m but increases from 1100 to 1800 m

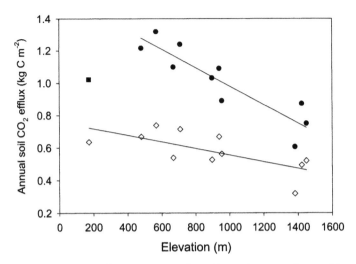

FIGURE 6.10   Relationships of growing season (May–September, open diamonds) and annual (solid circles) soil CO_2 efflux to elevation gradients (Redrawn with permission from Ecosystems: Kane *et al.* 2003).

along an altitudinal and thermal gradient in the Italian Alps (Rodeghiero and Cescatti 2005). In an 18-month experiment with laboratory incubation of soils from an altitudinal gradient in northern Arizona, microbial respiration increases from 372 to $534\,g\,C\,m^{-2}\,yr^{-1}$, with increasing elevation from 1900 to 2300 m, possibly due to differences in soil carbon pool sizes (Conant *et al.* 2000). Decomposition rate constant, k, decreases with elevations in a logarithmical function (Silver 1998). Litterfall nitrogen and phosphorus, together with elevation, can explain 83% of the variability in the k values. Overall, combined effects of multiple factors—temperature, soil moisture, length of growing season, frost-free days, and snow-free days along an altitudinal gradient—may contribute to the decreasing trend of soil respiration with altitudes. But the confounded effects of soil respiration on altitudinal patterns are difficult to unravel for understanding mechanisms of multifactor interactions.

TOPOGRAPHY

Microclimates at different topographic locations can influence soil respiration with different microsite factors, such as soil temperature (Kang *et al.* 2000), soil water content (Western *et al.* 1998), incident solar radiation (Kang *et al.* 2002), evapotranspiration (Running *et al.* 1987), and subsurface water redis-

TABLE 6.4   Estimated annual forest floor CO2 efflux and soil characteristics of four topographic locations (valley, NE slope, SW slope, and ridge-top positions) on the Walker Branch Watershed in Tennessee (Hanson *et al.* 1993)

Topographic Location	Annual CO_2 Efflux $(g\,C\,m^{-2}\,yr^{-1})$	Fine Roots $(mg\,cm^{-3})$	Soil Carbon (%)	Soil Nitrogen (%)	Forest Litter $(g\,m^{-2})$
Valleys	736	3.7 ± 2.4	3.5 ± 1.3	0.21 ± 0.07	519 ± 180
NE slopes	818	7.7 ± 3.8	2.8 ± 0.9	0.20 ± 0.06	606 ± 193
SW slopes	845	11.9 ± 3.8	2.8 ± 0.9	0.15 ± 0.06	623 ± 229
Ridge tops	927	12.5 ± 7.5	2.9 ± 0.8	0.16 ± 0.04	767 ± 231

Modified with permission from Tree Physiology: Hanson *et al.* 1993.

tribution (White *et al.* 1998). Soil respiration and soil moisture are significantly greater on north-facing slopes than on south-facing slopes in six temperate mixed hardwood forest slopes in Korea, probably due to moisture limitations in the south-facing slopes (Kang *et al.* 2003). In a white oak forest in Missouri, higher CO_2 efflux rates at low-slope positions are attributable to the greater soil water, litter mass, and roots than at high-slope positions (Garrett and Cox 1973). Hanson *et al.* (1993) chose four topographically distinct locations (valley bottom, ridge top, northeast-facing slopes, and southwest-facing slopes) to examine spatial patterns of soil CO_2 efflux in an upland oak forest in Tennessee. The estimated annual CO_2 efflux is lower in the valley bottom than in upslope and ridge-top locations, resulting from low fine-root density and high coarse fraction percentage (Table 6.4). Overall, there are no consistent patterns of soil respiration along topographic gradients among the studies, although north-facing and south-facing slopes have significantly different soil temperature, soil moisture, and/or vegetation cover.

Responses to Disturbances

Since the Industrial Revolution, human activities have altered many facets of the earth's system, inducing climatic changes causing substantial perturbations to ecosystems. These anthropogenic perturbations, together with natural disturbances, have influenced various processes of CO_2 production and transport in soil. Whereas Chapter 5 focused on soil respiration as regulated by individual environmental and biological factors, this chapter describes changes in soil respiration in response to disturbances. The disturbances affect soil respiration as external forcing variables via either natural events or manipulative experiments. Those variables include rising atmospheric CO_2 concentration, climatic warming, changes in precipitation frequency and intensity, substrate reduction or addition, nitrogen deposition and fertilization, and agricultural cultivation. This chapter also evaluates the interactive effects of multifactor disturbances on soil respiration.

7.1. ELEVATED CO_2 CONCENTRATION

Soil respiration usually increases when ecosystems are exposed to elevated CO_2. For example, when a 15-year-old stand of loblolly pine in North Carolina is exposed to Free-Air CO_2 Enrichment (FACE), soil respiration increases by 22% in the first five years of the experiment (Fig. 7.1, King *et al.* 2004). Similarly, when a grassland community in California is exposed to elevated CO_2 for three years, the flux of CO_2 from the soil surface increases from 323 to $440\,g\,C\,m^{-2}$ year^{-1} (Luo *et al.* 1996). Soil respiration increases by 12 to 40.6% in a sweetgum forest in Tennessee and developing popular forests in Wisconsin and Tuscany, Italy, at elevated CO_2 in comparison with that found at ambient CO_2 (King *et al.* 2004).

Zak *et al.* (2000) synthesized 47 published studies on responses of soil carbon and nitrogen cycling to elevated CO_2. The synthesis includes pot experiments with monoculture of 14 graminoid, 8 herbaceous, and 18 woody plant species and field experiments in intact annual grasslands, tallgrass prairie, and alpine pastures. In experiments with monoculture of grasses and intact grasslands, soil respiration varies from a 10% decline with *Lolium perenne* to a 162% increase with *Bromus hordeaceus* at elevated CO_2 compared with that found at ambient CO_2. The mean response of grasses and grassland ecosystems is a 51% increase with high variability (coefficient of variation = 100%). A few studies of herbaceous species show higher rates of soil respira-

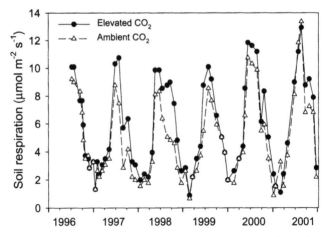

FIGURE 7.1 Soil respiration rates at ambient and elevated CO_2 from the Duke Forest FACE experiment. Open and closed symbols are ambient and elevated CO_2 plots respectively (Redrawn with permission from Global Change Biology: King *et al.* 2004).

tion at elevated than at ambient CO_2. Soil respiration with woody plants always increases under elevated CO_2 (42 ± 24.1%), with a range of 5 to 93%.

Increased soil respiration at elevated CO_2 results mainly from changes in substrate supply to the rhizosphere. Rising atmospheric CO_2 stimulates plant photosynthesis and growth. Recent reviews indicate that increases in CO_2 concentration by 200 to 350 ppm usually stimulate photosynthesis by 40 to 60% (Ceulemans and Mousseau 1994, Medlyn *et al.* 1999) and aboveground biomass growth by 22.4%, averaged over 186 paired observations (Fig. 7.2a, Luo *et al.* 2006). Increased photosynthetic carbon fixation and plant biomass growth result in delivery of more carbon substrate to belowground at elevated CO_2 than at ambient CO_2. Increased carbon substrate stimulates root and soil carbon processes, such as root biomass, specific root respiration, root turnover rates, litter production, litter decomposition, root exudation, soil priming, and microbial activity.

Elevated CO_2 stimulates belowground biomass growth by 31.6%, averaged over 168 paired observations (Fig. 7.2b), fine-root production by up to 96% (Allen *et al.* 2000, Tingey *et al.* 2000, King *et al.* 2001), and fine-root turnover (Higgins *et al.* 2002). Fine-root respiration for maintenance and growth contributes 28 to 70% to the total soil CO_2 efflux (Ryan *et al.* 1996). Thus, seasonal increases in soil CO_2 efflux at elevated CO_2 are closely related to the increase in fine-root production and biomass (Thomas *et al.* 2000). Increased root production and turnover rates result in higher heterotrophic respiration at elevated than at ambient CO_2. In addition, dead fine roots contribute to SOM during litter decomposition.

Several studies indicate that elevated CO_2 results in decreases in specific respiration rates of roots (Callaway *et al.* 1994, Crookshanks *et al.* 1998, George *et al.* 2003). Among the three components, maintenance respiration is by far the largest, accounting for 92% and 86% of the total fine root respiration at the loblolly pine and sweetgum forests respectively (George *et al.* 2003), while respiration due to root growth and nitrogen uptake and metabolism is minor. The root-specific maintenance respiration decreases by 24% in the loblolly pine forest and does not significantly vary in the sweetgum forest at elevated CO_2 (George *et al.* 2003). The CO_2-induced changes in specific root respiration are generally associated with decreases in fine-root nitrogen concentration and increases in storage carbon content (Cotrufo *et al.* 1998, Callaway *et al.* 1994, Crookshanks *et al.* 1998). However, the decrease in the specific respiration rates is usually overridden by a substantial increase in fine-root production, resulting in an increase in the total fine-root respiration.

Elevated CO_2 usually stimulates litter production and has little effect on specific rates of litter decomposition. Litter biomass increases by 20.6%, averaged over 14 paired observations (Fig. 7.2d). In general, elevated CO_2 has no

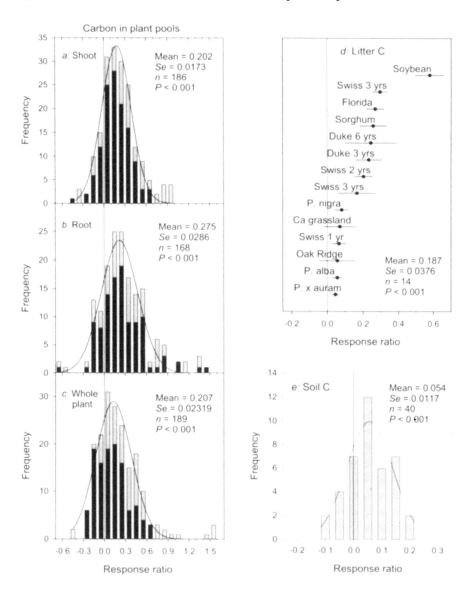

effect on specific rates of leaf litter decomposition within plant species (Finzi *et al.* 2001, Norby *et al.* 2001, Allard *et al.* 2004), although N concentration in green leaves decreases under elevated CO_2 (Cotrufo *et al.* 1998). Thus, increased litter production at elevated CO_2 increases the total amount of

FIGURE 7.2 Changes in carbon input into ecosystems via plant (a–c), litter (d), and soil (e) at elevated CO$_2$ in comparison with those at ambient CO$_2$ as indicated by frequency distributions of response ratios (RR). $RR = \ln(\bar{X}_e) - \ln(\bar{X}_a)$, where $\bar{X}_e$ and $\bar{X}_a$ are measured C contents either in plant, litter, and soil at elevated and ambient CO$_2$ respectively. In panels a to c, the solid part of bars indicates data points from ground-area-based measurements, while the gray part of bars indicates data points from plant-based measurements. In panel d, data are from the open-top chamber (OTC) experiments in Auburn, Alabama, for "Soybean" and "Sorghum" (Torbert et al. 2000); the OTC experiment in a grassland in Switzerland for "Swiss 1 yr, 2 yrs, and 3 yrs" (Leadley et al. 1999) and "Swiss 3 yrs" (Niklaus et al. 2001); the OTC experiment in an oak woodland in Florida for "Florida" (Johnson et al. 2003); the FACE experiment in the Duke loblolly pine forest in North Carolina for "Duke 3 yrs" (Schlesinger and Lichter 2001) and "Duke 6 yrs" (Lichter et al. 2005); three pure stands in the FACE experiment in Italy for "P. nigra," "P. alba," and "P. x auram" (full spelling for auram is auramericana.) (Calfapietra et al. 2003); an OTC experiment in California grassland for "CA grassland" (Higgins et al. 2002); the FACE experiment in the sweetgum forest in Oak Ridge, Tennessee, for "Oak Ridge" (Johnson et al. 2004). Each panel presents mean, standard error (Se), sample size (n), and probability (P). The solid line is the fitted normal distribution to frequency data. The vertical line is drawn at RR = 0. RR can be converted to percentage changes by $(e^{RR} - 1) \times 100\%$. The RR means of 0.202, 0.275, 0.207, 0.187, and 0.054 are equivalent to 22.4, 31.6, 23.0, 20.6, and 5.6% increases in carbon contents in shoot, root, whole plant, litter, and soil pools respectively at elevated CO$_2$ in comparison with those at ambient CO$_2$ (Luo et al. 2006).

---

substrate available for heterotrophic respiration, thereby contributing to increased soil CO$_2$ efflux.

Root exudation and rhizodeposition can be an important pathway to deliver carbon substrate from plants to soil. Elevated CO$_2$ increases carbon allocation to roots (Norby et al. 1987) and potentially increases root exudation, leading to stimulation of microbial respiration in the rhizosphere. Increased rates of carbon exudation into the rhizosphere under elevated CO$_2$ have been reported mostly in pot studies (Rouhier et al. 1994, Cheng and Johnson 1998, Cheng 1999). It is technically challenging to quantify root exudation and its priming effects in the field.

Due to increased litter production and carbon allocation to root growth and turnover, soil carbon content increases by 5.6% at elevated CO$_2$, averaged over 40 paired observations (Fig. 7.2e). Increased carbon substrate in soil stimulates microbial growth and respiration. However, a laboratory study could not detect significant changes in microbial biomass and specific rates of microbial respiration in root-free soil collected from three of the four forest FACE sites in North Carolina, Tennessee, Wisconsin, and Italy (Zak et al. 2003). Mycorrhizal colonization under elevated CO$_2$ increased from 0% to 78%, depending on tree species and type of mycorrhizae, in a developing, mixed plantation of Populus sp. in Tuscany, Italy (King et al. 2004). Mycorrhizal density increased at 34 weeks after pulse ^{14}C labeling at elevated CO$_2$ (Norby et al. 1987).

Elevated CO_2 can affect soil moisture dynamics that in turn regulates soil respiration. Elevated CO_2 increases soil moisture as a result of decreased plant transpiration (Clifford *et al.* 1993, Field *et al.* 1995, Hungate *et al.* 1997). The increased soil moisture can prolong ecosystem photosynthesis into dry seasons (Field *et al.* 1995), enhance bacterial motility and accessibility to substrates (Hamdi 1971), stimulate protozoan grazing and associated nitrogen mineralization (Kuikman *et al.* 1991), increase substrate diffusion (Davidson *et al.* 1990), and cause the higher gross mineralization in elevated CO_2 (Hungate *et al.* 1997). Thus, any one or more combinations of these mechanisms would increase soil respiration, especially in water-limited ecosystems. In a Colorado grassland, for example, elevated CO_2 increases soil respiration rates by ~25% in a moist growing season and by ~85% in a dry season (Pendall *et al.* 2003), partially due to alleviation of water stress on photosynthesis and respiration.

7.2. CLIMATIC WARMING

As discussed in Chapter 5, soil respiration is generally sensitive to temperature. As a consequence, most modeling studies assume that the increase in soil respiration per 10°C rise in temperature—Q_{10}—is about 2.0. With the temperature sensitivity of soil respiration, almost all global biogeochemical models predict a loss of carbon from soils as a result of global warming (Schimel *et al.* 1994, McGuire *et al.* 1995, Cox *et al.* 2000).

When natural ecosystems are exposed to experimental warming, soil CO_2 efflux generally increases (Peterjohn *et al.* 1993, Hobbie 1996, Rustad and Fernandez 1998, Melillo *et al.* 2002, Zhou *et al.* 2006). A meta-analysis of data collected at 17 sites from four broadly defined biomes (high tundra, low tundra, grassland, and forest) shows that soil respiration under experimental warming increases at 11 sites, decreases at one site, and does not change at five sites (Rustad *et al.* 2001). The weighted mean increase in soil respiration in response to warming is 20%, which corresponds to a mean increase of $26\,mg\,C\,m^{-2}\,hr^{-1}$ (Fig. 7.3). The relative simulation of soil respiration per degree of warming decreases with increasing temperature.

Warming-induced increases in soil respiration likely result from changes in multiple processes (Shaver *et al.* 2000), since warming affects almost all physical, chemical, and biological processes in an ecosystem. Global warming extends the length of the growing season (Lucht *et al.* 2002, Norby *et al.* 2003), alters plant phenology (Price and Waser 1998, Chmielewski and Rötzer 2001, Dunne *et al.* 2003, Fang *et al.* 2003), stimulates plant growth (Wan *et al.* 2005), increases mineralization and soil nitrogen availability (Rustad *et al.* 2001, Shaw and Harte 2001, Melillo *et al.* 2002), reduces soil water content

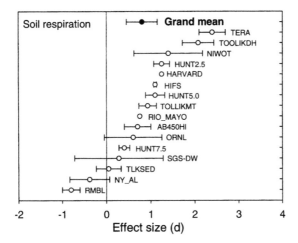

FIGURE 7.3   Mean effect sizes (d: open cycle) and 95% confidence intervals from individual sites are included in the meta-analysis for soil respiration. Abbreviation of sites are TERA = TERA trees, OR, USA; TOOLIKDH = Toolik Lake-dry heath study, AK, USA: NIWOT = Niwot Ridge, CO, USA; HUNT2.5 = Huntington Wildlife Forest, 2.5°C study, NY, USA; HARVARD = Harvard Forest, MA, USA; HIFS = Howland Forest, ME, USA; HUNT5.0 = Huntington Wildlife Forest, 5.0°Cstudy, NY, USA; TOOLLIKMT = Toolik Lake-moist tussock study, AK, USA; RIO_MAYO = Rio Mayo, Argentina: AB450H = Abisko Nature Reserve, e.s.l. 450 m, high heat study; ORNL = Oak Ridge National Laboratory, TN, USA; HUNT7.5 = Huntington Wildlife Forest, 7.5°C study, NY, USA; SGS-DW = Shortgrass Steppe-day-time warming; TLKSED = Toolik Lake-wet sedge study, AK, USA; NY_AL = Ny Alesund, Norway; RMBL = Rocky Mountain Biological Laboratory, CO, USA. Redrawn with permission from Oecologia: Rustad *et al.* (2001).

(Harte *et al.* 1995, Wan *et al.* 2002), and shifts species composition and community structure (Harte and Shaw 1995, Saleska *et al.* 2002, Weltzin *et al.* 2003). All the processes can directly and indirectly affect soil respiration on different time-scales. For example, experimental warming in a North American grassland significantly stimulates growth of aboveground biomass by 19, 36.4, and 14% in spring and 49.3, 34.2, and 9.6% in autumn in 2000, 2001, and 2002 respectively (Wan *et al.* 2005). During these three years, annual mean soil respiration correlates positively with the total aboveground biomass across different plots. The warming-induced percentage changes in annual mean soil respiration also correlate positively with the warming-induced changes in the aboveground biomass with a slope of 0.39, suggesting that annual mean soil respiration increases by approximately 39% for one unit increase in the aboveground biomass under warming compared with that under control.

Responses of soil respiration to warming differ with location. The magnitude of the response of soil respiration to soil warming is greater in cold, high-latitude ecosystems than in warm, temperate areas (Kirshbaum 1995, Parton et al. 1995, Houghton et al. 1996). Recent climatic warming has likely caused a great loss of carbon in tundra and boreal soils (Oechel et al. 1995, Goulden et al. 1998). The variability is also shown in a meta-analysis that effect sizes of experimental warming on soil respiration are much greater at forest sites than that at the grassland sites (Rustad et al. 2001). The site differences in responses of soil respiration to warming are likely related to soil organic C contents, vegetation types, and variability in climatic conditions.

There is a trend that the magnitude of respiratory response to warming decreases over time (Rustad et al. 2001, Melillio et al. 2002). In a soil-warming experiment with heating cables in the Harvard Forest in New England, the yearly efflux of CO_2 from the heated plots is approximately 40% higher than that in the control plots in the first year of the experiment (Fig. 7.4). The warming effects gradually disappear after the six-year warming treatment. The decline trend in the warming effects on soil CO_2 efflux is attributable to acclimatization (Luo et al. 2001a, Melillo et al. 2002), depletion of substrate (Kirschbaum 2004), extension of growing seasons (Dunne et al. 2003, Bowdish 2002), stimulated plant productivity (Wan et al. 2005), and fluctuation of environmental factors such as drought (Peterjohn et al. 1994, Rustad and Fernandez 1998).

Acclimation is usually referred to as a phenomenon whereby, in response to a change in temperature, the rate of respiration is initially altered (i.e., either increased or decreased) and then gradually adjusted toward the original value prior to the change in temperature. For example, in response to an increase in temperature, the rate of respiration is initially stimulated. But the stimulating effect declines upon acclimation of the system in question to the high temperature, so that the rate of respiration at the high temperature approaches the rate at the original temperature. Conversely, the rate of respiration is initially lowered in response to a treatment of low temperature and then gradually increases upon acclimation to the original rate before the treatment. The adjustment in respiration rates during acclimation can result from many processes, such as depletion of substrate, changes in enzymatic activities and/or composition, and shifts in microbial community. The acclimation of soil respiration to warming is regulated by soil clay content, soil water content, and substrate quality and quantity. Respiratory acclimation to warming results in decreased temperature sensitivity as indicated by lowered Q_{10} values (Luo et al. 2001a). For example, in a field-warming experiment in a tallgrass prairie in central Oklahoma, the Q_{10} value decreased from 2.70 in the unwarmed plots to 2.43 in the warmed plots (Fig. 7.5).

The decrease in sensitivity of soil respiration to warming over time is likely due to accelerated decomposition, potentially leading to depletion of labile soil carbon pool (Kirschbaum 2004, Pajari 1995, Strömgren 2001, Niinistö et

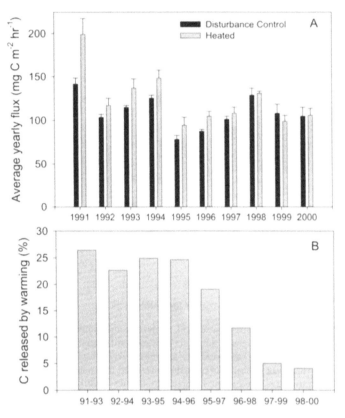

FIGURE 7.4 (A) Average yearly fluxes of CO_2 from the heated and disturbance control plots. Measurements were made from April through November from 1991 through 2000. Error bars represent the standard error of the mean ($n = 6$ plots) between plots of the same treatment. (B) Percentage increase in the amount of carbon released from the heated plots relative to the disturbance control plots. The data are presented as three-year running means from 1991 through 2000 (Redrawn with permission from Science: Melillo *et al.* 2002).

al. 2004). Many modeling studies have demonstrated that warming stimulates oxidation of SOM; depletes carbon substrate in soil pools, particularly in the labile carbon pools; and then results in decreases in temperature sensitivity of soil respiration (Elisasson *et al.* 2005, Gu *et al.* 2004). However, experimental warming in an Oklahoma grassland significantly increases labile carbon and nitrogen contents in soil pools (Tedla 2004). The increases in the labile carbon and nitrogen pools are attributable to stimulation of plant growth and inputs of organic carbon into soil under warming (Wan *et al.* 2005), although warming may directly accelerate decomposition.

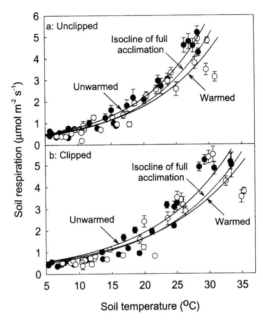

FIGURE 7.5 The relationships between soil respiration and temperature in unwarmed (solid circles) and warmed (open circles) treatments with standard errors. Panel *a* is for the unclipped subplots and panel *b* for the clipped subplots. Response curves of soil respiration to temperature in the warmed treatment are below that in the unwarmed treatment, indicating that warming reduced temperature sensitivity. The theoretical isoclines of full acclimation are defined as identical rates of respiration in the warm-exposed plots to that in the unwarmed control plots. Soil temperature used in this analysis was measured at a depth of 5 cm and was 1.5°C higher in the warmed than the unwarmed plots without clipping and 1.9°C higher with clipping (Luo *et al.* 2001).

Warming causes a shift in the soil microbial community structure toward more fungi (Zhang *et al.* 2005), likely contributing to decreases in the sensitivity of soil CO_2 efflux to temperature. Fungi are more tolerant of high soil temperature and dry environments than are bacteria, due to their filamentous nature (Holland and Coleman 1987). Warming also increased soil microbial biomass carbon and nitrogen contents in a North American grassland (Tedla 2004) and in a dwarf shrub dominated tree-line heath and a high latitude fellfield at Abisko Swedish Lapland due to increased organic carbon inputs. The stimulation of microbial biomass C and N resulting from high organic input to soil has been reported in Australia, the United Kingdom, and Denmark (Sparling 1992, Degens 1998, Michelsen *et al.* 1999). Changes in microbial biomass may also contribute to alterations in the temperature sensitivity of soil respiration.

A decrease in soil moisture under warming possibly reduces root and microbial activity, affecting the sensitivity of soil respiration to warming (Peterjohn *et al.* 1994, Rustad and Fernandez 1998). Experimental warming decreases moisture contents in litter and soil. The latter counterbalances the positive effect of elevated temperature on litter decomposition and soil respiration (McHale *et al.* 1998, Emmett *et al.* 2004).

7.3. CHANGES IN PRECIPITATION FREQUENCY AND INTENSITY

Changes in precipitation frequency and intensity have greatest impact on soil respiration in xeric ecosystems or dry seasons of mesic ecosystems. It has been observed that soil respiration in arid or semiarid areas shows dynamic changes within a raining cycle. The rate of respiration in dry soil usually bursts to a very high level after rainfall and then declines as the soil dries (Fig. 7.6). The increments in respiration caused by rainfall events are inversely related to the rate of respiration before the rain (Xu *et al.* 2004). Irrigation in arid lands usually releases drought stress and therefore stimulates soil respiration rates. For example, irrigation with or without fertilization equally stimulates soil respiration in a Saskatchewan grassland (de Jong *et al.* 1974).

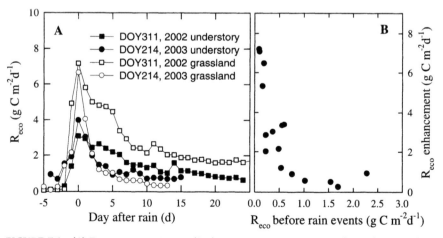

FIGURE 7.6   (A) Ecosystem respiration (R_{eco}) response to rain events in the understory of the savanna woodland and the grassland. (B) Enhancements of R_{eco} (the difference in respiration after and before rain events), which are inversely related to R_{eco} before rain events (Redrawn with permission from Global Biogeochemical Cycles: Xu *et al.* 2004).

Addition of water in a manipulative experiment also stimulates soil CO_2 efflux in grassland (Liu et al. 2002a).

Alteration of rainfall amounts and temporal variability results in changes in soil water content and then affects soil CO_2 efflux. A reduction in rainfall amounts usually results in lowered soil respiration. Similarly, prolonged water deficits between periods of rainfall also reduce soil CO_2 efflux as a result of increased plant and microbial stress (Bremer et al. 1998). In the Konza prairie, a 70% reduction in the natural rainfall quantity decreases soil respiration by 8% (Harper et al. 2005). A 50% increase in the length of dry intervals between rainfalls reduces soil respiration by 13% (Fig. 7.7). When both the rainfall amounts and rainfall intervals are altered, soil respiration decreases by 20%. The changes in soil CO_2 efflux are accompanied by changes in plant productivity.

In the Walker Branch throughfall displacement experiment, treatments with either an increase or decrease of throughfall by 33% did not significantly affect soil respiration in the forest (Hanson et al. 2003). The throughfall treatments do not significantly affect either litter quality or litter decomposition rates.

Wetland drainage increases soil aeration and stimulates respiration rates by releasing oxygen limitation to soil organisms. Soil respiration rates in

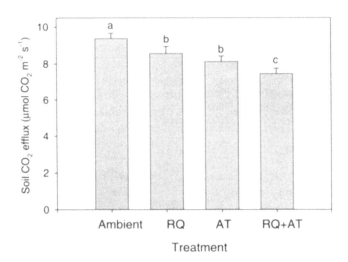

FIGURE 7.7   Mean plus one standard error of soil CO_2 of efflux (μmol $CO_2 m^{-2} s^{-1}$) during growing seasons for each experimental treatment across the four years of a study in Kansas. The treatments are abbreviated as follows: ambient rainfall (Ambient), reduced rainfall quantity (RQ), altered rainfall timing (AT), and reduced quantity + altered timing (RQ + AT) (Redrawn with permission from Global Change Biology: Harper et al. 2005).

northern peatlands, for example, correlate positively with depth to the water table (Fig. 7.8, Moore and Knowles 1989, Luken and Billings 1985). Rates of soil respiration increase after peatland drainage in Finland (Silvola *et al.* 1985). Such a response is most profound in poorly decomposed peats. Virgin mores accumulate soil carbon but become CO_2 sources in the atmosphere after drainage (Silvola 1986). Increased rates of soil respiration from drained peatlands result in loss of organic carbon from many wetland soils of the world.

Water manipulation also affects temperature sensitivity of soil respiration due to the interactive effects of changes in soil water content and temperature. Soil CO_2 efflux is more sensitive to soil water content than to soil temperature during prolonged drying cycles in a tallgrass prairie (Bremer *et al.* 1998). Increased rainfall variability reduces Q_{10} values, because both low and high soil water contents occur more frequently in the altered rainfall timing treatment (Harper *et al.* 2005). The temperature sensitivity of soil respiration also varies with years and other attributes of ecosystems. The Q_{10} values are low in the wet years and high in the dry years in a multiyear study of a grassland and a beech-spruce forest in Germany (Dörr and Münnich 1987).

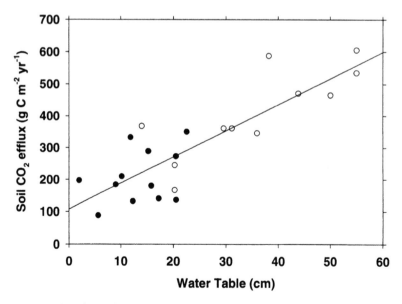

FIGURE 7.8   The relationship between annual average CO_2 effluxes and average depth of the water tables in boreal peatlands of Finland. Solid symbols are the virgin sites and open symbols are the drained sites (Modified with permission from Journal of Ecology: Silvola *et al.* 1996).

But some studies found that the Q_{10} values are lower in the well-drained sites than the in wetter sites (Davidson *et al.* 1998, Xu and Qi 2001a, Reichstein *et al.* 2003). Davidson *et al.* (1998) attribute the variable responses of Q_{10} to site-specific moisture conditions and/or rainfall distributions within a single year. Complex interactive effects of soil water and temperature on CO_2/O_2 diffusion, root and microbial activities could result in the diverse responses of the temperature sensitivity of soil CO_2 efflux to water availability.

7.4. DISTURBANCES AND MANIPULATIONS OF SUBSTRATE SUPPLY

Many of the natural disturbances and experimental manipulation result in changes in substrate supply to root and microbial respiration. The disturbances include fire or burning; harvesting, thinning, or girdling of forests; grazing, clipping, and shading in grasslands; and litter removal or addition. In general, soil respiration decreases with reduction in substrate supply and increases with addition of substrate supply.

FIRE OR BURNING

Wildfire is one of the primary regulators of carbon uptake and release on landscape scales. In general, fire reduces soil respiration. The magnitude of reduction in soil respiration depends on the severity of the fire and the time that elapses after five (Weber 1990, O'Neill *et al.* 2002). Soil respiration in burned forests, for example, is significantly lower than in intact forests; and the decrease in soil respiration is greater in severely burned forests than in mildly burned forests (Sawamoto *et al.* 2000). Soil respiration decreases by 6%, 5%, and 22% for the controlled burning, removal of red straw, and total litter removal respectively in a *Pinus palustris* forest, in contrast to those in the control (Reinke *et al.* 1981). Although the soils become significantly warmer after fire, losses of vegetation, litter, and surface SOM result in significant decreases in soil CO_2 efflux in the burned areas in comparison with that in the control in three stands—black spruce, white spruce, and aspen (O'Neill *et al.* 2002). Burning also dampens seasonal fluctuations in CO_2 efflux and lowers Q_{10} values because of reduced root activity. Even so, fire thaws the permafrost soil and thickens active soil layers, enhancing decomposition and net loss of stored carbon from the frozen ecosystems (O'Neill *et al.* 2003, Zhuang *et al.* 2003). However, burning stimulates soil respiration in a tallgrass prairie in Kansas (Tate and Striegl 1993). The measured soil

CO_2 efflux during the 200-day sampling period is 15.7, 14.5, 13.9, and 10.3 g CO_2 $m^{-2} d^{-1}$ for burned prairie, unburned prairie, wheat, and sorghum respectively.

FOREST HARVESTING, THINNING, AND GIRDLING

Forest harvesting can have a dramatic impact on soil physical and chemical properties due to tree removal and soil modification by harvesting equipment (Pritchett and Fisher 1987). Due to biomass removal, forest harvesting usually increases soil heating, water evaporation at the soil surface, and diurnal fluctuations of soil surface temperature. Forest harvesting also leaves a large amount of forest litter and dying tree roots that decompose easily (Startsev et al. 1997). All the changes in physical properties and biological attributes potentially affect soil respiration. Forest harvesting by clear-cutting, for example, stimulates (Gordon et al. 1987, Hendrickson et al. 1989), suppresses (Nakane et al. 1986, Mattson and Smith 1993), or has no effect on (Edwards and Ross-Todd 1983, O'Connell 1987, Toland and Zak 1994, Edmonds et al. 2000) soil respiration, depending on harvest methods, forest types, speed of regeneration, and climate conditions (Table 7.1).

In a northern spruce forest and a *Pinus elliottii* plantation in Florida, clear-cutting plots release more CO_2 than do uncut plots in the first year following the treatment, due to the increased soil temperature and decomposition of logging debris and fine roots (Lytle and Cronan 1998, Ewel et al. 1987a). Soil respiration increases distinctively after clear-cutting in the white spruce forests of interior Alaska, especially in summer (Gordon et al. 1987). Rates of soil respiration and magnitude of increases after clear-cutting depend not only on contents and decomposition rates of SOM but also on amounts of logging debris and harvest methods (Ewel et al. 1987a). Soil respiration rates increase in the first two years after partial cutting, but this increase disappear in the third year in a Japanese cedar forest (Ohashi et al. 1999).

In other studies, forest clear-cutting has been found to reduce soil respiration. A comparative study in Saskatchewan, conducted in the 1994 growing season, shows that tree harvesting in a mature jack pine stand reduces soil CO_2 efflux from 22.5 to 9.1 mol CO_2 m^{-2} (Striegl and Wickland 1998). The undisturbed forest site is a net sink of 3.9 mol CO_2 m^{-2}, while the clear-cut site is the net source of 9.1 mol CO_2 m^{-2}. Reduction of soil respiration is attributed to disruption of carbon supply from the canopy to the rhizosphere. In years following the clear-cutting, soil respiration increases with time as new trees and herbaceous plants are established (Weber 1990, Gordon et al. 1987, Hendreickson et al. 1989, Striegl and Wickland 2001). In a study of a jack pine forest, the clear-cutting results in a >50% reduction in soil respira-

TABLE 7.1   Direct comparison of soil respiration rates in uncut and clear-cut forests

Forest Type, Location	Soil Respiration Rate ($g\,m^{-2}\,yr^{-1}$)			Yr	Reference
	Control	Clear-cut	Difference		
Pinus elloittii, Florida	1300	2600	1300 (100%)	1st	Ewel *et al.* (1987b)
Eucalyptus, Victoria, Australia	830	1060	230 (27.7%)	2nd	Ellis (1969)
Liquidambar and Quercus, Texas, USA	493	712	219 (44.4%)	1st	Londo *et al.* (1999)
Acer rubrum, Maine	645	765	120 (18.6%)	4–6	Fernandez *et al.* (1993)
Picea rubens, Maine (182 d)	379	441	124 (16.3%)	1st	Lytle and Cronan (1998)
Quercus nigra, Mississippi	514	620	106 (20.7%)	1st	Schilling *et al.* (1999)
Picea glauca, Alaska	440	530	90 (20.5%)	2–4	Gordon *et al.* (1987)
Pinus densiflora, Japan	1255	676	−579 (−46.1%)	1st	Nakane *et al.* (1983)
Acer and Betula, Ontario, Canada	369	240	−285 (34.9%)	1st (165d)	Laporte *et al.* (2003)
Pinus banksianai, Prince Albert	270	109	−161 (−59.6%)	(growing season)	Striegl and Wickland (1998)
Populus trenuloides, Ottawa, Canada	355	299	−56 (−15.8%)	2nd	Weber 1990
Populus trenuloides, Ottawa, Canada	320	303	−17 (−5.3%)	1st	Weber 1990
Populus trenuloides, Ottawa, Canada	328	320	−8 (−2.4%)	3rd	Weber 1990
Quercus-Carya, Tennessee	529	488	−41 (−7.8%)	1st	Edwards and Ross-Todd (1983)
Acer and Quercus, Michigan	487	467	−20 (−4.2%)	1st	Toland and Zak (1994)
Acer and Tilia, Michigan	469	474	5 (1.1%)	1st	Toland and Zak (1994)
Quercus and Acer, Virginia	171	171	0	0.5–23 (summer)	Mattson and Smith (1993)

tion compared with that of a mature forest in the first growing season after the treatment (Striegl and Wickland 2001). However, soil respiration is higher by about 40% in an 8-year-old stand but lower by about 25% at a 20-year stand than that of a mature forest during the growing season. As the forest grows to more than 20 years old, soil respiration may continue to decrease as the rate of tree growth slows and pioneer grasses, annuals, and small shrubs are replaced by lichen (Fig. 7.9). The dynamics of soil respiration during forest succession after clear-cutting are attributable to changes in vegetation and its associated carbon supply.

Studies of conifer forests in Oregon (Vermes and Myrold 1992), northern hardwood forests in Michigan (Toland and Zak 1994), and fir forests in western Washington (Edmonds *et al.* 2000) show no apparent effects of clear-cutting on soil respiration. This is likely because the enhancement of microbial respiration offsets the decrease in root activity after clear-cutting. A comparative study indicates that soil C content is lower by 30% in a nearby logged area where open spaces have been invaded by dense shrub than in an old-growth forest reserve (Wang *et al.* 1999). Measured soil respiration in the two contrasting forests does not differ much in summer. The respiration rate in the logged site in winter is about 50% of that in the forested site.

Forest thinning partially removes trees from a stand to reduce competition, improve tree productivity, and reduce wildfire risk. Like forest har-

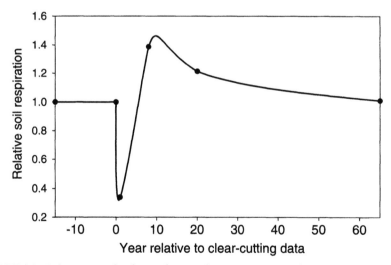

FIGURE 7.9 Relative rate of jack pine forest soil respiration versus forest age. A value of 1.0 represents mature forest (Redrawn with permission from Canadian Journal of Forst Research: Striegl and Wickland 2001).

vesting, thinning decreases stand density and leaf area, increases light and nutrient availability, and alters soil thermal and moisture regimes. In addition, mechanical thinning compacts soil, causing a decrease in soil aeration and restricting root growth and microbial activities (Poff 1996). Thinning-induced changes in those processes inevitably affect soil respiration.

Like clear-cutting, forest thinning also produces diverse effects on soil respiration. Forest thinning with 30% removal of biomass in the Sierra Nevada Mountains in California decreases total soil respiration at a given temperature and water content, does not change the sensitivity of soil respiration to temperature or to water, yet increases the spatial homogeneity of respiration (Tang *et al.* 2005b). This decrease in soil respiration is likely due to the decrease in root density and carbon substrate supply after thinning. Soil respiration is not significantly affected by the thinning treatment in an old-growth, mixed conifer forest in California, possibly due to decomposition of increased litter inputs that offset the reduction in root respiration (Ma 2003). However, thinning increases soil respiration by 40% in a Japanese cedar (*Cryptomeria japonica*) stand three to four years after the treatment but have no effect in year 5 (Ohashi *et al.* 1999). Soil respiration increases by 43% with selective thinning in the mixed conifer forest and by 14% in the hardwood forest (Concilio *et al.* 2005). Similarly, increases in soil respiration after forest thinning have been observed in other studies (Gordon *et al.* 1987, Hendrickson *et al.* 1989, Misson *et al.* 2005). Increases in soil respiration may result from increased soil temperature and moisture (Gordon *et al.* 1987), decomposition of increased dead roots or aboveground litter layer inputs (Rustad *et al.* 2000), and changed litter quality from fresh leaves of logging slash (Fonte and Schowalter 2004).

Girdling instantaneously terminates the flow of photosynthates from the tree canopy through the phloem to the roots and rhizosphere, while water transport in the reverse direction through the xylem is not affected for days. Thus, the tree girdling reduces substrate supply but does not immediately affect soil environemnts such as moisture and temperature. It does not physically displace roots or soil organisms, nor does it sever roots or fungal hyphae. The tree girdling is an ideal approach to study effects of substrate supply from the aboveground photosynthesis on soil respiration. In a large-scale girdling experiment with nine plots, each containing about 120 trees, girdling reduces soil respiration by up to 37% within five days and about 54% within one to two months relative to respiration on ungirdled control plots (Högberg *et al.* 2001). In the second year after girdling, differences in soil respiration between the girdled and ungirdled plots are smaller than in the first year (Bhupinderpal-Singh *et al.* 2003).

GRAZING, CLIPPING, AND SHADING IN GRASSLANDS

A considerable portion of CO_2 released via soil respiration is derived from recently fixed carbon by plant photosynthesis. Thus, soil respiration is very responsive to changes in carbon supply caused by grazing, clipping, and shading in grasslands (Craine et al. 1999, Craine and Wedin 2002, Wan and Luo 2003). Grazing affects soil respiration directly or indirectly through many processes. For example, grazing removes live biomass periodically during the growing season, regulates plant community composition, alters plant canopy structure, changes chemical composition of litter input into the soil (Bremer et al. 1998, LeCain et al. 2000, Wilsey et al. 2002), adds urinary and fecal input into soil (Augustine and McNaughton 1998, Sirotnak and Huntly 2000), induces defensive chemicals in plants (Bryant et al. 1991), causes an increase or decrease in plant root exudation (Bargdett et al. 1998), and affects soil microclimate. Generally, grazing reduces soil respiration (Ohtonen and Väre 1998, Johnson and Matchett 2001, Stark et al. 2003, Cao et al. 2004) due to reduced root biomass (Johnson and Matchett 2001) and decreased supply of labile C substrate to microbes and roots (Stark et al. 2003). However, soil respiration and microbial metabolic activity are enhanced by reindeer grazing in the suboceania tundra heaths (Stark et al. 2002) because of increased rates of nutrient cycling. Reindeer grazing also increases the proportion of graminoids that allocate more carbohydrate than forbs for fine-root growth. The urine and feces produced by mammalian herbivores stimulate soil microbial processes too.

Clipping is often used in manipulative experiments to mimic mowing for hay in grasslands, which is a common land use practice in many regions. Clipping reduces soil CO_2 efflux by 19% to 49% in grassland ecosystems (Bremer et al. 1998, Craine et al. 1999, Wan and Luo 2003). Methods of clipping and durations of study affect responses of soil respiration to clipping. Wan and Luo (2003) kept clipping aboveground biomass to maintain bare ground in the clipped plots during the whole study period of one year. The repeated clipping leads to a 33% decrease in annual mean soil CO_2 efflux (Fig. 5.1). Bremer et al. (1998) studied soil respiration in three clipping treatments (early-season clipping, full-season clipping, and no clipping) and adjacent grazed and ungrazed pastures at three separate sites. Clipping reduceds soil respiration by 21 to 49% on the second day after clipping, even with higher soil temperatures in the clipped plots than in the control plots. Daily soil respiration is 20 to 37% less in the grazed pastures than in ungrazed pastures, because of reduced canopy photosynthesis and lowered carbon allocation to the rhizosphere. However, clipping once a year for four years has no significant effects on soil CO_2 efflux in a grassland in the central United States (Zhou *et al.* 2006). Long-term clipping reduces carbon and nitrogen contents in both labile and recalcitrant soil pools, obviously due to partial removal of plant biomass that could otherwise have been returned to the soil (Almendinger 1990, Janzen *et al.* 1992, Rühlmann 1999, Ghani *et al.* 2003). Recalcitrant carbon pools in soil decreased by 2 to 12% in clipped plots in comparison with those in unclipped plots in long-term field experiments conducted at several sites across many ecosystems (Rühlmann 1999, Tedla 2004).

Shading also decreases the supply of carbon substrate to roots and root-associated processes. As a consequence, soil respiration decreases by 40% under shading at two-day experiments in a tallgrass prairie in the northern U.S. Great Plains (Craine *et al.* 1999). Year-round shading in a tallgrass prairie of the southern Great Plains reduces soil respiration on all the time-scales (diurnal, transient, and annual) irrespective of the minor concurrent changes in soil temperature and moisture. Annual mean soil respiration decreases significantly, by 23 and 43% for the shading and shading plus clipping treatments respectively (Fig. 5.1, Wan and Luo 2003).

LITTER REMOVAL AND ADDITION

A significant fraction of soil respiration is attributable to the decomposition of plant litter (Bowden *et al.* 1993, Lin *et al.* 1999, Sulzman *et al.* 2005). Thus, soil respiration usually decreases with litter removal and increases with litter addition (Boone *et al.* 1998, Jonasson *et al.* 2004). Complete removal of aboveground litter reduces soil respiration by up to 25%, and double litter increases it by approximately 20% (Fig. 7.10). The litter addition or removal also affects temperature sensitivity of soil respiration (Table 5.1).

7.5. NITROGEN DEPOSITION AND FERTILIZATION

Responses of soil respiration to nitrogen fertilization and deposition are extremely variable depending on fertilizer types, loading levels, and site conditions. Fertilization increases soil respiration in a central North Carolina forest (Gallardo and Schlesinger 1994), a temperate forest in Germany (Brume and Besse 1992), pine forests in Russia (Repnevskaya 1967), spruce forests in Norway (Borken *et al.* 2002), and a grassland in minnesota (Fig. 7.11 Craine *et al.* 2000). The stimulation of soil respiration by nitrogen fertilization results from increased fine-root biomass in fertilized plots (Reich *et al.* 2001, Craine *et al.* 2002). However, nitrogen fertilization depresses soil CO_2 efflux in abandoned agricultural fields in Canada (Kowalenko *et al.* 1978), in a native

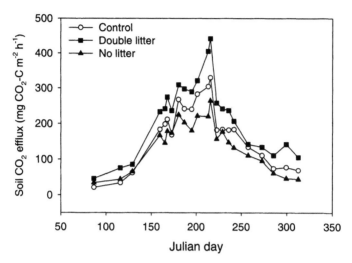

FIGURE 7.10   Soil CO_2 efflux for Harvard Forest litter manipulation plots. Measurements are made over one year from 16 June 1994 to 14 June 1995. Control = normal litter input, no litter = aboveground litter excluded from plots annually, double litter = aboveground litter doubled annually (Modified with permission from Nature: Boone *et al.* 1998).

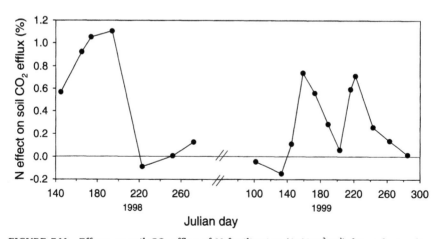

FIGURE 7.11   Effects on soil CO_2 efflux of N fertilization ($4\,g\,N\,m^{-2}\,yr^{-1}$) for each sampling period in 1998 and 1999 (Modified with permission from New Phytologist: Craine *et al.* 2001).

grassland of Saskatchewan (de Jong *et al.* 1974), and in 11 year-old loblolly pine (*Pinus taeda*) plantations in North Carolina (Maier and Kress 2000). The long-term fertilization (17 years) using NH_4NO_3 with a rate of $74\,kg\,N\,ha^{-1}\,yr^{-1}$ in *Pinus sylvestris* forests in the United Kingdom depresses soil respiration by 30 to 40% (Persson *et al.* 1989). Both the autotrophic and heterotrophic components of soil respiration are significantly lower (by approximately 40%) in fertilized than nonfertilized plots in a large-scale girdling experiment with a 40-year-old Norway spruce, although aboveground production in the non-girdled stands is about three times higher in fertilized than nonfertilized plots (Olsson *et al.* 2005). Phosphorus fertilization increases stem growth of trees but reduces soil CO_2 efflux by approximately 8% in a mature *Eucalyptus pauciflora* forest (Keith *et al.* 1997). Fertilized and nonfertilized barley fields in Sweden have similar soil respiration rates (Paustian *et al.* 1990). Fertilization does not affect CO_2 efflux in a mature slash pine plantation in Florida (Castro *et al.* 1994) and in lobolly pine plantations in North Carolina (Oren *et al.* 2001).

    Variable responses of soil respiration to fertilization are also observed in wet ecosystems. Drained peat-bog forests showed no changes, increases, or decreases in soil respiration rates in response to fertilization at different sites (Silvola *et al.* 1985). Soil respiration decreases with fertilization in the flood-plain alder and white spruce sites and increases in the birch/aspen site in Alaska (Gulledge and Schimel 2000).

    Nitrogen addition to ecosystems potentially affects a number of processes of soil respiration. Nitrogen fertilization can enhance plant dark respiration, stimulate specific rates of root respiration, and increase root biomass (Mitchell *et al.* 1995, Ibrahim *et al.* 1997, Griffin *et al.* 1997, Lutze *et al.* 2000). However, fertilization could reduce belowground carbon allocation and negatively affect both root and rhizosphere microbial respiration (Franklin *et al.* 2003, Giardian *et al.* 2003, 2004, Olsson *et al.* 2005). Nitrogen effects on decomposition of litter and SOM are also highly variable, being either positive (Van Vuuren and Van Der Eerden 1992, Boxman *et al.* 1995, Magill and Aber 2000, Hobbie 2000), negative (Koopmans *et al.* 1997, Resh *et al.* 2002), or unaffected (Gundersen 1998, Hoosbeek *et al.* 2002). Decomposition of cellu-loses or other more labile compounds in litter and SOM are stimulated by nitrogen addition, whereas decomposition of lignin or other recalcitrant com-pounds of litter and SOM are inhibited by nitrogen addition (see Chapter 5). As a consequence, the net effects of nitrogen fertilization on soil respiration vary with sites, soil types, and vegetation covers. No clear patterns have emerged from available data. Short- and long-term effects of fertilization may also differ as vegetation adapts to new nutrient regimes. In short, mechanisms that regulate responses of soil respiration to nutrient addition are poorly understood.

7.6. AGRICULTURAL CULTIVATION

Cultivation disturbs soil and usually improves soil aeration and moisture conditions. As a consequence, environments for decomposition of SOM improve, resulting in increases in soil respiration. Cultivation also disrupts soil aggregates, exposing stable, adsorbed organic matter to microbial activity (Elliotts 1986, Six *et al.* 1998). In the short term, therefore, soil respiration is generally stimulated by cultivation disturbance. For example, newly cropped plots generated by slash-and-burn release more CO_2 than an uncut forest plot in Thailand (Tulaphitak *et al.* 1983). Soil CO_2 efflux from wheat is greater than that from the native grassland vegetation in Missouri (Buyanovsky *et al.* 1987) and Saskatchewan (de Jong *et al.* 1974). Losses of carbon from cultivated soils may be as large as $0.8 \, Pg \, C \, yr^{-1}$ globally (McGuire *et al.* 2001).

The loss of organic matter in soil means depleted substrate for soil respiration over time. Thus, the long-term cultivation usually results in decreases in soil respiration. In southern Queensland, for example, the concentration of soil carbon decreases by up to 70% after more than 40 years of cultivation at the Langlands-Logie site (Fig. 7.12). After 22 years of conversion of an annual grassland to a lemon orchard in central California, soil carbon content decreases by 26% and annual soil respiration by 11% with litter and 31% without litter (Wang *et al.* 1999). In addition, cultivation is usually accompa-

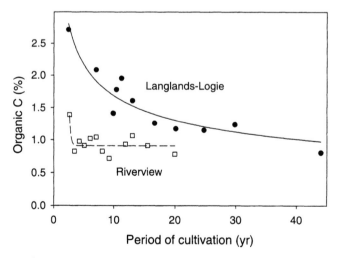

FIGURE 7.12   Change in soil C with period of cultivation at two sites in southern Queensland (Data from Dalal and Mayer 1986).

nied by a harvest of biomass. Inputs of plant litter to soils in crop fields are lower than native vegetation after it is converted to agricultural fields, contributing to depletion of soil carbon stocks.

SOM is lost less when "no tillage" agriculture is practiced on lands that have been cultivated for a long time. No-tillage cropping on previously cultivated lands increases SOM (Kern and Johnson 1993, Dao 1998) and enhances carbon storage in temperate regions and subhumid and humid tropics (Paustian et al. 1997). For example, no-tillage practice for 11 years increases organic carbon content in a silt loam (0 to 5 cm) soil in Oklahoma by 65% compared with a moldboard plow treatment (Dao 1998). The increased storage of carbon in soil is usually associated with reduced rates of soil CO_2 efflux during the conversion to no-tillage cropping from conventional tillage (Curtin et al. 2000, Al-Kaisi and Yin 2005).

When cultivation is supplemented with other practices, soil respiration may be affected in different ways. Addition of straw to soil or on the surface substantially increases soil respiration (Table 7.2). Soil subjected to moist-dry cycles from 90% field capacity to below the permanent wilting point before watering releases 36 to 62% less CO_2 than soil with continuous watering every two or three days to 90% of field capacity (Curtin et al. 1998).

7.7. INTERACTIVE AND RELATIVE EFFECTS OF MULTIPLE FACTORS

Natural disturbances and anthropogenic perturbations often involve simultaneous changes in multiple factors, which could potentially have complex interactive influences on soil respiration. The complex interactive effects of two or more variables are usually not predictable from the effects of individ-

TABLE 7.2 Total amount of CO_2-C released (gm^{-2}) within 77 d as influenced by straw addition, placement method, and moisture regime

Moisture Regime	No Straw	Incorporated Straw		Surface Straw	
		Fresh	Weathered	Fresh	Weathered
Continuously moist	24.7 a	68.1 b	76.9 c	40.1 d	42.5 d
Moist-dry cycle	12.5 a	42.6 b	42.3 b	17.6 c	15.4 c

Note: Two types of straw are either incorporated into or placed on the soil surface at a rate equivalent to 2800 kg ha^{-1}. Fresh straw is collected shortly after harvest. Weathered straw is the standing stubble that has been in the field for a year (Curtin et al. 1998).

ual variables in terms of directions and magnitudes. Thus, it is critical to examine interactive effects on soil respiration with multifactor manipulation experiments (Beier 2004, Norby and Luo 2004).

Zhou *et al.* (2006) conducted two experiments—one long-term with a 2°C increase and one short-term with a 4.4°C increase—to investigate main and interactive effects of the three factors (i.e., warming, clipping, and doubled precipitation) on soil respiration and its temperature sensitivity in a tallgrass prairie of the U.S. Great Plains. While the main effects of warming and doubled precipitation are significant, interactive effects among the factors are not statistically significant either for soil respiration or their temperature sensitivities, except for the warming × clipping interaction (Table 7.3). Similarly, interactive effects of elevated CO_2 and temperature are not statistically significant in the *Acer* stand (Edwards and Norby 1998) and in a boreal forest with Scots pine (*Pinus sylvestris L.*) (Niinistö *et al.* 2004). In addition, elevated CO_2 and warming have no interactive effects on three components of soil respiration—rhizosphere respiration, litter decomposition, and SOM oxidation—except SOM oxidation in 1994 and rhizosphere respiration in 1995 (Lin *et al.* 2001).

The interactions are significant neither between elevated CO_2, nitrogen supply, and plant diversity (Craine *et al.* 2001) nor between elevated CO_2 and O_3 (Kasurinen *et al.* 2004) in influencing soil CO_2 efflux. However, there is a strong interactive effect on root respiration between elevated temperature and soil drying for the Concord grape grown in a greenhouse (Huang *et al.* 2005) and for citrus (Bryla *et al.* 2001). Decomposition of "old" organic carbon is stimulated more by elevated CO_2 and warming together than by elevated CO_2 alone, but this interaction is strongly mediated by nitrogen supply in a warming-CO_2-nitrogen experiment in tunnels with ryegrass swards (Loiseau and Soussana 1999).

In addition, Johnson *et al.* (2000) evaluated the relative importance of chronic warming, nitrogen, and phosphorus fertilization in influencing gross ecosystem photosynthesis, ecosystem respiration, and net ecosystem productivity in wet sedge tundra at Toolik Lake, Alaska. The fertilization with both nitrogen and phosphorus increases ecosystem respiration two- to fourfold in comparison with that in the control. The fertilized plots consistently released more CO_2 than the warmed or control plots. The stimulated respiration from fertilized plots occurs in spite of the fact that the depth of thawed soil is reduced by ~30% in these plots. Nutrient fertilization strongly affects plant cover and results in a fivefold increase in biomass and leaf area (Shaver *et al.* 1998), which in turn regulates seasonal and diurnal CO_2 exchanges. The increase in respiratory CO_2 exchanges is related to changes at the canopy level. However, warming of the Arctic wet sedge ecosystem does not significantly affect ecosystem respiration over the entire season. Soil temperatures in the greenhouse are as much as 8°C higher than the control plots early in the season and 2°C higher later in the season. Increased temperature might cause early canopy development and lengthen the growing season, rather than directly affect instantaneous rates of photosynthesis.

Regardless of the presence or absence of interactions at particular sites of experiments, multifactor experiments provide the opportunity to investigate two or more variables simultaneously in influencing ecosystem processes under the same climatic and edaphic conditions. Such experiments can illustrate areas of uncertainty and offer data to test whether models are appropriately characterizing interactions (Norby and Luo 2004).

PART IV

Approaches

Methods of Measurements and Estimations

There is nothing more important than accurate measurements of CO_2 effluxes in the development of the science of soil respiration. Without accurate measurements, we would not have high confidence in collected data, could not objectively evaluate relative magnitudes of soil respiration among ecosystems, and might not use data to probe mechanisms and to understand the processes of soil respiration. Also dependent on accurate measurements are partitioning of measured soil respiration into different source components, estimation of belowground allocation, and development of models to predict or simulate soil respiration in novel environments. This chapter first presents methodological challenges in measuring soil respiration, then describes measurement methods, and finally evaluates their advantages and disadvantages.

8.1. METHODOLOGICAL CHALLENGES AND CLASSIFICATION OF MEASUREMENT METHODS

Accurate measurements of soil CO_2 efflux are extraordinarily challenging due to the very properties of CO_2 transport in a porous medium of soil. Transport of CO_2 takes place under the influence of both concentration gradients (diffusion flow) and pressure gradients (mass flow). First, as discussed in Chapter 4, the CO_2 concentration in soil is usually many times greater than that in ambient air with a steep gradient. Any measurement methods that disturb the soil CO_2 concentration and/or distort the gradient would result in serious errors. Second, the CO_2 transport from deep soil layers to the surface is driven primarily by diffusion along steep gradients. At the soil surface, CO_2 release is strongly influenced by changes in atmospheric pressure and pressure fluctuation caused by gusts or wind. Since soil is a porous medium, particularly at the soil surface where porosity is usually the highest, small changes in driving forces or mechanisms of CO_2 transport would alter the releases of CO_2 from soil. Third, soil respiration is extremely heterogeneous over time and space (see Chapter 6). It is highly challenging to sample representative spots at representative times and accurately quantify spatial and temporal variability in soil respiration.

To cope with the challenges in measuring soil respiration, scientists have conducted extensive research in the past several decades to develop a variety of measurement methods (Chapter 1). Most commonly used are chamber methods (Fig. 8.1), which provide direct measurements of CO_2 efflux at the soil surface. Depending on the presence or absence of air circulation through chamber, chamber techniques can be categorized as either dynamic or static methods. The dynamic chamber methods allow air to circulate between the chamber and a measurement sensor, which is usually an infrared gas analyzer

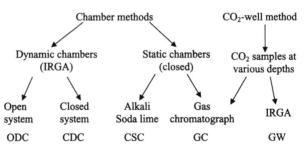

FIGURE 8.1   Classification of direct methods of measuring soil respiration.

(IRGA), to measure CO_2 concentration in the chamber over time. Presently, the most commonly used method in laboratory and field measurements is the closed dynamic chamber (CDC) method, which operates in a fully enclosed mode on soil surface and measures changes in CO_2 concentration in the chamber over a short time. Some scientists employ the open dynamic chamber (ODC) method to measure soil CO_2 efflux. This method operates in a continuously ventilated, quasi-steady-state mode to measure differential changes in CO_2 concentration as air passes over the soil surface. The closed static chamber (CSC) method isolates an amount of atmosphere from the environment during a measurement period as alkali solution or soda lime is used to trap CO_2. A rate of soil efflux is then estimated from the trapped CO_2. With a static chamber, CO_2 concentration can also be measured from air samples at two or more different times during enclosure using syringe samples, which are analyzed with either a gas chromatograph (GC) or IRGA to estimate the rate of soil CO_2 efflux.

The soil respiration can be also estimated from gradients of CO_2 concentration along a soil vertical profile using the gas well (GW) method. Recently, many studies indirectly estimated soil respiration from measurements of net ecosystem exchange (NEE) of carbon made by micrometerological methods such as eddy covariance (Baldocchi et al. 1986, Wohlfahrt et al. 2005) and Bowen-ratio/energy balance (BREB) (Dugas 1993, Gilmanov et al. 2005). The measured NEE is ecosystem respiration at night or the difference between canopy photosynthesis and ecosystem respiration during daytime. The measured NEE is partitioned into photosynthesis, aboveground respiration, and soil respiration.

8.2. CLOSED DYNAMIC CHAMBER (CDC) METHOD

The CDC method is to use a closed chamber to cover an area of ground surface and meanwhile allow air to circulate in a loop between the chamber and a CO_2-detecting sensor (IRGA) during the measurements (Fig. 8.2 and Appendix). Once a closed chamber covers the soil surface, the CO_2 concentration in the chamber rises, due to release of CO_2 from beneath the soil surface (Table 8.1). The rate of CO_2 increase is proportional to the soil CO_2 efflux. To determine the respiration rate, we usually use an IRGA to measure the increase in chamber CO_2 concentration over time. With two CO_2 concentration values measured at the starting and ending points respectively during a short time, the increment in the amount of CO_2 in the chamber can be used to estimate the rate of soil CO_2 efflux (F) with the following equation (Field et al. 1989):

$$F = \frac{(c_f - c_i)V}{\Delta t A} \qquad (8.1)$$

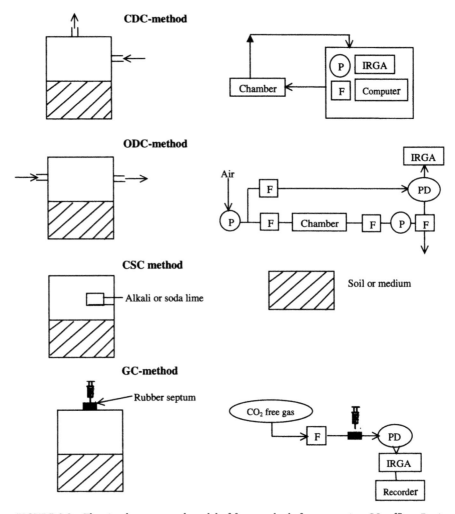

FIGURE 8.2   The simple conceptual model of four methods for measuring CO_2 efflux. P: air pump; F: flow meter; PD: perma pure drier; IRGA: infrared gas analyzer (Modified with permission from Applied Soil Ecology: Bekku *et al.* 1997b).

where c_i is the initial CO_2 concentration, c_f is the final CO_2 concentration, and V is the system volume, including chamber and tube volumes, Δt is the time between the two CO_2 measurement points, and A is the soil surface area covered by the chamber. When multiple data points are taken during one measurement, a gradual increase in the CO_2 concentration in the chamber

TABLE 8.1 Operating principles, advantages, and disadvantages of various measurements estimation methods for soil respiration

Method	Abbreviation	Operating Principle	Advantage	Disadvantage	Comments
Closed dynamic chamber	CDC	Temporal gradient by building up CO_2 in chamber	1. Commercially available and easy to use. 2. IRGA calibration less important due to non-steady state. 3. Short measurement time and flexible for spatial sampling with a portable system.	1. Builds up CO_2 concentration in chamber that distorts the gradient for diffusion. 2. Labor-intensive, with a portable system to sample temporal variation.	Most of the commercially available systems are based on the principles of this method.
Open dynamic chamber	ODC	Differential CO_2 at inlet and outlet	1. High accuracy if artifacts removed. 2. Steady-state measurement. 3. Allows continuous measurements and high temporal resolution.	1. Sensitive to pressure differences inside and outside the chamber. 2. Takes time to reach steady state in chamber. 3. Needs power supply. 4. Requires differential gas analyzer and mass flow controller.	Most of the ODCs are homemade and run continuously.
Closed static chamber (alkali or soda-lime trapping)	CSC	Stored or absorbed by base solutions or soda lime	1. Inexpensive. 2. Potential to integrate the diurnal change. 3. Easy operation in the field and fast laboratory preparation. 4. Off-site analysis of samples.	1. Less accurate due to effects of CO_2 building up on diffusion process. 2. Long enclosure/exposure times cause change in microenvironments in chamber.	

TABLE 8.1—(*Cont'd*)

Method	Abbreviation	Operating Principle	Advantage	Disadvantage	Comments
	CSC			3. Edge effects, especially in small, shallow chambers.	
Gas chromatograph	GC	Discrete temporal gradient by building up CO_2 in chamber	1. Parallel analyses of other trace gases and isotopic composition. 2. Easy to use and samples can be stored.	1. Labor-intensive to sample temporal variation. 2. Needs a trajectory of headspace CO_2 building up to estimate respiration correctly. 3. Requires a GC in the lab.	
Gas-well	GW	Spatial gradient by diffusion	Estimation of source depths of CO_2 production.	Difficulty in estimation of soil and air diffusivity.	
Eddy-flux	EF	CO_2 mixing ratio in eddies	1. Nonintrusive. 2. Measured under natural turbulent conditions. 3. Sampling a large surface area to represent spatial heterogeneity.	1. Errors inherent in NEE measurements due to fetch requirements and nighttime atmospheric inversion. 2. Difficult to partition NEE into photosynthesis, aboveground, and soil respiration.	Data of NEE are widely available from networks of flux measurements

can be fitted by a linear regression equation with a slope of b. From the slope b, the respiration rate is estimated by:

$$F = \frac{bV}{A} \tag{8.2}$$

If the enclosure time of the CDC system is long enough to alter the CO_2 gradient, equation 8.2 is no longer applicable. Chamber enclosure could increase CO_2 concentration in the upper part of the soil profile. Thus, fluxes calculated from fitting a linear equation to data of CO_2 concentrations within the chamber are less than those expected under the natural condition outside the chamber, because a proportion of the CO_2 produced is stored within the soil profile while the chamber is in place. The discrepancy caused by this effect increases with air-filled porosity and decreases with the height of the chamber (Conen and Smith 2000, Table 8.1). To correct the depression of CO_2 releases from soil by high CO_2 concentrations in the chamber, a nonlinear regression equation is required (Davidson et al. 2002b).

For field measurements of soil respiration, a collar that exactly matches the size of the chamber is usually installed to a certain depth in the soil to reduce CO_2 leaking. The bottom edge of the soil chamber is sharpened. A foam gasket around the flange of the soil chamber provides a seal between the chamber and the collar. Pressure equilibrium between the air in the chamber and the surrounding air is maintained by a tube or relief vent. Air is mixed in the chamber using a diaphragm air-sampling pump that circulates air through the chamber at a certain flow rate, depending on chamber design. Chamber air is usually withdrawn at the top of the soil chamber, passes through an IRGA for continuous measurements of CO_2 concentration, and reenters the chamber through an air-dispersion ring at the bottom. Chamber CO_2 concentration should not be allowed to build up too far above ambient CO_2 concentration, or the flux will be underestimated because soil CO_2 efflux decreases with chamber CO_2 concentration (Fig. 8.3). The best estimate of the flux is obtained when concentration inside the chamber is equal to that outside. Thus, the system design should make measurements of CO_2 efflux around ambient CO_2 concentration. The commercial products are usually designed to scrub the chamber concentration to just below an ambient target and then measure CO_2 concentration as it rises to slightly above the ambient. Soil CO_2 efflux can be obtained in about 1 to 15 minutes, depending on the system design and the magnitude of the soil CO_2 efflux.

Most of the commercially available instruments for measurement of soil CO_2 efflux are built according to the principles of the CDC method (see Appendix). The soil respiration system developed by PP Systems in Hitchin, U.K., consists of the soil respiration chamber and either the Environmental Gas Monitor or Differential CO_2/H_2O Infrared Gas Analyzers. The portable

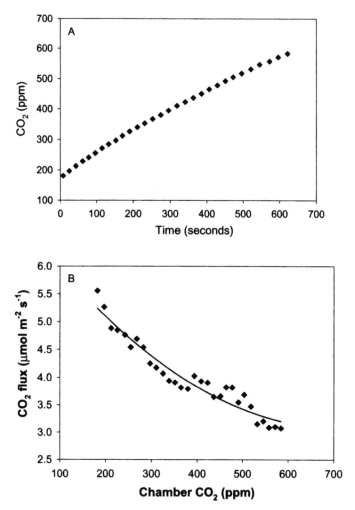

FIGURE 8.3 Panel A: chamber CO_2 concentration varies with duration when the closed chamber covers soil surface. Panel B: soil CO_2 efflux dependency on chamber CO_2 concentration. (Redrawn with permission from Chemical Geology: Welks *et al.* 2001).

CDC systems developed by the Li-Cor BioSciences in Lincoln, Nebraska, combine the Li-Cor 6200 gas analyzer with the Li 6000-09 chamber or the Li-Cor 6400 gas analyzer with the Li 6400-09 soil chamber. A newly developed, fully automated system, the Li-Cor 8100 is also based on principles of the CDC method and can repeatedly measure soil CO_2 efflux at one spot over time. The system includes the analyzer control unit, which houses the system electronics, the IRGA, and the movable chamber. The portable soil respiration measurement system, SRC-1000 and SRC-2000, developed by Dynamax in Houston, Texas, consists of a console programming unit and a soil respiration chamber.

As an example, the Li-Cor 6400 system with 6400-09 soil chamber is further described here. The Li-6400-09 soil respiration chamber is equipped with a pressure relief vent. The standard chamber with a diameter of 95.5 mm and a volume of 991 cm^3 is placed on a PVC collar (diameter 103 mm, height 50 mm) installed to a soil depth of 20 to 30 mm. Air is circulated from the chamber to the IRGA and back by a mixing fan. Before each cycle of flux measurement, air in the chamber headspace is scrubbed down 10 to 20 ppm below the ambient CO_2 concentration and then allowed to rise as a consequence of CO_2 efflux. During this period, at least five datum points of CO_2 concentrations are taken. This procedure can be repeated a few more times for each measurement. A measurement cycle usually lasts one to two minutes in grasslands and forests or two to five minutes in soil with very low rates of soil respiration. The efflux is calculated by fitting a nonlinear curve to measured CO_2 concentrations in the chamber over time.

8.3. OPEN DYNAMIC CHAMBER (ODC) METHOD

The ODC method uses a differential mode to estimate CO_2 effluxes in contrast to the closed dynamic system that uses changes in CO_2 concentration over a period of time (Fig. 8.2). With the ODC method, ambient air flows from an inlet through a chamber to an outlet (Fang and Moncrieff 1998, Iritz et al. 1997, Table 8.1). The air leaving the chamber is enriched in CO_2 concentration relative to the air entering the chamber, due to CO_2 release from respiration at the soil surface. Assuming that the rates of respiration and air flow through the chamber are constant, the soil respiration can be estimated by:

$$F = \frac{u_o c_o - u_e c_e}{A} \tag{8.3}$$

where c_o is the CO_2 concentration in the air leaving the chamber, c_e is the CO_2 concentration in the air entering the chamber, u_e is the rate of air flow entering the chamber, u_o is the rate of air flow leaving the chamber, which differs from air flow entering the chamber because soil respiration adds CO_2, and A is the soil surface area covered by the chamber.

The open system with differential mode has been extensively used in study (Witkamp and Frank 1969, Edwards and Sollins 1973, Kanemasu et al. 1974, Denmead 1979, Fang and Moncrieff 1996, Rayment and Jarvis 1997, Lund et al. 1999, Pumpanen et al. 2001). For example, Edward and Riggs (2003) have developed a movable-lip chamber with the open system. A chamber is permanently installed at soil surface with a movable lip. The lip is open most of the time. When a measurement starts, the lip closes over the chamber in response to a control signal. It remains closed for a period of several minutes while the measurement is made. During the measurement, the IRGA operates in differential mode when equivalent flow rates of reference gas (ambient air) and sample gas (air exiting chamber) are maintained with mass flow controllers. A large mixing bottle is usually used to buffer frequent changes in ambient CO_2 concentration. Once the measurement is taken, the lip opens again to allow normal drying and wetting of the soil and litterfalling into the soil surface between measurements. The movable-lip, ODC designed by Edward and Riggs (2003) has been adapted by Dynamax, Inc. to be a commercial instrument called SRC-MV5 (see Appendix).

With the ODC method, the CO_2 efflux is obtained from the difference in the amounts of CO_2 between the inlet air and the outlet air of the chamber (Equation 8.3). A difference between the inflow and the outflow rates can cause a pressure difference between the chamber and the ambient air and thus can generate additional air flow between the chamber and the soil. Even a pressure difference of 1 Pascal (Pa) can cause substantial errors in CO_2 efflux measurements (De Jong et al. 1979; Fang and Moncrieff 1996, 1998; Lund et al. 1999, Table 8.1). Therefore, the design of an ODC system requires a minimal pressure difference between the chamber interior and the atmosphere to eliminate any mass flow of air into or out of the chamber. In practice, it is inevitable that the chamber is leaky to some extent during a measurement due to the porous nature of soil and pressure differences between the inside and outside of the chamber. In the past, air seals were usually achieved by maintaining a slight positive pressure within the chamber, ensuring that ambient air did not enter the chamber and dilute the air inside (Šesták et al. 1971). Air seals may equally well be created with a slight negative pressure within the chamber, drawing in ambient air and ensuring that no chamber air is lost (Rayment and Jarvis 1997). The ODC system, Dynamax SRC-MV5, uses specially designed inlet and outlet fittings to ensure that there is no internal pressure gradient in the chamber. Also, accurate measurements of air flow rates through the chamber are critical for the calculation of soil respiration rates (Rayment and Jarvis 1997).

8.4. CLOSED STATIC CHAMBER (CSC) METHODS

The CSC methods cover an area of soil surface with a chamber having a chemical absorbent inside to absorb CO_2 molecules within a certain time (Fig. 8.2 and Table 8.1). The chemical absorbents for CO_2 trapping include alkali (NaOH or KOH) solution and soda lime, which consists of NaOH and $Ca(OH)_2$. The alkali solution method is probably the oldest method of soil respiration measurement (Lundegårdh 1927), while the soda-lime method is probably the most frequently used static technique because it is inexpensive and easy to use (Monteith et al. 1964, Edwards 1982, Jensen et al. 1996, Grogan 1998). Since the chamber is closed without air flow except CO_2 releases from soil, this method is sometimes also called the non-steady-state or non-through-flow chamber technique.

ALKALI TRAPPING

Soil respiration is determined using alkali traps by absorbing CO_2 released from the soil into a sealed headspace chamber for a specific period of time using NaOH or KOH solutions. At the end of the adsorption period, the total mass of CO_2 in the alkali traps is determined by titrating the NaOH or KOH solutions with a dilute HCl to a set pH value. The rate of soil respiration (F) is calculated using the total amount of CO_2 trapped over an absorption period (Δt_{abs}):

$$F = \frac{C_{trap} - C_{blank}}{\Delta t_{abs} A} \tag{8.4}$$

where C_{trap} is the amount of CO_2 trapped in the enclosure, C_{blank} is the amount of CO_2 in a blank control solution that is used to account for any bias caused by contamination of the alkali solution, and A is the area of the surface covered by the chamber.

The estimated rate of soil respiration using this technique varies with different solution strengths, volumes, chamber sizes, absorption times, and absorption areas (Kirita 1971, Gupta and Singh 1977). An increase in the normality of NaOH from 0.25 to 0.75 N has no effect on CO_2 absorption capability when sufficient volumes (>30 ml) of NaOH are used (Gupta and Singh 1977). An increase in the absorption area of up to 19.9% of the total surface area of the ground enclosed has no effect on CO_2 absorption at 0.25 and 0.5 N alkali concentrations either. An increase in the volume of NaOH beyond 30 ml has no effect on the measured rate of soil respiration at the concentrations tested in the range of 0.5 to 2 N (Minderman and Vulvo 1973). However, the rate of CO_2 efflux determined by the static chamber method is very sensitive to adsorption times, exhibiting a power decrease with time (Fig. 8.4). The efflux rates from a minicosm study decrease with absorption time from 20.3 mg CO_2 m^{-2}h^{-1} for absorption time of 1 h to 3.7 mg CO_2 m^{-2}h^{-1} for an absorption time of 48 h at temperature of 5°C (Kabwe et al. 2002). Similarly, the flux rates from the mesocosm decrease from 276 mg CO_2 m^{-2}h^{-1} for the absorption time of 1 h to about 24 mg CO_2 m^{-2}h^{-1} for the absorption time of 110 h. The CO_2 flux rates with the alkali-trapping technique reported in the literature are obtained mostly under long absorption times, typically over 24 h.

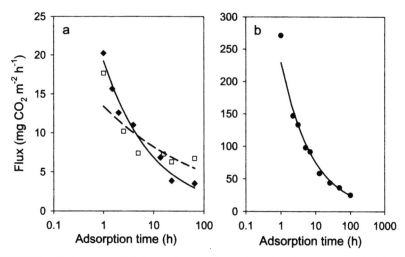

FIGURE 8.4 Variations in CO_2 fluxes from various adsorption times measured with static chambers (alkali traps) for (a) the low temperature (□) and high temperature (♦) minicosms and (b) mesocosm (•) (Redrawn with permission from Journal of Hydrology: Kabwe *et al.* 2002).

After reviewing the literature on measurements made with the CSC methods, Rochette and Hutchinson (2003) made recommendations for optimizing the design of the measurement procedure. Their recommendations include (1) that the optimal strength of the alkali solution is ≈0.5 to 1.0 M; (2) that the alkali trap should have a total capacity approximately three times greater than the amount of CO_2 expected to be released during the deployment period; (3) that a 20% ratio of exposed alkali trap area to emitting soil surface area provides good absorption efficiency in many situations, but can be altered when needed to keep headspace CO_2 concentration as close as possible to the ambient level; (4) that the chamber should be nonvented and should have good seals that minimize CO_2 exchange between the chamber and its surroundings; and (5) that the deployment period should be at least 12 and preferably 24 h to minimize measurement bias due to the initial non-steady-state condition, as well as bias due to chamber-induced temperature disturbances.

SODA-LIME TRAPPING

The soda-lime technique has been used for more than 40 years to measure CO_2 effluxes from soil under field conditions (e.g., Monteith *et al.* 1964). Soda lime is a mixture of sodium and calcium hydroxides that reacts with CO_2 to form carbonates. The amount of CO_2 adsorbed by soda lime in a chamber over the soil surface is determined by the gain in soda-lime dry weight during the sampling period. The increase in weight is directly related to the absorption of CO_2 with a correction factor. Protocols for its use are described in detail by Zibilske (1994). In brief, oven-dried (105°C) soda lime (1.5 to 2.0 mesh) is put in an open jar and placed on the soil surface beneath a closed chamber. Blanks that are necessary for CO_2 flux calculations are sealed in cylinders. Soda-lime traps are removed after 24 hours, oven-dried, and reweighed to determine the amount of CO_2 absorbed.

The CO_2 adsorption rate of soda lime is rarely in equilibrium with the efflux rates to be measured at the soil surface, leading to potential errors in measurements. The method tends to overestimate soil CO_2 efflux in its low range and underestimate it in its high range compared with dynamic methods (Yim et al. 2002). The technique can potentially underestimate soil surface CO_2 effluxes by 10 to 100% (Norman et al. 1992, Rochette et al. 1992, Haynes and Gower 1995, Nay et al. 1994). Thus, it becomes necessary to use calibration curves to compensate for this error (Edwards 1982, Grogan 1998). Usually, larger errors occur for chambers that are not well designed to match the rates of soil respiration they are intended to measure (Hutchinson and Rochette 2003).

Healy et al. (1996) numerically evaluated the accuracy of measurements by the static chamber. Enclosure with a static chamber on the soil surface slows down CO_2 efflux in comparison with that in the absence of the chamber, primarily resulting from distortion of the soil CO_2 concentration gradient. As a consequence, the CO_2 concentration gradient decreases in the vertical component and increases in the radial component, thus decreasing the rate of diffusion in the vertical direction. To improve the accuracy of measurements, the CSC method should be designed to mix air in the chamber headspace thoroughly, minimize deployment time, maximize the height and radius of the chamber, and push the rim of the chamber into the soil to avoid leaking.

When serious design deficiencies are avoided, the CSC methods offer simple, inexpensive means to obtain multiple, reliable, time-integrated estimates of soil respiration, particularly at remote locations (Table 8.1). The measurements with the soda-lime or alkali trapping can provide a single, integrated estimate of soil respiration over a daily time-scale that incorporates the effects of diurnal fluctuation in abiotic variables on CO_2 efflux. The methods are robust and economical, making them appropriate for a large number of repeated field measurements that are necessary to account for enormous spatial heterogeneity in soil surface CO_2 effluxes.

8.5. GAS CHROMATOGRAPH (GC)

In addition to being continuously measured with an IRGA on site, gas samples can be taken from the field with syringes and brought back to the laboratory for analysis with a GC or IRGA. A variant of this method is to place an IRGA such as LiCor-7500 in the closed chamber without air circulation. The procedure of taking gas samples is similar to the CSC methods. Chambers are either newly covered on an area of ground surface or permanently installed with removable lids. The lids are opaque, to eliminate CO_2 fixation by plants in the chamber during measurements. The lids are fitted with rubber septa for syringe sampling (Fig. 8.2). The chamber headspace is sampled by syringe soon after sealing the lip and at intervals every a few minutes for a short time (Gulledge and Schimel 2000). Gas samples are usually taken with 10 mL glass syringes and stored in the sealed syringes until analysis. As samples are extracted with the needle, compensation air is simultaneously drawn into the chamber through a pressure equilibrium tube.

Gas samples in the sealed syringes are analyzed for CO_2 or O_2 concentrations (or other trace gases) using a GC (Gulledge and Schimel 2000, Knoepp and Vose 2002, Abnee et al. 2004) or IRGA (Bekku et al. 1995, 1997b). A GC is a device used to separate components in a gas sample. When it is injected into a gas stream, a gas sample is swept through the packed column or the open tubular column (e.g., stainless steel Porapak_N column) with a thermal conductivity detector (TCD) plumbed in series. The ultrasonic detector, which is more sensitive than a TCD, is also used for CO_2 analysis (Blackmer and Bremner 1977). Some molecule components of air samples are slowed down more than others, so that different components exit the column sequentially.

After the sample is pulled out of the flask with a syringe, the syringe is inserted into the injector with a finger pressed on the plunger to counteract the pressure within the GC. Injections should be done quickly. The plunger is quickly depressed to withdraw the syringe needle. The output from the detector (in minivolts) is transformed to soil air CO_2 concentration that is measured by comparing integrated peak areas of samples with standard gases. Once data of CO_2 concentration are obtained from the GC, the soil CO_2 efflux can be estimated with either Equation 8.1 for two-point measurements, Equation 8.2, or some forms of nonlinear equations for multiple-point measurements.

The GC method can potentially underestimate the rate of soil CO_2 fluxes in comparison with other methods by up to 45% (Knoepp and Vose 2002). The measurement period also significantly affects the flux rates due to decreased CO_2 releases from soil with increased CO_2 concentration inside the chamber. When the measurement period increases from 10 to 30 minutes, the flux rates are underestimated by 15% on coarse and dry fine sands and by 10% on wet fine sands (Pumpanen *et al.* 2004). The advantage of the GC method is that the fluxes of several gas species (e.g., CH_4, CO_2, NO_x) can be measured simultaneously from the same gas samples (Table 8.1).

8.6. CHAMBER DESIGN AND DEPLOYMENT

CHAMBER DESIGN

To accurately measure CO_2 efflux rates at the soil surface, the chamber methods have to be designed to account for several factors (Table 8.1). Although some of these factors have been mentioned in the above sections, here we provide detailed discussion on them. First, the release of CO_2 at the soil surface is regulated primarily by the concentration gradient between the soil and the ambient atmosphere. Building up CO_2 concentration in the chamber, particularly with the closed-chamber methods, will reduce the CO_2 concentration gradient and then depress the CO_2 release, leading to underestimation of soil CO_2 efflux (Healy *et al.* 1996). Second, since soil is a porous medium, a small pressure differentiation between the inside and outside of the chamber can alter air flow into and out of soil and thus substantially affect soil CO_2 efflux (Kanemasu *et al.* 1974, Fang and Moncrieff 1996). The mass flow controller that regulates pressure with the ODC method therefore has to be carefully selected and adjusted to maintain balanced pressure. With the closed-chamber methods, building up CO_2 concentration can alter pressure in the chamber, causing a divergence of flux away from the chamber toward the outside of the chamber (Norman *et al.* 1997). Third, air in the chamber headspace has to be thoroughly mixed so that the chamber CO_2 concentration can be sampled correctly. The air mixing needs to be achieved without causing localized pressure gradients. Fourth, when a closed chamber is placed on a moist soil surface on dry, sunny days, air temperature and water vapor in the chamber rapidly increase. As a consequence, the air CO_2 partial pressure proportionally decreases, possibly resulting in underestimation of the CO_2 efflux. In this case, a dilution factor is needed to correct the humidity effect.

To avoid disturbance of soil each time when a measurement is made, soil collars need to be permanently installed at the very beginning of a study. Soil CO_2 concentration in subsurface layers is usually several times higher than that at the surface. Disturbance of soil will release a large amount of CO_2 from soil and cause overestimation of CO_2 efflux. Ideal soil collars are large enough to cover bare surface spots within a canopy. Thus, soil collars can be much bigger for measurements in forests than for those in grasslands. In cases where soil collars could not be placed on soil surface without plants, plants have to be clipped one or a few days before measurements are made to eliminate plant respiration. Soil collars may also have "edge effects" due to altered soil physical properties or plant growth. Because collars are usually located between impermeable areas such as rocks or larger roots near the surface, measured efflux from small chambers is likely to be larger than flux rates averaged over a large area.

Soil CO_2 efflux can be measured accurately only by a system that does not alter either soil respiratory activity, the CO_2 concentration gradient, the pressure, or air motion near the surface. In summary, chamber designs must consider the following principles for reliable measurements:

1. Minimize changes in natural microclimate within chamber.
2. Minimize disturbances of soil.
3. Do not cause change in pressures within a chamber.
4. Do not build up or deplete CO_2 enough to cause substantial changes in the gradient of CO_2 concentration or leak CO_2 into or out of the chamber.
5. Measure water vapor pressure with a correction factor.
6. Have relatively stable intake CO_2 concentration for an open dynamic chamber.

Commercially available instruments have been designed mostly with these principles in mind. For example, the LiCor-6400-09 soil respiration chamber has a pressure equilibration tube, air-mixing fan, and automatic program for scrubbing CO_2 in chamber to avoid its building up. Other CDC systems, such as soil respiration system made by PP Systems, SRC1000 and 2000 by Dynamax, and the Li-Cor 8100, are similarly designed. The commercially available instrument, Dynamax Model SRC-MV5 and PP systems model CFX-2 is the ODC system (see Appendix).

CHAMBER DEPLOYMENT

Even with a well-designed chamber and carefully selected spots for soil collar installation, accuracy of measurements may still depend on deployment of chambers, since all the chamber methods have to deal with spatial and temporal variability in soil respiration (Table 8.1). To cope with the variability, measurement chambers that have been developed to measure soil surface respiration are usually deployed in three ways: manual measurements with a portable-chamber system, automatic measurements with one movable-lip chamber system, and automatic measurements with a multiple-chambers system.

The portable-chamber system, such as LiCor-6400-09 (directly linked to the LiCor-6400 IRGA), can be taken to different locations to take measurements at spots with different experimental treatments or spatial variations. The portable-chamber system usually requires preinstalled collars to reduce soil disturbances. It usually requires personal attendance to collect data and therefore has low temporal resolution. To sample representative soil CO_2 efflux, measurements are made at a certain time of day (e.g., 1000 to 1500). To avoid variability caused by rain events, measurements are usually not taken immediately after rains. Since soil respiration can vary dramatically with soil moisture after rains, particularly in arid and semiarid lands (Lee et al. 2002, Liu et al. 2002a, Xu et al. 2004), it is very difficult to have representative measurements of soil respiration within the wetting-drying cycles.

A movable-lip chamber is usually installed permanently at the soil surface. The measurement of soil respiration can be made with either the CDC (Goulden and Crill 1997, King and Harrison 2002) or ODC methods (Rayment and Jarvis 1997, Edwards and Riggs 2003). Since the lip is open most of the time, the system allows normal drying and wetting of the soil and litterfall into the soil surface between measurements. The movable-lip chamber system provides a high temporal resolution of measurement of CO_2 efflux. It can be operated continuously for long periods while the soil microclimate naturally fluctuates over diurnal, seasonal, and interannual time-scales. The measured soil respiration with the movable-lip chamber system is highly comparable to that measured by a portable-chamber system at individual points (Edward and Riggs 2003). Cumulative soil respiration over several weeks is lower with the movable-lip chamber than with the portable chamber. While the movable-lip chamber can adequately provide high temporal resolution, it is expensive to have many chambers in different locations to quantify spatial variability.

A cluster of chambers that connect to an automatically sampling IRGA system can record spatial variability on a local scale with high resolution of temporal variability. The multichamber system can use either a CDC or ODC design. In general, such a system comprises an IRGA and several parallel channels, each linked to a chamber and the sample and reference gas units. One gas unit consists of a pump, a mechanical flow controller, and a magnetic valve. In addition, the gas unit can allow both overpressure and underpressure to be applied to the chambers. Behind the magnetic valve, the air stream passes through an electronic flow meter and a gas cooling unit to an IRGA (Kutsch et al. 2001). Commercially available products usually allow researchers to choose the number of channels to be used for measurements. For example, the soil respiration and integrated measurement systems from Dynamax, Inc. offer four choices: 4, 8, 12, and 24 channels. To date, most of the multiple chambers are built by researchers themselves according to their own needs (e.g. Low et al. 2001, Sabre et al. 2003, Liang et al. 2004, 2005).

   Chamber measurements of soil respiration usually yield systematic errors whenever air mixing in the chamber headspace differs from that at the soil surface prior to the chamber deployment. Due to turbulence fluctuation, the predeployment air fluxes at the soil surface are rarely at a steady state (see Chapter 4). Since gusts and wind cause random variation in the predeployment air movement at the soil surface, it is impossible to design a chamber technique and/or sampling scheme enabling air mixing in the chamber headspace to mimic precisely that prior to chamber deployment. Thus, it seems inevitable that measurement errors occur in individual observations, particularly when a chamber significantly alters atmospheric mixing processes near the soil surface (Hutchinson et al. 2000). The errors may be averaged out with many observations.

   Based on an assessment by Davidson et al. (2002b) of artifacts, biases, and uncertainties in chamber-based measurements of soil respiration, distortion of diffusion gradients causes underestimation of effluxes by less than 15% in most cases. This underestimation can be partially corrected for with curve fitting and/or can be minimized by using brief measurement periods. Underpressurization or overpressurization of the chamber induced by flow restrictions in air circulation designs can cause significant errors, which can be avoided with properly sized chamber vents and unrestricted flows.

8.7. GAS-WELL (GW) METHOD

The GW method samples CO_2 and O_2 concentrations at two or more depths along a vertical profile of soil. The method usually requires a permanent installation of CO_2 sampling tubes (e.g., stainless steel tubes) in midway of each horizon in the litter, organic matter, and mineral soil layers. The ends of the tubes have several holes to allow air to pass through the tubes and to be collected in syringes. These air samples in syringes are then injected into an IRGA through a mixing chamber or GC in the laboratory to determine CO_2 concentrations of the samples. An automated sampling system (Fig. 8.5) has been developed by Hirsch et al. (2002, 2004) to measure CO_2 concentrations at several depths in the soil. At each depth, air is withdrawn from the soil air-filled pore space by a diaphragm pump through a microporous Teflon tube 25 cm in length into a solenoid manifold, which selects sampling channels from different depths. After entering the sampling system, the air is dried, filtered, and transported to an IRGA to measure CO_2 concentration with a specific flow rate. The air from different channels is alternately sampled for one to two minutes each, once an hour.

   Measured CO_2 concentrations at different depths usually form a gradient of CO_2 concentrations through the soil profile (see Chapter 4). The gradient,

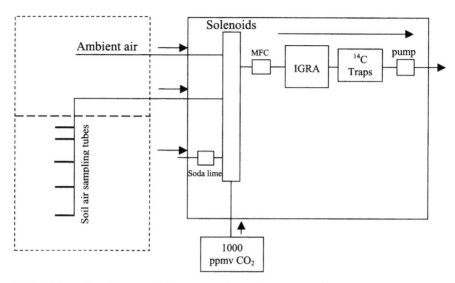

FIGURE 8.5 Flow diagram of the automated sampling system of the GW method. At each depth, air is withdrawn from the soil air-filled pore space by pump through tubing into a solenoid manifold. The air is dried and filtered. Flow is controlled by a mass-flow controller (MFC) for measuring CO_2 concentration using IRGA. The IRGA is zeroed with soda lime and calibrated from a calibration tank of 1000 ppmv CO_2 once an hour. After exiting the IRGA, the sample air passes through a second solenoid manifold where stainless steel molecular sieve is to trap ^{14}C for isotope measurement (Redrawn with permission from Journal of Geophysical Research: Hirsch *et al.* 2003).

together with diffusion of gas, is used to calculate soil respiration in each layer (de Jong and Schappert 1972). The GW method assumes that diffusion is the major mechanism by which gases move vertically in soils; it is described by the equation:

$$F = -D_s \frac{dc}{dz} \qquad (8.5)$$

where F is flux of gas in unit of g CO_2 cm^{-2}s^{-1}, D_s is diffusion constant in soil in unit of cm^{2}s^{-1}, c is concentration of gas (g CO_2 cm^{-3} air), and z is depth (cm). The diffusivity coefficient, D_s, varies with soil porosity and tortuosity (Dörr and Münnich 1990). The negative sign in Equation 8.5 indicates that the flux flows in the direction from high CO_2 to low CO_2. Equation 8.5 can be modified to incorporate a source term of CO_2 production for the conservation of matter:

$$\frac{dG}{dt} = -\frac{dF}{dz} + S \qquad (8.6)$$

where G is the amount of gas ($g\,cm^{-3}$ of soil), t is time, and S is respiratory CO_2 production in layer z ($g\,cm^{-3}$ of soil). The amount of CO_2 per cm^3 of soil can be calculated by:

$$G = cV_A \qquad (8.7)$$

where V_A is air-filled pore space in $cm^3\ cm^{-3}$ of soil. Equations 8.6 and 8.7 combined give:

$$\frac{dc}{dt} = \frac{1}{V_A}\left[-\frac{dF}{dz} + S\right] \qquad (8.8)$$

If the diffusion is considered to be a steady-state process, concentration, c, is constant with time. Then Equation 8.8 reduces to

$$\frac{dF}{dz} = S \qquad (8.9)$$

Either Equation 8.8 or 8.9 may be used to calculate the amount of CO_2 respired when combined with Equation 8.5.

The GW method has been used to estimate soil CO_2 production in different soil layers and surface CO_2 efflux at different sites, for example, in eastern Nova Scotia (Risk *et al.* 2002a, b), in an old-growth neotropical rainforest, La Selva, in Costa Rica (Schwendenmann *et al.* 2003), and in other ecosystems (Vose *et al.* 1995, Kabwe *et al.* 2002). The GW method with an automated sampling system is used to measure the seasonal cycle of CO_2 production and isotope ^{14}C in different soil depths at a northern old black spruce site in northern Manitoba (Hirsch *et al.* 2002). Deep soil respiration is sensitive to soil thaw. Much of the CO_2 produced in deep layers results from decomposition of old organic matter that is fixed from the atmosphere centuries ago, rather than root respiration. The daily cycle in the top 20 cm of the boreal forest litter layer is very strong, with a small surface CO_2 gradient and low concentrations during the day and a large surface gradient and high concentrations at night (Hirsch *et al.* 2004).

The GW method is based on a few assumptions that may influence the accuracy of estimated CO_2 efflux. For example, the method assumes that the gradient of CO_2 concentration in the soil surface layer can be approximated by the gradient in deep soil layers, since it is very difficult to measure concentration gradients at the soil surface. This assumption works only if the mass of the gas in question is conserved. However, most fine roots in ecosystems, particularly in forests and hot deserts, are distributed and thus generate great sources of CO_2 in the top layers of soil. The source strengths of CO_2 production that vary with each segment (as defined by the depth of the gas wells) along a soil profile must be taken into consideration when the GW method is used to estimate soil respiration. Data on source strengths of CO_2 production within these segments of the soil are rarely available. In addition, the gradient of the CO_2 concentration in the soil surface layers is strongly affected by soil moisture as shown in a boreal forest (Billings *et al.* 1998) and gusts (Hirsch *et al.* 2004).

The GW method is highly dependent on soil and air diffusivity (D_s), which are very difficult to estimate. There are many algorithms for D_s (reviewed by Mattson 1995, Johnson *et al.* 1994, and Moldrup *et al.* 1996), but all involve effective porosity (air-filled pore space). Because the diffusivity of CO_2 in air is many times greater than in water, water effectively restricts CO_2 diffusion from soils by reducing effective pore space. The effects of moisture content on D_s are complicated by the pressure of dead-end pores and changes in the size distribution of gas-filled pores. Yet most models of soil respiration using the GW method make a simple assumption that D_s is reduced in proportion to the reduction in air-filled pore space. Reviews by Colin and Rasmuson (1988), Mattson (1995), Moldrup *et al.* (1996), and Šimůnek and Suarez (1993) provide details of various models for D_s and its changes with soil moisture content. Nevertheless, all models predict that adding water to soils will reduce D_s. If irrigation has no instantaneous effect on CO_2 production, its net effect is first to drive the high CO_2-concentrated air out of the soil and then to have a temporary reduction in soil CO_2 efflux until a new steady state is achieved. Thus, addition of water can cause either increases (e.g., deJong *et al.* 1974, Wiant 1967) or decreases (Buchmann *et al.* 1997, Kowalenko *et al.* 1978) in soil CO_2 efflux, due to changes in CO_2 concentration in soil airspace and gas diffusivity.

The efflux of CO_2 by processes other than diffusion, such as gusts, convection, and atmospheric pressure fluctuations (see Chapter 4), can affect the accuracy of the GW method (de Jong 1972, Hirsch *et al.* 2004). Such events are excluded from chambered methods by the chambers themselves and are ignored in the GW method. In the presence of advective flows in the soil induced by pressure changes above the surface, Equation 8.5 has to be modified (Schery *et al.* 1984) to be:

$$F = -D_s \frac{dc}{dz} + vc \qquad (8.10)$$

where v is the advective velocity (i.e., mass flow of air through the soil).

8.8. MISCELLANEOUS INDIRECT METHODS

Soil respiration has also been estimated by a variety of indirect methods. Those methods usually measure ecosystem respiration or NEE of carbon, which is the difference between canopy photosynthesis and ecosystem respiration during daytime and the ecosystem respiration at night. From measured NEE or ecosystem respiration, soil respiration may be derived.

The commonly used methods of measuring NEE are eddy covariance and (Bowen-ratio/energy balance) BREB. The basic concept of these micrometeorological methods is that gas transport from the soil surface is accomplished by eddies that displace air parcels from the soil to the measurement height. The eddy-covariance technique ascertains the net exchange rate of CO_2 across the interface between the atmosphere and a plant canopy by measuring the covariance between fluctuations in vertical wind velocity and CO_2 mixing ratio (Baldocchi et al. 2003). The BREB method is based on a surface energy balance that assumes similarity between the turbulent exchange coefficients of sensible heat, latent heat, CO_2, and momentum to compute net CO_2 fluxes from flux-gradient relationships among water vapor, CO_2, and heat (Denmead 1969, Baldocchi et al. 1981, Dugas et al. 1997, Gilmanov et al. 2005). The accuracy of CO_2 fluxes calculated using the BREB method is influenced by the assumed equality of the turbulent exchange coefficients and measurement errors of input variables, such as net radiation, temperature, and humidity gradients (Dugas et al. 1997). Other micrometeorological methods include the aerodynamic (Lemon 1969, Takagi et al. 2003), eddy accumulation (Pattey et al. 1992, 1993; Katul et al. 1996; Baker 2000), mass balance (Denmead et al. 1996, 1998), dual tracer (Denmead 1995), and surface renewal methods (Paw et al. 1995; Spano et al. 1997, 2000).

Eddy covariance and BREB systems are nonintrusive micrometeorological methods that impose minimal influences on microenvironments of the soil surface compared with chamber-based methods (Dugas 1993). Those methods can measure CO_2 efflux continuously over long periods and integrate large surface areas (Baldocchi 1997) so that the spatial heterogeneity is integrated under "natural" turbulent conditions Table 8.1. The successful applications of these techniques depend on several conditions. An extensive, homogeneous upwind fetch and atmospheric steady-state conditions are prerequisites (Baldocchi and Meyers 1991). The micrometeorological methods are usually not suited to small-scale measurements (Jensen 1996), and the implementation is expensive compared with other ways (Le Dantec et al. 1999). Eddy-covariance methods make it difficult to measure understory fluxes when turbulence is low and the footprint is difficult to identify. Correction of the nighttime fluxes is also needed when both storage during stable conditions and advection of the carbon flux exist on the site Table 8.1.

It is still very difficult to partition measured NEE into soil respiration, aboveground plant respiration, and canopy photosynthesis. The ecosystem respiration can be derived from nighttime eddy flux measurements above the canopy or by analysis of the daytime measurements (Falge et al. 2003). Distinction between respiration from soil and from aboveground plant parts is not possible without using empirical estimates or other supplemental measurements. Correlation of the eddy-covariance flux with chamber measurements can be used for correction and estimation of soil CO_2 effluxes over a larger area (Subke and Tenhunen 2004).

Other indirect methods for estimation of soil respiration include Lagrangian analysis of canopy carbon source and sink profiles (Katul *et al.* 1997), nocturnal measurements of CO_2 concentration profiles in planetary boundary layer (Denmead *et al.* 1996), and carbon balance based on litterfall-soil respiration ratio (Raich and Naderhoffer 1989, Davidson *et al.* 2002a). For example, Katul *et al.* (1997) used the Lagrangian dispersion model to infer soil respiration from canopy CO_2 profiles that the near-ground air is a CO_2 source.

8.9. METHOD COMPARISON

Performances of different measurement methods have been compared in a number of studies (Bekku *et al.* 1997b, Norman *et al.* 1997, Le Dantec *et al.* 1999, Janssens *et al.* 2000, Davidson *et al.* 2002b, Yim *et al.* 2002, Liang *et al.* 2004, Pumpanen *et al.* 2004). Comparison of measurement systems is usually conducted with either known rates of CO_2 effluxes from a surface against which all the systems can be compared with or repeated measurements by several systems, one after another, at one location Table 8.1. With known effluxes from the surface of a simulated soil, Nay *et al.* (1994) evaluated the methods using CSC and CDC. According to Edwards (1982), the CSC with the soda-lime absorbent overestimates CO_2 efflux in its low range and underestimates it in the high range. According to Norman *et al.* (1992), the CDC method with IRGA consistently underestimates efflux rates by 15% (Fig. 8.6).

Knoepp and Vose (2002) evaluated three chamber methods (i.e., CSC with NaOH or soda lime, GC, and ODC) using sand-filled cylinders to simulate a soil system and three concentrations of standard CO_2 gas to represent low, medium, and high soil CO_2 flux rates. Flux rates measured with the ODC method equal the actual CO_2 flux at all three CO_2 concentrations. The other two methods all underestimate soil CO_2 efflux in different levels. Nonetheless, the flux rates measured with soda lime and GC correlate well with the rates measured with the ODC method (Fig. 8.7). The correlations can be used to standardize data collected with different methods and then allow comparisons of data from different studies.

Against known CO_2 fluxes ranging from 0.32 to 10.01 μmol CO_2 m^{-2}s^{-1}, Pumpanen *et al.* (2004) compared 20 chambers from different research groups for measurement of soil CO_2 efflux. The 20 chambers each belong to one of the three chamber methods (i.e., CSC, CDC and ODC). The measured flux rates by the CSC method range from underestimation by 35% to

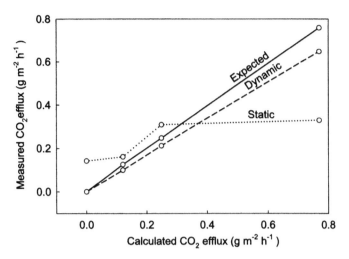

FIGURE 8.6 CO_2 efflux measured by chamber method compared with CO_2 efflux calculated by Fick's law as known effluxes (Redrawn with permission from Ecology: Nay *et al.* 1994).

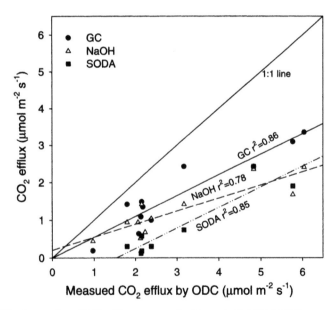

FIGURE 8.7 Regression of CO_2 efflux measured with static 2.0 M NaOH base trap (NaOH), static soda-lime trap (SODA), and closed-chamber system using GC analysis of changes in headspace CO_2 concentration (GC) against measured CO_2 efflux by the ODC IRGA system (Knoepp and Vose 2002).

overestimation by 6%. With the CDC method, the rates range from underestimation by 21% to overestimation by 33%, depending on chamber types and the methods of mixing air within the chamber headspaces. The ODCs work almost equally well in all sand types and overestimate the fluxes on average by 2 to 4%.

With known and constant CO_2 fluxes injected into the bottom of the minicosm, Kabwe et al. (2002) assessed three techniques (closed dynamic chambers, static chambers, and gradient calculations from GW measurements) in determining soil CO_2 efflux rates. The dynamic closed-chamber technique yields accurate measurements of fluxes over a range of CO_2 effluxes observed from natural unsaturated media. The concentration gradient method estimates efflux rates reasonably well, but generates uncertainties due to both the concentration gradient and the gaseous diffusion coefficient in the soil air. The static-chamber method underestimates the flux rates at high CO_2 effluxes and with adsorption times >24 h. When the adsorption time is 1 h for the mesocosm, the static-chamber method yields an estimate of CO_2 effluxes relatively comparable to the other two methods.

Many individual investigators have compared methods by making repeated measurements of different systems at the same site. Such comparison studies usually demonstrate relative differences among different measurement methods. For example, Janssens et al. (2000) conducted an in situ comparison of four measurement systems—the static chamber with soda lime, the eddy-covariance methods, one CDC system from PP Systems, and the CDC with the LiCor 6200. Among the four systems, PP Systems systematically measured the highest flux rates. The measured flux rates are lower by 10, 36, and 46% with the LiCor 6200, the soda-lime, and the eddy-covariance methods respectively than with PP Systems. The measured rates are well correlated among three chamber methods, but not with the eddy-covariance method. Norman et al. (1997) also compared four methods for measuring soil CO_2 efflux (CSC, CDC, ODC, and eddy covariance). Systematic differences exist among the four methods. The rates measured with the four methods can all be brought into reasonable agreement using correlation factors from 0.93 to 1.45. Variability due to spatial heterogeneity contributes to 15% uncertainty in measured CO_2 flux rates. Many other comparison studies have been done, such as that by Le Dantec et al. (1999), Rayment (2000), Lankreijer et al. (2003), and Liang et al. (2004, 2006).

From comparison studies, there is no universal consensus established yet on which method is the best and can be used as a standard for soil respiration measurement. In spite of that, several comparison studies do suggest that the ODC method has emerged as the most reliable one, although it is highly complicated in terms of controlling the pressure inside the chamber and requires substantial technical investment.

CHAPTER 9

Separation of Source Components of Soil Respiration

Multiple sources contribute to the respiratory releases of CO_2 at the soil surface (see Chapter 3). Each of the source components involves different biological and ecological processes and likely responds differently to environmental change. Accurate partitioning of observed soil respiration to various source components is a critical step toward mechanistic understanding of soil respiration itself and its responses to environmental change. In the past several decades, scientists have developed a rich array of methods to quantify different components (Turpin 1920, Anderson 1973, Hanson *et al.* 2000). Those methods can be categorized into roughly three groups: experimental manipulation of components, isotope tracing, and inference analysis. In each group, there are several methods of partitioning soil respiration. Each of the methods utilizes special characteristics of respiratory processes to quantify one or more components. Figure 9.1 summarizes those methods in terms of component partitioning of soil respiration. For example, trenching and clipping are designed to study root-derived carbon processes. Bomb ^{14}C tracer potentially characterizes carbon processes with distinctive residence times of each component. This chapter describes most of the methods shown in Figure 9.1. However, methods with glucose addition, ^{13}C or ^{14}C labeled litter, and soil incubation are designed to study one source component of soil respiration and will not be discussed in this book.

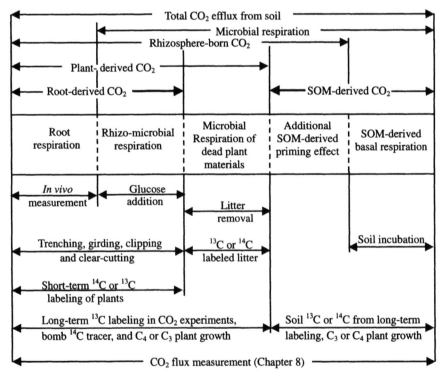

FIGURE 9.1 A conceptual scheme of soil respiration, showing compartments, sources, and component separation methods of soil respiration.

9.1. EXPERIMENTAL MANIPULATION METHODS

Experimental manipulation physically alters one or more source components of soil respiration to quantify their relative contributions. The manipulative experiments that have been conducted include direct measurements of each component, root exclusion, severing substrate supply to the rhizosphere, and litter removal.

DIRECT COMPONENT MEASUREMENTS AND INTEGRATION

This method is to measure a specific rate of CO_2 efflux from each component (i.e., roots, litter, and SOM) and their respective masses. The root respiration is usually measured from freshly cut roots (Edwards and Sollins 1973). Litter is removed from the ground surface and placed in a cuvette for measurement. CO_2 efflux from the same soil after the roots are removed is generally incubated and measured in the laboratory (Lamade et al. 1996, Thierron and Laudelout 1996). The measured specific rate of CO_2 release from each component is multiplied with the corresponding mass to estimate respiration rates for each component. Summarization of each component yields the total soil CO_2 efflux. The estimated soil CO_2 efflux should be compared with an in situ measurement of total efflux rates to validate the partitioning. In reality, however, scientists often measure in situ total soil CO_2 efflux and the litter and root components to estimate other components, which are difficult to measure or isolate, by subtraction.

This method of component measurements and integration is relatively simple and conceptually straightforward. However, in vitro analysis of root tissue usually involves digging out of the soil, severing from the plant, and washing soil out of the roots before respiration measurements are taken (Vose and Ryan 2002). This procedure causes severe root damage and drastically alters the rhizosphere environment such as symbiotic mycorrhizae, O_2, and CO_2 concentrations (Hanson et al. 2000). This method also involves soil disturbance that damages soil structure and results in a rate significantly different from the respiration rate in natural ecosystems (Nakane et al. 1996, Ohashi and Satio 1998). Removal of litter alters moisture content and gas diffusivity. To minimize disturbance effects on component measurements, the severed roots should be analyzed before desiccation or physiological death occurs. Adequate time is required to allow disturbed soil to equilibrate after disturbance in experiments.

ROOT EXCLUSION

The root exclusion method is used to estimate root respiration indirectly by comparing measured CO_2 efflux rates at soil surface with or without living roots. This method first removes roots in the soil and then measures soil CO_2 efflux rates without roots. In some careful studies, soil is usually placed back in the reverse order of removal, and further root growth is prevented by barriers after root removal (Hanson *et al.* 2000). Thus, root respiration is estimated by subtracting the measured CO_2 efflux rate from soils without root from that with roots. Results of studies using this method indicate that root contributions to the total soil respiration range from 45 to 60% in a 29-year-old mixed forest plantation in Connecticut (Wiant 1967) and from 54 to 78% in a study of pine seedlings planted in large buried pots (Edwards 1991). This technique can avoid the contribution of dead roots to CO_2 production compared with trenching, as discussed below, and allow the measurement of root biomass in the study plots. However, using the root removal technique in natural ecosystems is time-consuming and significantly disturbs soil structure. Environmental variables, such as soil temperature and moisture, are altered by root removal (Wiant 1967, Thierron and Laudelout 1996), resulting in changes in respiration rates.

SEVERING SUBSTRATE SUPPLY TO THE RHIZOSPHERE

Several methods have been used to sever carbon supply to roots and rhizosphere. Among them are trenching, clear-cutting in forests, clipping and shading in grasslands, tree girdling, and litter removal.

Trenching

Trenching cuts carbon supply from trees to blocks of soil so as to estimate relative contributions of autotrophic and heterotrophic respiration to the total soil respiration. Trenching can be implemented in several ways. For example, Bowden *et al.* (1993) dug trenches to a depth of 70 to 100 cm (20 cm below the rooting depth) around the plots (3×3 m) with a protection belt of 0.5 m outside the plots in an 80-year-old hardwood stand in the Harvard Forest, New England. The trenches are backfilled after lining with corrugated fiberglass sheets to prevent root ingrowth. A similar trenching experiment is done in a 40-year-old balsam fir (*Abies balsamea*) forest in New Brunswick (Lavigne *et al.* 2004) and in 17- and 40-year-old larch plantations in northeastern China (Jiang *et al.* 2005). Trenching can be implemented by inserting root barriers into soil to cut off root growth and carbon supply without digging soil. Buchmann (2000) inserted PVC collars (10 cm deep, 10 cm internal diameter) to exclude root growth in Norway spruce stands in Bavaria, Germany. Similarly, Wan *et al.* (2005) inserted PVC tubes of 80 cm^{-2} in area and 70 cm in depth into grassland soil in the central U.S. Great Plains to separate heterotrophic from autotrophic respiration.

Measurements of CO_2 efflux at the soil surface in the untrenched plots where roots can normally grow are taken to quantify total soil respiration. Observed CO_2 efflux in the trenched plots without the presence of live roots is the heterotrophic respiration from microbial decomposition of litter and SOM. The difference in observed CO_2 effluxes between the trenched and untrenched plots is an estimate of autotrophic respiration. Trenching studies demonstrate that the root contribution to the total soil respiration is 33% or 123 g C m^{-2} yr^{-1} in the Harvard Forest (Bowden *et al.* 1993), 20 to 30% in the spruce stand (Buchmann 2000), and 38% in the American grassland (Wan *et al.* 2005). Trenching in the balsam fir forest does not affect the temperature sensitivity of soil respiration but decreases the baseline respiration by 40 to 50% in comparison with that in the control plots (Table 9.1, Lavigne *et al.* 2004).

Trenching severs roots. Dead roots usually decompose faster than SOM, possibly resulting in pulse releases of CO_2 after trenching. Thus, a simple subtraction of measured CO_2 efflux between the trenched and untrenched plots may underestimate the root contribution to the soil respiration. Because trenching also restricts plant water uptake, soil moisture content is higher in trenched than in untrenched plots (Hart and Sollins 1998). Altered soil moisture content likely affects heterotrophic respiration rates.

Clear-cutting in forests, clipping and shading in grasslands

Clear-cutting in forests and clipping in grasslands share the same features by cutting and clearing the aboveground parts of vegetation to create vegetation-

TABLE 9.1 Parameter values of b, c, and d by fitting $R_s = ce^{d\varphi_s}e^{b(T_s-10)}$ (R_s = soil respiration rates, φ_s = soil water potential, and T_s = soil temperature) to data observed in the trenching experiment in the Balsam fir (Adapted with permission from *Tree Physiology*: Lavigne *et al.* 2004)

Treatment	Season	b	c	d	n	r^2
Untrenched	Spring	0.095	5.61	15.28	30	0.77
		(0.010)	(0.28)	(2.57)		
Untrenched	Autumn	0.050	5.30	0.11	28	0.70
		(0.012)	(0.26)	(0.02)		
Trenched	Spring	0.092	2.94	5.22	38	0.71
		(0.010)	(0.17)	(2.83)		
Trenched	Autumn	0.028	2.69	0.30	28	0.49
		(0.009)	(0.15)	(0.15)		

free soils. As a consequence, live roots and carbohydrate supply to the soil from aboveground is reduced, and resultant soil respiration decreases (Brumme 1995, Striegl and Wickland 1998). Shading in grasslands blocks light to reduce carbohydrate supply to root systems (Craine *et al.* 1999, Wan and Luo 2003). Clear-cutting creates gaps in forest stands and forms root-free patches when the forest gap sizes range from several square meters at the minimum to tens of square meters. In grasslands, clipping of areas of one or a few square meters is adequate to study root contribution to the total soil respiration.

Ohashi *et al.* (2000), for example, cut four trees and created a gap of 2.5 m × 2.5 m in a 10-year-old Japanese cedar (*Cryptomeria japonica*) in southwest Japan in March 1996. Four types of measurement plots are set up at the center of the gap, at 0.8 m (edge of the gap), at 1.6 m (edge of the surrounding stand, and at 6.0 m (in the forest as control) from the center of the gap (Fig. 9.2). Measured soil respiration does not differ among the four plots in the first year. In the second year, soil respiration measured at the center of gap decreases by approximately 50% compared with that in the control. The root respiration that is estimated from the differences between soil respiration in

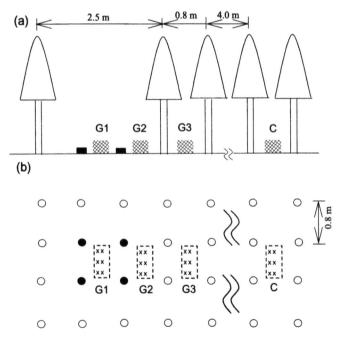

FIGURE 9.2 Location of measurement plots, (a) side view, (b) plan view. Dashed rectangles are for measurement plots, (×) measurement point; (●) felled tree; (○) living tree (Redrawn with permission from *Ecological Research*: Ohashi *et al.* 2000).

the center of the gap and that in the control correlates with soil surface temperature. The correlation illustrates a seasonal trend of higher proportional rates of root respiration in the summer than in the winter.

Clipping and shading are used to manipulate substrate supply to soil respiration in a tallgrass prairie of the U.S. Great Plains (Wan and Luo 2003). Reduced substrate supply significantly decreases soil respiration by 33, 23, and 43% for the clipping, shading, and clipping plus shading treatments respectively (Fig. 5.1). Root and rhizosphere respiration, respiration from decomposition of aboveground litter, and respiration from oxidation of SOM and dead roots contribute 30, 14, and 56% respectively to annual mean soil respiration. Similarly, two days after clipping in a Kansas tallgrass prairie, soil respiration decreases by 21 to 49%, despite the fact that clipping increases soil temperature (Bremer *et al.* 1998). The rate of rhizosphere respiration in planted barrel medic (*Medicago truncatula* Gaertn. Cv. Paraggio) decreases immediately after defoliation (Crawford *et al.* 2000). In a Minnesota grassland, two days of shading causes a 40% reduction in soil respiration, while clipping reduces soil respiration by 19% (Craine *et al.* 1999).

Several biological and environmental factors can confound estimation of root contributions to soil respiration with the clear-cutting, clipping, and shading methods. The forest cutting and grassland clipping may temporarily increase soil respiration due to accelerated decomposition of dead roots and/ or stored carbohydrate (Toland and Zak 1994). Accelerated decomposition of dead roots occurs in a tropical forest (Tulaphitak *et al.* 1985), a hardwood forest (Londo *et al.* 1999), and a northern mixed forest (Hendrickson *et al.* 1989). It may take a long time for microorganisms to decompose dead roots fully. The relative decomposition rate of dead roots is 0.13 year^{-1} in a Japanese plantation (Nakane 1995). Dead root decomposition contributes $50 \, g \, C \, m^{-2} \, yr^{-1}$ to the soil respiration in the second year of the cutting experiment (Ohashi *et al.* 2000). In addition, forest cutting or grassland clipping may stimulate growth of roots of the remaining plants.

The death of live roots may decrease rhizospheric microbes and microbial respiration, leading to an overestimation of root respiration per se. Decomposition of dead roots may change soil nutritional environments, affecting microbial respiration indirectly. Elimination of rhizosphere activity changes microbial community composition and alters uses of soil carbon substrates.

Clear-cutting, clipping, and shading potentially alter soil temperature and moisture. As a result of the removal of a substantial portion of the canopy, the treatment plots receive more incoming shortwave radiation during the daytime but trap less long-wave radiation at night than the control plots. Temperature is higher by day and lower at night, and upper layers of litter and soil become drier in the treatment plots than in the control plots. The absence of roots, however, can decrease plant water uptake and transpiration, resulting in increases in soil moisture. Changes in temperature and moisture affect respiration rates, compromising the estimation of root contributions to soil respiration. To minimize the changes in environmental conditions, Nakane *et al.* (1983, 1996) used a frame box covered with nets in clear-cut areas to maintain similar environments as in the controls. Ohashi *et al.* (2000) used the small gaps that do not result in much change in environmental conditions. Wan and Luo (2003) used correction functions to account for the effects of altered temperature and moisture on soil respiration.

Tree girdling

Girdling of trees is an approach first presented by Högberg *et al.* (2001) to separate autotrophic respiration from heterotrophic respiration in a boreal Scots pine forest in northern Sweden. Girdling strips the stem bark to the depth of the current xylem at the breast height in order to discontinue the supply of current photosynthates from the tree canopy through the phloem to the roots and their mycorrhizal fungi, while water is allowed to transport upward through the xylem without physically disturbing the delicate root-microbe-soil system. Forest girdling reduces soil respiration by about 50% within one to three months in comparison with nongirdled control plots (Fig. 9.3, Högberg *et al.* 2001, Subke *et al.* 2004). Högberg *et al.* (2001) found that root activity contributes up to 56% of soil respiration during the first summer. In the second year after girdling, estimated root contribution increases to 65% of the soil respiration, presumably due to depletion of starch reserves of girdled tree roots (Bhupinderpal-Singh *et al.* 2003). As consequence, the second-year estimate of root contribution may be more reasonable than the first-year estimate. A significant advantage of the girdling technique is that roots are not killed instantly but rather gradually transformed into root litter, which is available for microbial respiration. In addition, the soil water status is affected less by the girdling treatment than by soil trenching, which cuts off plant uptake of water. However, the soil respiration measured in the girdled plots includes the respiration of roots of understory plants that are not manipulated. Also, a part of root death may stimulate respiration levels of heterotrophic organisms. These processes likely lead to underestimation of root contributions to the soil respiration.

Litter Removal

Litter removal is an approach to determine the contribution of litter decomposition to soil respiration. Removal of existing litter and/or exclusion of litterfall as a result of placing litter traps over the litter treatment plots can eliminate microbial respiration due to litter decomposition. The litter contri-

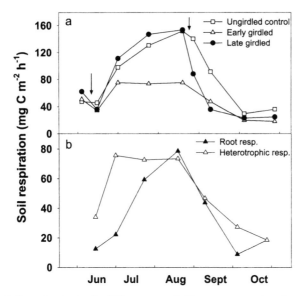

FIGURE 9.3   Soil respiration in the different tree-girdling treatments in a Scots pine forest at
Åheden. a: Respiratory soil CO_2 efflux from ungirdled control, early girdled, and late girdled
plots. b: Calculated root respiration (respiration on control plots minus that on early girdled
plots) and heterotrophic respiration (respiration on early girdled plots). Redrawn with permission from Nature: Högberg *et al.* (2001).

bution to soil respiration is estimated by subtracting CO_2 efflux rates measured in the plots with litter removal from the rates in the control plots. The
litter removal manipulation is usually conducted together with root exclusion. In a Mediterranean mixed oak forest ecosystem in Italy, a litter removal
and root exclusion experiment showed that aboveground litter decomposition, root respiration, and belowground SOM decomposition account for 21.9,
23.3, and 54.8% respectively of the annual soil respiration (Rey *et al.* 2002).
The contribution of aboveground litter to the total soil respiration is larger
in spring and autumn than in the summer, in accordance with the seasonal
pattern of litterfall. The contribution of root respiration is largest in autumn
prior to leaf litterfall (Fig. 9.4). Removal of aboveground litter in a grassland
decreases soil respiration by 14% (Wan and Luo 2003).

9.2. ISOTOPE METHODS

Isotopes are often used to trace the fates and transformations of an element
as it goes through ecological processes without environmental disturbance

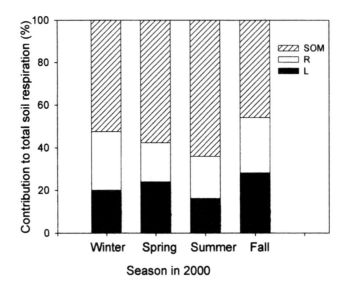

FIGURE 9.4   Relative contribution of aboveground litter (L), root respiration (R), and below-ground decomposition (SOM) to the total soil respiration over the year 2000 in four seasons (Redrawn with permission from *Global Change Biology*: Rey *et al*. 2002).

(Coleman and Fry 1991). Isotopes used in soil respiration studies are primarily radioactive carbon-14 (^{14}C), stable carbon-13 (^{13}C), and occasionally ^{18}O (Lin *et al*. 1999, Trumbore 2000). The use of isotope tracers requires that (1) different source components of soil respiration have different isotopic values and (2) there is no significant fractionation of isotopes during processes of carbon from source assimilation to output where isotope samples are taken. Fundamental principles of isotopes and their applications to ecological research are described by Coleman and Fry (1991), Dawson *et al*. (2002), and Flanagan *et al*. (2005). This section focuses on applications of isotope methods to partitioning of soil respiration.

Four isotope methods have been commonly applied to partitioning of soil respiration. The first method is to use differences in natural abundance of isotopes (mainly ^{13}C) created through different fractionation by C_3 and C_4 plants. The second method is to use depleted ^{13}C signals in pure CO_2 sources that fumigate CO_2 experiments to partition CO_2 efflux from old versus recently formed soil carbon components. The third method is to use "bomb ^{14}C", created by nuclear bomb explosions, to examine carbon dynamics from roots and different fractions of SOM. The fourth method is to create different source values of isotopes by adding a trace amount of isotopes to plants or ecosystems in labeling experiments (Table 9.2).

TABLE 9.2 Summary of different isotopic methods in partitioning study of soil respiration

Labeling Method	Isotope Sources	Source Concentration	Labeling	Study Sites
Growing C_3 plant on C_4 soil or C_4 plant on C_3 soil	Natural abundance	Constant	Continuous	Field or greenhouse
CO_2 experiment	Depleted ^{13}C	Constant	Continuous	Field or greenhouse
Bomb ^{14}C	Enriched ^{14}C	Varying with time	Continuous	Field
Labeling experiment	Enriched ^{14}C or ^{13}C	Constant	Pulse or continuous	Greenhouse or growth chamber

GROWING C_3 PLANTS ON C_4 SOIL OR C_4 PLANTS ON C_3 SOIL

Plants with the C_3 photosynthetic pathway (i.e., C_3 plants) produce carbohydrate with a $\delta^{13}C$ value of ~27‰, whereas photosynthate from C_4 plants has a $\delta^{13}C$ value of ~13‰. C_3 plants are more depleted in ^{13}C relative to C_4 plants, due to physical and enzymatic discrimination against ^{13}C molecules during C_3 photosynthesis (O'Leary 1988). Long-term occupancy of either C_3 or C_4 plants in an ecosystem leaves isotope signatures in SOM. Thus, the isotope value of SOM is usually close to that of the dominant plants in the ecosystem, being ~27‰ for a C_3 plant-dominant ecosystem (hereafter called C_3 soil) and ~13‰ for a C_4 plant-dominant ecosystem (hereafter called C_4 soil). In C_3 and C_4 mixed grasslands, soil isotope values are between those for the C_3 and C_4 soils.

When an ecosystem experiences a shift in vegetation from C_3 to C_4 plants (e.g., growing C_4 crops after deforestation-removal of C_3 tree plants in tropical regions) or vice versa (e.g., C_3 tree encroachment into C_4 grasslands), the $\delta^{13}C$ value of root and rhizosphere respiration is different from that of microbial respiration of old SOM (Rochette et al. 1999). Taking advantage of differences in $\delta^{13}C$ values between C_3 and C_4 plants and between C_3 and C_4 soils, researchers often grow C_4 plants in C_3 soil or C_3 plants in C_4 soil to partition soil CO_2 efflux into sources of old versus recently formed carbon (Schonwitz et al. 1986, Wedin et al. 1995, Cheng 1996).

For example, Rochette et al. (1999) grew maize, a C_4 species, on a soil where spring wheat and perennial forage used to grow. Measured $\delta^{13}C$ values of SOM and maize roots are −25.0 and −13.7‰, respectively. Measured $\delta^{13}C$

values of the total soil respired CO_2 are ~−24‰ in the first 40 days after planting, increase linearly from day 40 to 70, and peak at ~−18‰ from day 70 to 100 after planting. Those $\delta^{13}C$ values are used in a two-source mixing model to estimate the fractional contribution of root respiration, f, to soil respiration (Robinson and Scrimgeour 1995):

$$\delta^{13}C_{R\text{-soil}} = f\delta^{13}C_{R\text{-root}} + (1 - f)\delta^{13}C_{R\text{-SOM}} \qquad (9.1)$$

where $\delta^{13}C_{R\text{-soil}}$, $\delta^{13}C_{R\text{-root}}$, and $\delta^{13}C_{R\text{-SOM}}$ are isotope ^{13}C values of the soil respiration, roots, and SOM respectively. Rearrangement of the above equation gives:

$$f = \frac{\delta^{13}C_{R\text{-soil}} - \delta^{13}C_{R\text{-SOM}}}{\delta^{13}C_{R\text{-root}} - \delta^{13}C_{R\text{-SOM}}} \qquad (9.2)$$

With the measured $\delta^{13}C$ values, we can solve the above equation to estimate f. The estimated root contribution to soil respiration varies with time, as indicated by variation in the $\delta^{13}C$ values of the soil-respired CO_2 (Fig. 9.5a, Rochette et al. 1999). Root and root-associated microbial respiration in the rhizosphere contributes up to 45% of soil respiration during the most productive part of the growing season. The estimated root contribution from the isotope method is comparable to that with the root exclusion technique (Fig. 9.5b).

Another approach to partitioning of ecosystem and soil respiration is based on ^{13}C enrichment in microbial (largely fungal) biomass. The $\delta^{13}C$ values in microbial biomass can be up to 5‰ higher than that in plant organic matter (Tu and Dawson 2005). The enrichment in ^{13}C signatures from microbial respiration can result from (1) temporal lags in ^{13}C movement though various ecosystem pools, (2) metabolic fractionation, (3) heterotrophic CO_2 fixation in roots and microbes, (4) selective uses of compounds with different ^{13}C values as substrate for respiration, and (5) kinetic fractionation during respiration. Tu and Dawson (2005) used the stable carbon isotope signatures to partition ecosystem respiration into three components: 25% from aboveground respiration, 33% from root respiration, and 42% from microbial decomposition of SOM from a redwood forest near Occidental, California.

Similarly, agricultural displacement of native ecosystems, crop rotation, forest-to-pasture conversions (Sanderman et al. 2003), shrub expansion in arid lands (Connin et al. 1997), and woody encroachment all potentially generate isotope disequilibrium, offering the possibility of studying components of soil respiration. However, such transition ecosystems are usually limited in distribution, and isotope signatures disappear over time after the conversion occurs.

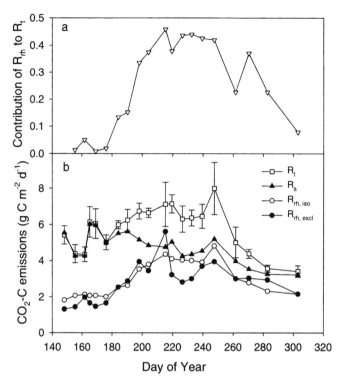

FIGURE 9.5 (a) Contribution of maize rhizosphere respiration (R_{rh}) to total soil respiration (R_t) in a maize crop during the 1996 growing season; (b) total soil (R_t), rhizosphere (R_{rh}), and SOM (R_s) respiration in a maize crop during the 1996 growing season. Estimates of R_{rh} are obtained by the ^{13}C isotopic technique ($R_{rh, iso}$) and root-exclusion technique ($R_{rh,excl}$). Vertical bars indicate ±SD (Redrawn with permission from *Soil Science Society of America Journal*: Rochette *et al.* 1999).

CO₂ ENRICHMENT EXPERIMENTS

Many CO_2 experiments have been conducted in natural ecosystems using open-top chambers (OTC) and free-air CO_2 enrichment (FACE) facilities in the past two decades. Those CO_2 experiments are designed primarily to study impacts of rising atmospheric CO_2 concentration on plants and ecosystems. Since they release pure CO_2 from commercial sources, those experiments also function as a continuous isotope labeling with depleted ^{13}C (Pataki *et al.* 2003). The CO_2 experiments with depleted ^{13}C usually result in a $\delta^{13}C$ value of approximately −40‰ in newly synthesized carbohydrate at elevated CO_2, whereas carbohydrate from pretreatment photosynthesis under ambient CO_2 has a $\delta^{13}C$ value of ~−27‰ (Fig. 9.6). Thus, the different isotopic values of

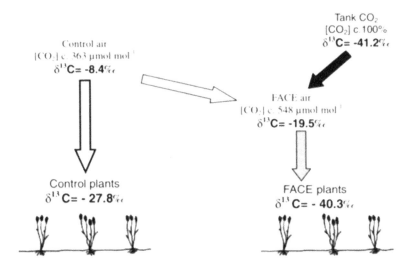

FIGURE 9.6 $\delta^{13}C$ values of bulk air, tank CO_2 from commercial sources, mixed air in the elevated CO_2 plots, and plants. The CO_2 experiments release pure CO_2 from commercial sources to increase its concentration in treatment plots. The commercial CO_2 is usually generated from fossil fuels with depleted ^{13}C, whereas air CO_2 has a $\delta^{13}C$ value of —8‰. The released pure CO_2 is mixed with air, resulting in a ^{13}C value of —19‰ in elevated CO_2 plots. Photosynthetic fractionation of the depleted $^{13}CO_2$ leads to a $\delta^{13}C$ value of approximately −40‰ in newly synthesized carbohydrate at elevated CO_2. Photosynthate at ambient CO_2 has a $\delta^{13}C$ value of —27‰. Open arrows, background air; solid arrows, pure commercial CO_2 from tank; gray arrows, mixed background (Modified with permission from *New Phycologist*: Leavitt *et al.* 2001).

carbohydrate synthesized before versus after CO_2 treatments create an opportunity to partition observed soil respiration into autotrophic and heterotrophic components.

The isotopic study in the CO_2 experiments involves measurements of $\delta^{13}C$ values of CO_2 respired from the rhizosphere ($\delta^{13}C_{R\text{-root}}$), CO_2 respired from root-free soil ($\delta^{13}C_{R\text{-SOM}}$), and CO_2 from soil surface efflux ($\delta^{13}C_{R\text{-soil}}$). In practice, we often measure the $\delta^{13}C$ value of newly produced roots or leaves as the estimate of the $\delta^{13}C$ value of CO_2 respired from the rhizosphere, because there is no fractionation during respiration (Cheng 1996, Lin and Ehleringer 1997). The $\delta^{13}C$ value of SOM is often measured from laboratory incubation of root-free soil collected from the CO_2 experiments (Andrew *et al.* 1999, Pendall *et al.* 2003). After one year of fumigation with ^{13}C-depleted CO_2 at the Duke FACE site, for example, $\delta^{13}C$ values of CO_2 are −39.3‰ from the rhizosphere, −25.7‰ from the root-free soil, and −32‰ from the soil respiration (Andrew *et al.* 1999). The three $\delta^{13}C$ values are fed into Equation 9.2 to obtain an estimate that root respiration contributes to 55% of soil respiration. Similarly, this approach has been applied to several other CO_2-enrichment experiments (Leavitt *et al.* 1994, 1996; Hungate *et al.* 1997; Nitschelm *et al.* 1997; Torbert *et al.* 1997; Van Kessel *et al.* 2000; Pendall *et al.* 2003) for estimation of relative contributions of different source components to the total soil respiration.

The isotopic partitioning approach works best when the differences in $\delta^{13}C$ values are greatest between source components that contribute to soil respiration. As the CO_2 experiments continue, the $\delta^{13}C$ value of root-free soil gradually increases and eventually approaches the value of rhizosphere respiration. Thus, the power of isotopic partitioning decreases. In addition, this approach is only applicable to respiration partitioning at elevated plots and not at ambient CO_2 plots. At ambient CO_2, the CO_2 source for photosynthesis has the identical $\delta^{13}C$ value before or after the CO_2 experiments. Thus, the $\delta^{13}C$ values of SOM and the rhizosphere C source are similar in ambient CO_2 plots, making it almost impossible to estimate relative source contributions to soil respiration. Several methods have been developed to remedy this situation. For example, at the OTC experiment site in the shortgrass steppes of northern Colorado, grazing has been reduced for 20 years prior to the experiment. A reduction in grazing pressure is accompanied by an increase in C_3 grass abundance (Milchunas *et al.* 1988). As a consequence, the $\delta^{13}C$ value of plant-derived carbon differs by 5‰ from that of SOM. Using that difference in the $\delta^{13}C$ values, Pendall *et al.* (2003) estimated that root respiration contributes up to 70% of the soil respiration at ambient CO_2 but only 25% at elevated CO_2. Similarly, a small but quantifiable difference between natural abundance ^{13}C in plants and SOC is used to estimate root contribution to soil respiration (Nitschelm *et al.* 1997). Other methods that have been used to estimate root contribution to soil respiration at ambient CO_2 in CO_2 enrichment experiments include (1) small subplots with soils from C_4 plant-dominated ecosystems within the ambient CO_2 plots where C_3 plants grow (Allison *et al.* 1983, Ineson *et al.* 1996, Cheng and Johnson 1998, Leavitt *et al.* 2001); (2) small subplots exposed to pulse pure ^{13}C labeling within ambient CO_2 plots (Hungate *et al.* 1997, Leavitt *et al.* 2001); and (3) CO_2 labeled ^{13}C or ^{14}C to fumigate entire control plots in chamber experiments (Lin *et al.* 1999, 2001).

A dual stable isotope approach has been applied by Lin *et al.* (1999, 2001) to partitioning of soil respiration into three components—rhizosphere respiration (root and root exudates), litter decomposition, and oxidation of SOM—under elevated CO_2 and elevated temperature in Douglas fir terracosms. Both soil CO_2 efflux rates and the ^{13}C and ^{18}O isotopic compositions of soil CO_2 efflux are measured. The measured $\delta^{13}C$ values of newly grown needles are ~−29‰ at ambient CO_2 and ~−35‰ at elevated CO_2, which are not affected by warming. The $\delta^{13}C$ values are ~−27‰ for litter and ~−24‰ for SOM.

Neither of them is affected by either warming or elevated CO_2 (Fig. 9.7). It is assumed that (1) the $\delta^{13}C$ value of CO_2 respired from the roots and rhizosphere ($\delta^{13}C_{R\text{-root}}$) equals the $\delta^{13}C$ value of the newly grown needles, (2) the $\delta^{13}C$ value of CO_2 respired from litter decomposition ($\delta^{13}C_{R\text{-litter}}$) equals the $\delta^{13}C$ value of litter, and (3) the $\delta^{13}C$ value of CO_2 respired from SOM oxidation ($\delta^{13}C_{R\text{-SOM}}$) equals the $\delta^{13}C$ value of SOM. Thus, those $\delta^{13}C$ values can be expressed by a three-source mixing model:

$$\delta^{13}C_{R\text{-soil}} = m\delta^{13}C_{R\text{-root}} + n\delta^{13}C_{R\text{-litter}} + (1 - m - n)\delta^{13}C_{R\text{-SOM}} \qquad (9.3)$$

where $\delta^{13}C_{R\text{-soil}}$ is the $\delta^{13}C$ value of soil respiration, m is the fraction of soil respiration attributable to root respiration, and n is the fraction of soil respiration due to litter decomposition.

In their study Lin et al. (1999, 2001) also measured $\delta^{18}O$ values of soil water at the top of the A horizon and litter water in the litter layer to estimate $\delta^{18}O$ values of soil CO_2 and litter-derived CO_2, respectively. The estimation is

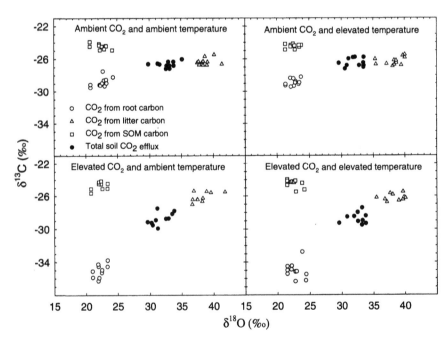

FIGURE 9.7 The carbon and oxygen isotope ratios of total soil CO_2 efflux (closed symbols) and its three major carbon sources (open symbols) in the terracosms under different CO_2 and temperature treatments (Redrawn with permission from *Global Chang Biology*: Lin et al. 1999).

based on assumptions that the $\delta^{18}O$ value of CO_2 released from decomposition of litter is in equilibrium with the litter water and that CO_2 released from soil (including both root respiration and SOM decomposition) reaches isotopic equilibrium with soil water in the top 0 to 5 cm layer (Ciais *et al.* 1997, Tans 1998). Their estimated $\delta^{18}O$ values released are similar among the temperature and CO_2 treatments. Thus, those $\delta^{18}O$ values can be expressed by a two-source mixing model:

$$\delta^{18}O_{R\text{-soil}} = n\delta^{18}O_{R\text{-litter}} + (1 - n)\delta^{18}O_{R\text{-topsoil}} \qquad (9.4)$$

where $\delta^{18}O_{R\text{-soil}}$ is the $\delta^{18}O$ value of soil respiration, $\delta^{18}O_{R\text{-litter}}$ is the $\delta^{18}O$ value of CO_2 released from litter decomposition, and $\delta^{18}O_{R\text{-topsoil}}$ is the $\delta^{18}O$ value of CO_2 from both root respiration and oxidation of SOM. The obtained three ^{18}O values (i.e., $\delta^{18}O_{R\text{-soil}}$, $\delta^{18}O_{R\text{-litter}}$, and $\delta^{18}O_{R\text{-topsoil}}$) are used to estimate the relative contribution of litter decomposition to the overall soil CO_2 efflux (i.e., n in Equation 9.4). With the estimated n and $\delta^{13}C$ data, the fraction of soil respiration attributable to root respiration (i.e., m in Equation 9.3) can be estimated to separate contributions of roots from that of oxidation of SOM. In most cases, litter decomposition is the dominant component of soil CO_2 efflux followed by rhizosphere respiration and SOM oxidation in their terracosms study (Lin *et al.* 1999, 2001). Both elevated CO_2 and warming stimulate rhizosphere respiration and litter decomposition. The oxidation of SOM is stimulated only by increased temperature. Release of newly fixed carbon via root respiration is the most responsive to elevated CO_2 while SOM oxidation is most responsive to increased temperature.

The isotopic methods may incur uncertainty in source partitioning of soil respiration due to assumptions about calculations of isotopic signals for different CO_2 sources. First, the CO_2 from rhizosphere respiration in most studies is assumed to have the same $\delta^{13}C$ value as that of newly grown parts in plants. If the $\delta^{13}C$ of the newly grown parts is more negative than that of the active roots and root exudates, the contribution of rhizosphere respiration to the soil CO_2 efflux is underestimated. Second, the $\delta^{13}C$ of CO_2 from SOM is assumed to be the same as that of bulk SOM. Bulk SOM is made up of several fractions, which may decompose at different rates and have different isotopic composition (Bird and Pousai 1997). The actual $\delta^{13}C$ of CO_2 from SOM oxidation is likely to be different from that of bulk SOM. If the carbon that contributes to the soil CO_2 efflux has more negative $\delta^{13}C$ values than bulk SOM, the relative contribution from SOM oxidation to the soil CO_2 efflux is underestimated. Third, the partitioning of the soil CO_2 efflux, particularly the dual-isotope approach with both ^{13}C and ^{18}O, depends in part on an isotopic equilibrium of CO_2 with soil water when it diffuses through various soil layers to the atmosphere (Tans 1998). The diffusion isotope fractionation factor is largely unknown, but presumably depends on diffusive transfer from soil to the atmosphere and turbulent transfer in the litter layer.

BOMB ^{14}C TRACER

The testing of thermonuclear bombs from about 1955 to the middle of the 1970s has enriched isotope ^{14}C composition in atmospheric CO_2 by producing huge thermal neutron fluxes to induce the "bomb" ^{14}C. This "atom bomb" effect on atmospheric isotope composition is first identified by De Vries (1958). The amount of bomb ^{14}C in the atmosphere reaches a peak in 1963 in the northern hemisphere and in 1965 in the southern hemisphere. Based on samples of grapes grown in Russia (Burchuladze *et al.* 1989) from 1950 to 1977 and direct atmospheric measurements from 1977 to 1996 (Levin and Kromer 1997), Δ^{14}C, the difference in parts per mil (or ‰) between the ^{14}C/^{12}C ratio in the sample compared with that of a universal standard (oxalic acid I, decay-corrected to 1950), increased from 0 in 1954 to 893‰ in 1964; it then gradually declined to $+97 \pm 5$‰ in 1997 in the northern hemisphere (Fig. 9.8). The decline in the atmospheric ^{14}C results from exchange of atmospheric ^{14}C with terrestrial ecosystems and oceans. A positive Δ^{14}C value contains bomb-produced radiocarbon in the samples, whereas a negative Δ^{14}C value indicates that carbon in the reservoir has, on average, been isolated from exchange with atmospheric $^{14}CO_2$ for at least the past several hundred years.

Photosynthetic fixation of $^{14}CO_2$ acts as the global continuous labeling experiment, providing a unique opportunity of tracing C sources from rhizosphere- versus soil-respired CO_2 (Dörr and Münnich 1986). Rhizosphere-respired CO_2 can presumably reflect the ^{14}C signature of contemporary atmospheric CO_2 due to fast transfer of photosynthetically fixed carbon to the rhizosphere. The ^{14}C values in SOM represent the bomb ^{14}C that is incorporated into organic matter some time ago due to its long residence time. Gaudinski *et al.* (2000) simulated dynamics of Δ^{14}C, as driven by variation of ^{14}C in the atmospheric CO_2 through time, in homogeneous, steady-state C pools with residence times of 10, 50, or 100 years (Fig. 9.8). The Δ^{14}C values in the SOM pool, with residence times of 10 years, track more closely to the atmospheric Δ^{14}C patterns than those in the pools with residence times of 50 and 100 years. The distinctive patterns of Δ^{14}C values in rhizosphere carbon and SOM pools with different residence times offer the possibility of partitioning soil respiration into different sources.

Gaudinski *et al.* (2000) conducted a study in Harvard Forest, New England to partition soil respiration using the bomb ^{14}C. The amount of Δ^{14}C from

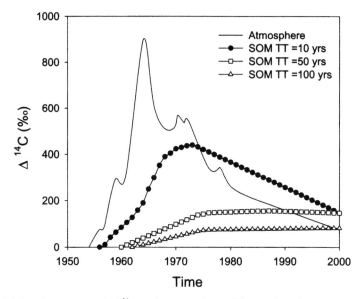

FIGURE 9.8 The time record of ^{14}C in the atmosphere of the northern hemisphere based on grapes grown in Russia (Burchuladze *et al.* 1989) and direct atmospheric measurements from 1977 to 1996 (Levin and Kromer 1997). Radiocarbon data are corrected for mass-dependent isotopic fractionation to $-25‰$ in ^{13}C. The ^{14}C content of a homogeneous, steady-state carbon pools with turnover times (TT) of 10, 50, or 100 years is compared with that of the atmosphere through time (Redrawn with permission from Biogeochemistry: Gaudinski *et al.* 2000).

soil-respired CO_2 can be partitioned into that derived from root respiration, root litter decomposition, leaf litter decomposition, and oxidation of SOM that resides in the soil for a long time. They measured $\Delta^{14}C$ values in leaf litter ($\Delta^{14}C_{LL}$), root litter ($\Delta^{14}C_{LR}$), humus and mineral carbon ($\Delta^{14}C_{H+M}$), and CO_2 respired at the soil surface ($\Delta^{14}C_R$). The $\Delta^{14}C$ value in CO_2 respired by roots ($\Delta^{14}C_{R-root}$) is assumed to equal that in the atmospheric CO_2 ($\Delta^{14}C_{atm}$). The $\Delta^{14}C$ values from these components are used in mass balance equations to determine the relative contribution of each component to the soil respiration as:

$$F_T = F_R + F_{LL} + F_{LR} + F_{H+M} \tag{9.5}$$

and

$$F_T\Delta^{14}C_R = F_R\Delta^{14}C_{R-root} + F_{LL}\Delta^{14}C_{LL} + F_{LR}\Delta^{14}C_{LR} + F_{H+M}\Delta^{14}C_{H+M} \tag{9.6}$$

where F_T is the annual soil respiration flux; F_R is the flux of CO_2 derived from recent carbon sources in root and rhizosphere; and F_{LL}, F_{LR}, and F_{H+M} are fluxes of CO_2 derived from leaf litter, root litter, and humus and mineral carbon sources respectively. Among all the parameters, $\Delta^{14}C_{LL}$, $\Delta^{14}C_{LR}$, $\Delta^{14}C_{H+M}$, $\Delta^{14}C_R$, and $\Delta^{14}C_{atm}$ (for $\Delta^{14}C_{R-root}$), are measured F_{LL} is constrained to be between 25 and $95\,gC\,m^{-2}yr^{-1}$, F_{H+M} is estimated from pool sizes of humified and mineral SOM and their turnover times. Thus, the above equations can be solved for the remaining unknowns, F_R and F_{LR}.

The bomb ^{14}C analysis by Gaudinski et al. (2000) indicates that approximately 59% of CO_2 produced annually in soil is derived from recent carbon fraction through root and rhizosphere respiration and 41% (34 to 51%) of CO_2 produced annually from decomposition of SOM with residence times greater than one year (Fig. 9.9). The decomposition of humus and mineral carbon fractions with residence times > 40 years contributes only 8% of the annual

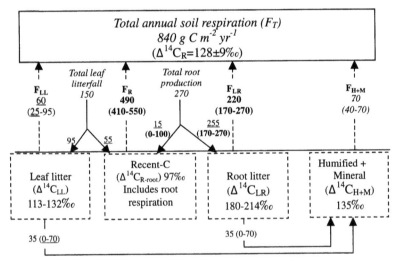

FIGURE 9.9 Results of isotopic mass balance approach to partitioning soil respiration into recent- versus pool-stored carbon sources. Solid arrows represent fluxes of organic carbon, while dashed arrows represent fluxes of CO_2. All units are in $gC\,m^{-2}yr^{-1}$ with the average (and range). Production of litter (leaf and root) is assumed to have the isotopic composition of the atmosphere (97‰) in 1996. Bold numbers represent direct results from the isotope mass balance model. Italicized numbers are independent measurements or calculated values used to constrain the model (see text for details), and underlined numbers are the resultant fluxes and transfers due to the model results and its constraints (Redrawn with permission from Biogeochemistry: Gaudinski et al. 2000).

respiration flux, with the remaining 33% (26 to 43%) from root and leaf litter decomposition.

LABELING EXPERIMENTS

A labeling experiment usually exposes the aboveground part of plants to a tracer (usually ^{14}C- or ^{13}C-labeled CO_2) inside a growth chamber or greenhouse (Fig. 9.10). Photosynthesis incorporates ^{14}C- or ^{13}C-labeled CO_2 into carbohydrate immediately following exposure. Over time, the labeled carbohydrate within labile carbon pools is used for respiration, incorporated into structural materials of plant tissues via growth, allocated to the rhizosphere, and built into SOM. To trace the fate of labeled carbon, samples of plant tissues, soil, and respired CO_2 are collected during and after exposure for the analysis of ^{14}C or ^{13}C. Relative amounts of radioactive ^{14}C or stable isotope ^{13}C are used to indicate partitioning of photosynthetically fixed carbon into different functional processes based on the mass conservation principle. When a labeling experiment is designed primarily to partition soil respiration into heterotrophic and autotrophic sources, the amounts of ^{14}C or ^{13}C in CO_2 respired from roots, SOM, and in total soil respiration are quantified. Then the mixing model in Equation 9.2 is applied to estimate the fraction of autotrophic versus heterotrophic respiration (Cheng and Johnson 1998).

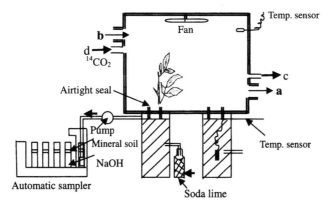

FIGURE 9.10 Equipment used for measuring root respiration of plants labeled with ^{14}CO$_2$: (a) and (b) connections with air mixing and temperature control equipment, (c) and (d) connections with ^{14}CO$_2$ and CO$_2$ regulating equipment (Redrawn with permission from Academic Press: Warembourg and Kummerow 1991).

Labeling experiments can be done in three ways: one-pulse labeling, repeated-pulse labeling, or continuous labeling during the growing season (Paterson *et al.* 1997). One-pulse labeling is the single addition of ^{14}C- or ^{13}C-labeled CO_2 for quantifying the distribution of labeled C within a plant and respired by plant tissues during a given period (Cheng *et al.* 1996). Repeated-pulse labeling has several additions of a tracer at different times during the growing season. This technique has been used successfully to approximate cumulative plant C budgets (Gregory and Atwell 1991) and cumulative belowground C input and rhizodeposition in barley (Jensen 1993) and in a temperate pasture ecosystem (Saggar and Hedley 2001). In the pasture experiment, ^{14}C-CO_2 losses are as high as 66 to 70% during summer, autumn, and winter but low (37 to 39%) during the spring.

Since pulse labeling is usually applied to small-stature plants in growth chambers or greenhouses, estimated autotrophic respiration from a labeling experiment is influenced by the stage of plant growth and the chase period. The latter is the elapsed time between pulse labeling, the final experimental measurements. The chase period should be long enough to allow plants to allocate the labeled carbon within plants and to belowground parts. Plant growth stage influences relative allocation of carbon to different plant parts over a growing season and thus alters the root contribution to soil respiration. Isotopes in pulse labeling are very dynamic due to rapid turnovers of carbon in some plant and rhizosphere pools (Meharg 1994). Thus, pulse labeling requires continuous or repeated measurements of the labeled isotopes in order to quantify carbon respired from fast turnover pools. Rhizosphere respiration rates may be overestimated because labeled isotopes are preferentially allocated to labile carbon pools (Paterson *et al.* 1997).

Continuous labeling is sequential uses of labeled carbon under laboratory or field conditions over time (Whipps and Lynch 1983, Merckx *et al.* 1985). This technique usually results in uniform labeling of all plant carbon pools, including labile metabolic substances and some plant structural components. Thus, continuous labeling offers information on cumulative carbon respired from roots and rhizosphere that has a different isotope signature from CO_2 produced during SOM decomposition. A continuous $^{14}CO_2$ labeling experiment with wheat and maize plants shows that rates of root respiration, rhizodeposition, and associated microbial respiration increase at the high nitrogen level in comparison with those at the low nitrogen level (Liljeroth *et al.* 1994). Expensive and cumbersome equipment for continuous labeling with tracer levels of ^{14}C makes field applications difficult, especially in forest communities. In addition, the radioactive ^{14}C labeling has environmental health restrictions and is mostly limited to short-term, laboratory experiments.

9.3. INFERENCE AND MODELING METHODS

REGRESSION EXTRAPOLATION AND MODELING ANALYSIS

Simple regression equations that relate root biomass to root and soil respiration or soil carbon content to microbial respiration are often used to estimate relative contributions of different components to soil respiration (Kucera and Kirkham 1971, Edwards and Sollins 1973, Pati *et al.* 1983, Katagiri 1988, Behara *et al.* 1990). Behara *et al.* (1990), for example, used a linear relationship between soil respiration and root biomass to estimate a value of 50.5% for the contribution of root respiration to the soil respiration. The fungal and bacterial contributions to soil respiration are estimated to be 44% and 5.5% respectively. Pati *et al.* (1983) estimated root, fungal, and bacterial contributions at 38%, 57%, and 5% respectively. However, the regression methods potentially generate substantial errors in estimated contributions of different source components to soil respiration, due to omission of many processes and difficulties in accurate measurements of root biomass.

Process-based models simulate processes of root and microbial respiration and can estimate relative contributions of different source components to soil respiration (Fang and Moncrieff 1999, Hui and Luo 2004). Most of the process-based models simulate root and microbial respirations by multiplying coefficients of specific rates of respiration with root biomass and content of organic matter respectively (see Chapter 10). The specific respiratory rates are regulated by environmental factors such as temperature, moisture, and CO_2 diffusion. Hui and Luo (2004) estimated that root respiration contributes 53.3% of the total soil respiration in the Duke loblolly pine forest in North Carolina. Most of soil CO_2 is produced in the top 30 cm of soil (Table 9.3).

TABLE 9.3 Contributions of root and microbial respiration to total soil respiration in different soil layers from a process-based model in the Duke Forest North Carolina (Hui and Luo 2004)

Layer	Thickness (m)	Root Respiration (%)	Microbial Respiration (%)
1	0.05	5.7	24.6
2	0.10	39.5	10.6
3	0.15	3.0	3.0
4	0.40	3.0	3.8
5	0.30	2.1	2.4
6	1.00	0.0	2.3
Total		53.3	46.7

DECONVOLUTION ANALYSIS

Deconvolution analysis utilizes characteristic response times of various carbon processes to a perturbation to separate components of soil respiration (Luo et al. 2001b). Soil respiration involves multiple processes, such as root exudation, root respiration, root turnover, decomposition of litter, and oxidation of SOM. Each of the processes has distinctive response times to perturbation, which are related to carbon residence times, that is, the time carbon remains in an ecosystem from entrance via photosynthesis to exit via respiration (Thompson and Randerson 1999). For example, belowground carbon cycling through the pathway of root exudation takes only a few weeks from photosynthesis to respiratory release (Cheng et al. 1994, Rouhier et al. 1996). In contrast, carbon cycling through the pathway of wood growth, death, and decomposition takes several decades from photosynthesis to respiratory release (Fig. 9.11). In response to either an increase in carbon influx (e.g., in an elevated CO_2 experiment) or a decrease in substrate supply (e.g., in a tree-girdling experiment), root exudation and root respiration change first, while SOM changes slowly.

Using the distinctive response times of various carbon processes, Luo et al. (2001b) developed the deconvolution approach to partitioning of soil respiration observed in the FACE experiment in the Duke Forest. The analysis assumes that a CO_2-induced change in soil respiration at elevated CO_2 is a

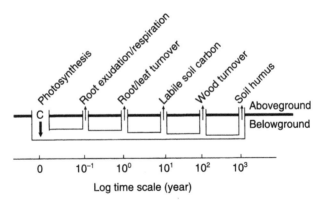

FIGURE 9.11 A schematic representation of rhizosphere C processes and their operational time-scales. In general, the C cycling from fixation to release takes weeks through the fast pathways of root exudation and root respiration, one year or longer through the pathway of root turnover (defined as growth, death, and decomposition), two to four years through needle turnover in the coniferous forest, decades through woody tissue turnover, and centuries or even millennia through the turnover of SOM in forests. (Luo et al. 2001b).

convolved response from all the CO_2 production processes in soil. The convolved response to elevated CO_2 depends on relative activities of those carbon processes. If the rapid carbon transfer pathways (e.g., root exudation, root respiration, and root turnover) contribute a substantial amount of carbon to soil respiration, the convolved response manifests a large and rapid increase in soil respiration after the CO_2 fumigation. In contrast, if the majority of carbon goes through the slow carbon pathways, the convolved response does not show up in the first few years after the CO_2 fumigation. Thus, the measured response of soil respiration to elevated CO_2 contains information about the relative importance of the CO_2 production processes.

At the Duke FACE experiment site, photosynthetic carbon influx into the ecosystem increases by 40% (Luo *et al.* 2001c). Elevation of CO_2 concentration did not result in a statistically significant difference in soil respiration in the first experimental year from August 1996 to July 1997 after the FACE experiment, but led to significant increases of 33.3% and 45.6% respectively in the second and third experimental years of the FACE experiment (Table 9.4). The increase of soil respiration during the first year was caused primarily by carbon released by root exudation and respiration, in the second year by root turnover in addition to root exudation and respiration, and in the third year by aboveground litterfall in addition to the above three pathways. By a qualitative comparison between responsive processes and observed increases, the deconvolution analysis suggests that the increases in root exudation and root respiration may be of minor importance in carbon transfer to the rhizosphere, whereas root turnover and aboveground litterfall are the major processes delivering carbon to soil.

TABLE 9.4   Observed CO_2 stimulation in soil respiration and associated mechanisms during the three experimental years of the FACE at the Duke Forest (Luo *et al.* 2001c)

Experimental Year	Period	Observed Change (%)	Possible Mechanisms
1	August 1996–July 1997	3.8	(1) root exudation and (2) root respiration
2	August 1997–July 1998	28.0	(1) root exudation, (2) root respiration, and (3) root turnover
3	August 1998–July 1999	45.6	(1) root exudation, (2) root respiration, (3) root turnover, and (4) aboveground litter

9.4. ESTIMATED RELATIVE CONTRIBUTIONS OF DIFFERENT SOURCE COMPONENTS

Many studies published in the literature generally suggest that root respiration contributes substantially to the soil CO_2 efflux. Two recent reviews both show that root contribution generally accounts for approximately 50% of the total soil respiration. Hanson *et al.* (2000) synthesized 50 studies published in the literature that either estimate root contribution to soil respiration or have sufficient data from which an estimate could be derived. The overall mean of root contribution to the total soil respiration is 48%, with a wide variation from less than 10% to greater than 90%. The low values of root contribution (i.e., <20%) are largely due to biases in measurement methods. Root contributions for sites dominated by forest vegetation account, on average, for 48.6% of soil respiration. The values of root contributions in the nonforest ecosystems are widely scattered throughout the entire range, with an overall average of 36.7%. Root contributions exhibit seasonality, usually being low during the dormant season, since root respiration depends on a supply of carbohydrates from canopy photosynthesis. Respiration usually increases dramatically during the active growing seasons.

Bond-Lamberty *et al.* (2004) synthesized published data from 53 different forest stands. The partitioning studies use a variety of methods, including root exclusion, comparison of unburned with recently burned stands, manipulation of photosynthate supply to roots and rhizosphere, root extraction, isotope labeling, and mass balance techniques. Their synthesis shows that either autotrophic or heterotrophic respiration correlates strongly with annual soil respiration across a wide range of forests (Fig. 9.12a). The root contributions to the soil respiration increase asymptotically with soil respiration itself (Fig. 9.12b). Low soil respiration is usually found in ecosystems with low production in which heterotrophic processes are likely dominant. And the autotrophic respiration accounts for a small fraction of soil respiration. As ecosystem production increases, so does the relative contribution of autotrophic respiration. For most of the ecosystems, the root contributions are within a range of 30 to 50% of soil respiration. Monte Carlo simulations show that the correlation between autotrophic and total soil respiration is not significantly affected by vegetation type, measurement method, mean annual temperature, precipitation, latitude, and soil drainage.

Our understanding of coarse partitioning of soil respiration into autotrophic and heterotrophic components has been considerably improved. Due to methodological difficulties, there is limited information on fine partitioning of soil respiration into components of surface or root litter, live roots, and various fractions of SOM. Moreover, complex interactions among soil compartments

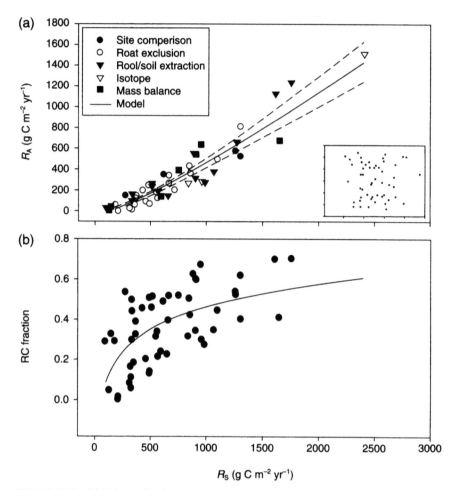

FIGURE 9.12 (a) Relationship between annual soil respiration (R_S) and its autotrophic (R_A) component, by study methods. For the 54 sites examined, $R_A^{0.5} = -7.97 + 0.93R_S^{0.5}$, $R^2 = 0.87$, $P < 0.001$. Dotted lines show model 95% confidence intervals; inset graph shows model residuals. (b) Root contribution (RC) to R_S. For the 53 sites examined, $RC = -0.66 + 0.16\ln(R_S)$, $R^2 = 0.38$, $P < 0.001$ (Adapted with permission from Global Change Biology: Bond-Laberty et al. 2004).

may result in positive or negative feedback on decomposition processes (Subke *et al.* 2004) and make fine partitioning of soil respiration very difficult. Although Bond-Lamberty *et al.* (2004) did not find any influences of many factors and different measurement methods on estimated contributions of autotrophic versus heterotrophic respiration, sources of variation caused by methods (e.g., disturbances in root exclusion study and assumptions in isotope calculations) and other factors (e.g., ecosystem type, dominant species, developmental stages, season of the year, and climatic conditions) are yet to be evaluated for estimating relative contributions to soil respiration.

Modeling Synthesis and Analysis

A model is derived either from experimental data and/or from process thinking. In this regard, modeling plays a critical role in synthesizing experimental results and analyzing processes of soil respiration. According to their origins, models that are used to study soil respiration are divided into two types: empirical and mechanistic. The empirical models use regression analysis to relate soil respiration to ecological variables such as temperature, soil moisture, precipitation, and carbon substrate. The mechanistic models, also called process-based models, are built upon our current understanding of environmental and biological processes that are involved in soil respiration. The process-based models can be further divided into a CO_2 production model and CO_2 production-transport model. The production model considers the processes that produce CO_2, whereas the production-transport model has vertical profiles of CO_2 production together with other variables, along which molecules of CO_2 are transferred to the soil surface. Models of soil and ecosystem respiration have also been applied to different temporal and spatial scales. In general, large-scale models are simpler than those applied to ecosystem scales where detailed processes can be examined. This chapter accordingly describes various approaches to modeling soil respiration.

10.1. EMPIRICAL MODELS

The empirical models are derived primarily from observed soil respiration as functions of environmental and biological variables. This type of model is usually simple in structure and does not identify fundamental processes that govern soil respiration. Regression analysis has been extensively applied to relationships of soil respiration to soil and air temperature, soil moisture, and precipitation. Many experimental studies also show that soil respiration is strongly regulated by substrate supply from canopy photosynthesis. Several empirical models have been developed to relate soil respiration to surrogate variables of substrate supply, for example, leaf area index (LAI). Moreover, soil respiration is interactively affected by multiple factors. Empirical models have also been developed to relate multiple factors to soil respiration.

TEMPERATURE-RESPIRATION MODELS

Respiration is fundamentally a cellular process and involves many biochemical reactions. Respiration models naturally borrow central principles from enzyme kinetics that describe relationships between enzyme activity and temperature (van't Hoff 1884, Arrhenius 1898). Since soil respiration observed in most of the studies usually increases in some form of accelerating rates with temperature (Fig. 5.9), the exponential equation as originally illustrated by van't Hoff (Equation 5.1) or the enzymatic reaction equation by Arrhenius (Equation 5.2) can well characterize the relationship between soil respiration and temperature. Besides these two models, many other empirical models have been developed in the literature to describe the relationship between soil respiration and temperature (Table 10.1). Most of them contain some forms of exponential and/or power functions. Except for the linear one for forest soil, all the models fit the data sets collected from a farmland and a mature sitka spruce plantation near Edinburgh, Scotland, with high determinant coefficients of 80 to 90% (Fang and Moncrieff 2001). Among them, the Arrhenius model has a sound theoretical basis and fits the data sets very well. The simple empirical equation, $R_s = a(T - T_{min})^b$, is more responsive to temperature changes at its low range than are the Arrhenius and exponential models.

While some forms of rate-accelerating equations usually fit data well (e.g., Figs. 5.9 and 7.5), the major controversy in studying responses of soil respiration to temperature arises from different views on the temperature sensitivity estimated from fitted equations. The exponential equation by van't Hoff gives one single Q_{10} value with Equation 5.4. When it fits 15 data sets, the exponential equation underestimates respiration at low temperature and overestimates at high temperature (Lloyd and Taylor 1994), indicating that

TABLE 10.1 Empirical equations commonly used to describe the relationships between soil respiration (R_s) and temperature (T)

Model	Properties	Reference
$R_s = a + bT$	Linear function	Rochette *et al.* (1991)
$R_s = ae^{bT}$	Exponential function	van't Hoff (1884)
$R_s = R_o Q_{10}^{\frac{T-T_o}{10}}$	Modified exponential function. R_o: respiration rate at temperature T_o Q_{10}: representing the relative increase (R/R_o) as temperature increases by 10°C	van't Hoff (1898)
$R_s = ae^{bT+cT^2}$	Second-order exponential function	O'Connell (1990)
$R_s = ae^{E/RT}$	Arrhenius function that accounts for activation energy (E) in chemical reaction. R: universal gas constant (8.314 J mol^{-1} K^{-1})	Arrhenius (1898)
$R_s = R_{10}e^{\left(\frac{E}{283.15R}\right)\left(1-\frac{283.15}{T}\right)}$	Modified Arrhenius function	Lloyd and Taylor (1994)
$R_s = a(T + 10)^b$	Varying power function to potentially account for more responsive R_S at low temperature	Kucera and Kirkham (1971)
$R_s = a(T - T_{min})^b$	Varying power function. T_{min} is a temperature when R_S equals zero.	Lomander *et al.* (1998)
$R_s = R_{10} + a(T - 10)^2$	Quadratic function. Period R_{10} is respiration rate at 10°C.	Holthausen and Caldwell (1980)
$R_s = \dfrac{(T - T_{min})^2}{(T_{ref} - T_{min})^2}$	Quadratic function with a hypothetical temperature (T_{min}) at which R_s equals zero.	Ratkowsky *et al.* (1982)
$R_s = \dfrac{1}{a + b^{-((T-10)/10)}}$	Logistic function with an "S" type response	Jenkinson (1990)
$R_s = \dfrac{1}{a + b^{-((T-10)/10)}} + c$	Logistic function with a minimum R_s at c and a maximum at ($1/a + c$)	Schlentner and Van Cleve (1985)
$R_s = (A_1 A_2 A_3)^z R_{max}$	$A_1 = (T_{max} - T)/(T_{max} - T_{opt})$ $A_2 = (T - T_{min})/(T_{opt} - T_{min})$ $A_3 = (T_{opt} - T_{min})/(T_{max} - T_{opt})$ z: shape parameter R_{max}: maximum measured R_s	Frank *et al.* (2002)

Note: a, b, and c are empirical coefficients to be estimated from regression analysis respiration may not follow a strict exponential relationship with constant temperature sensitivity in the full span of temperature (Fig. 10.1). The Arrhenius equation (Equation 5.2) is based on the kinetic theory and can account for decreasing activation energy with increasing temperature. The Arrhenius equation can be expressed in a form of exponential equation (Lloyd and Taylor 1994):

$$R_s = R_{ref} e^{E_0\left(\frac{1}{T_{ref}-T_0} - \frac{1}{T-T_0}\right)}$$

(10.1)

where R_{ref} is soil respiration at a reference temperature, T is the absolute temperature in degrees Kelvin (K), T_{ref} is a reference temperature in degrees Kelvin (K), E_0 is an activation-energy-type empirical coefficient, and T_0 is the low temperature limit for the soil respiration in Kelvin (K). R_{ref}, E_0, and T_0 can be empirically estimated from data sets.

Lloyd and Taylor (1994) found that Equation 10.1 provides an unbiased estimate of soil respiration rates across the entire temperature range for 15 data sets collected from a wide range of ecosystem types. The three-parameter model can account for declining temperature sensitivity of soil respiration as soil temperature increases (Fig. 10.1). The declining temperature sensitivity may result from diverse genera of microorganisms with different activation energies for decomposition of different chemical compounds in litter and

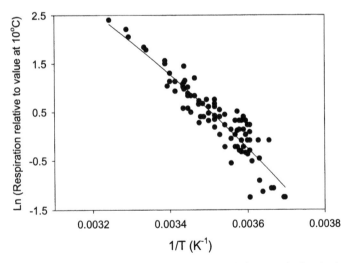

FIGURE 10.1 The natural logarithm of the respiration rate (relative to the fitted value at 10°C) expressed in relation to the reciprocal of the absolute temperature. The activation-energy-like parameter, E_0, (i.e., the slope of the relationship), changes inversely with temperature (Redrawn with permission from Functional Ecology: Lloyd and Taylor 1994).

SOM. Equation 10.1 has been found to well represent responses of soil respiration to temperature in other studies (Thierron and Laudelout 1996, Savage and Davidson 2001, Richardson and Hollinger 2005, Reichstein et al. 2005). However, some researchers found no distinct advantage in using the Arrhenius equation in comparison with other exponential-type models (Buchmann 2000). More important, it is impossible for soil respiration to keep increasing exponentially as temperature increases. It will eventually start to decline when temperature increases to a range beyond optimum (e.g., Figs. 5.7 and 5.8). None of the empirical models except the one by Frank et al. (2002) in Table 10.1 reflects negative impacts of high temperature.

Although some forms of rate-accelerating equations can well fit the temperature-respiration relationships, the derived empirical models (Equations 5.1, 5.2, and 10.1) do not reveal causality between temperature and respiration. This argument holds true particularly with data in the literature obtained mostly from seasonal measurements of temperature and respiration rates. Over a growing season, temperature is highly correlated with radiation. The latter determines carbon supply to the rhizosphere and strongly affects soil respiration. Other biotic and abiotic factors (e.g., phenology) covary with temperature and radiation over seasons to influence the seasonal course of soil respiration. Fitting a simple model to the data that reflect convolution of many processes and complex interactions of multiple factors could not reveal fundamental mechanisms underlying the respiration-temperature relationships (Davidson et al. 2006). To isolate the temperature effect from other variables, laboratory incubation has been used to study respiration at different temperatures, while other environmental factors are controlled at constant. The derived temperature sensitivity from the incubation studies is much less confounded than that from seasonal measurements. However, the incubation studies destroy soil structure and disconnect carbon flows from plants to the rhizosphere. To fundamentally improve our understanding of the temperature-soil respiration relationship, we have to conduct innovative experiments with environmentally controlled facilities to eliminate confounding effects by other factors. Such experiments need to be carried out at levels of whole ecosystems and individual components as well to quantify interactions of multiple processes.

MOISTURE-RESPIRATION MODELS

Unlike the patterns evident in the temperature-respiration relationship, no comparably consistent relationships between moisture and soil respiration have been identified across studies. Indeed, the nature and shape of the moisture-respiration relationship are largely unknown (Fig. 5.10). The inconsistent, variable responses of soil respiration to moisture result partly from complex mechanisms that are involved in moisture regulations of CO_2 production and transport processes and partly from fluctuation in moisture conditions in the field (see Chapter 5). Soil microorganisms as a community develop a suite of mechanisms to cope with water stresses so that moisture effects on microbial growth and death are minor (Harris 1981). Soil moisture affects microbial CO_2 production mainly through diffusion of oxygen, substrates, other gases, and solutes to or from sites of microbial activities at the level of soil aggregates. The function of gas and solute diffusion with soil water content is described by Equation 5.5 (Papendick and Campbell 1981). To examine further the idea that diffusion regulates microbial activities under different soil moisture levels, Skopp *et al.* (1990) conducted a laboratory experiment with soil incubation. Their results show that soil respiration increases with relative water content up to 0.7 and then declines (Fig. 10.2). The increase in soil respiration with water content at its low range is due to increased diffusion of substrate to sites where microbial organisms can use it. The decrease in soil respiration with water content at the high range is due to limitation of O_2 diffusion. The microbial respiration (R), which is limited by either substrate or O_2, can be described by:

$$R = Min \begin{cases} \alpha(\theta_v)^f \\ \beta(\varepsilon - \theta_v)^g \end{cases} \qquad (10.2)$$

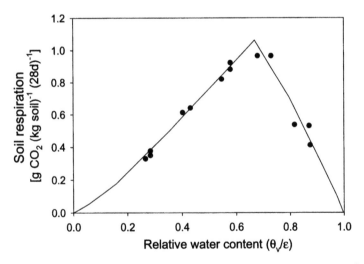

FIGURE 10.2   The relationship between soil respiration and relative water content with regression curve by Eq. 10.2 in the Yolo soil, California, USA (Redrawn with permission from *Soil Science Society of America Journal*: Skopp et al. 1990).

where θ_v is the volumetric soil water content ($m^3 m^{-3}$), ε is the total porosity ($m^3 m^{-3}$); α, β, and f and g are empirical coefficients that are estimated from experimental results and vary with soil types and other factors. Equation 10.2 well describes microbial respiration measured in the soil incubation study (Fig. 10.2).

Although microbial respiration in the laboratory incubation study shows an abrupt change from substrate limitation to O_2 limitation as soil water content increases, observed relationships between soil respiration and moisture in the field display different patterns (Table 10.2). For example, a natural logarithmic relationship is best to describe the relationship between normalized soil respiration and soil water potential in the Harvest Forest ecosystem (Fig. 10.3a, Davidson et al. 1998). A parabolic function can fit highly scattered data of soil respiration with volumetric soil moisture in the Texas grassland (Mielnick and Dugas 2000). Most of the empirical equations developed in the literature are obtained from measured soil respiration over moisture gradients or growing seasons. When soil moisture content is manipulated in the tallgrass prairie of Oklahoma with a relatively constant soil temperature over the experimental period, the relationship between soil CO_2 efflux and moisture is also widely scattered (Liu et al. 2002a). The scattered relationship is possibly caused by soil CO_2 degassing and other complex processes in the rhizosphere. Overall, soil CO_2 efflux increases with soil water availability and can be quantitatively described by an asymptotic equation (Fig. 10.3c). Quantitative relationships have not been developed to describe responses of soil respiration to water contents in relatively wet environments, such as water logging or wetlands.

Soil respiration also varies within a precipitation-drying cycle in the natural world. When soil is dry, a rain event can significantly enhance soil respiration. The effect of rain events on soil respiration is positively related to water input (i.e., amount of precipitation), is negatively related to water loss, and diminishes over time. This precipitation-drying pattern is modeled with a wetting index (I_w) in a Scots pine (*Pinus sylvestris* L.) stand in Belgium (Curiel Yuste et al. 2003) as:

$$I_w = \alpha + \log\left(\frac{\sqrt{P}}{VPD_a t^2} \right) \tag{10.3}$$

where α is a constant, P is the amount of precipitation during the last rain event (mm), t is time since the last rain event (h), and VPD_a is the mean vapor pressure deficit of the atmosphere at 1.5 m above the forest floor (kPa) averaged over the last 24 h. The equation uses the square root of the rain intensity to minimize its contribution and amplifies the contribution of time by using the square of the elapsed time, since the stimulating effect of rain on soil respira-

TABLE 10.2　Empirical equations commonly used to describe the relationships between soil respiration and soil water content (Rs = soil respiration)

Equation	Parameter	Measurement	Reference
$R_s = -a \times \ln(-\psi) + b$	Ψ = water potential	Lab incubations	Orchard and Cook (1983) Davidson et al. (2000)
$R_s = 383.63(\theta_v - 0.1)(0.7 - \theta_v)^{2.66}$	θ_v = % volumetric water content	Field fluxes in Texas	Mielnick and Dugas (2000)
$R_s = (a \times WF) + (b \times WF^2) + c$	WF = water-filled pore space	Lab incubations	Doran et al. (1991)
$R_s = \exp(-e^{(p - q\theta)})$	θ = % volumetric water content	Field fluxes in Belgium	Janssens et al. (2001)
$R = 0.664 \dfrac{W - 25.0}{7.88 + (W - 25.0)}$	W = % gravimetric water content	Field fluxes in central Oklahoma	Liu et al. (2002a)
$R_s = \text{Min} \begin{cases} \alpha(\theta_v)^f \\ \beta(\varepsilon - \theta_v)^g \end{cases}$	θ = % volumetric water content ε = water content at field capacity	Lab incubations	Skopp et al. (1990)
$R_s = \dfrac{(c_0 - c_b)D_0 k \theta^3}{s}$	c_0, c_b = the solute concentration at a cell surface and in bulk soil D_0 = diffusivity θ = % volumetric water content s = the diameter of a bacterial cell	Lab incubations	Papendick and Campbell (1981)
$I_w = \alpha + \log\left(\dfrac{\sqrt{P}}{VPD_a t^2}\right)$	I_w = rewetting index P = precipitation during the last rainfall (mm) t = time since the last rain (h) VPD_a = mean vapor pressure deficit	Field flux in a Scots pine (Pinus sylvestris L.) stand in Belgium	Curiel Yuste et al. (2003)

Note: a, b, c, k, f, g, p, q, α, and β are empirical coefficients to be estimated by regression analysis.

tion is highly ephemeral. Figure 10.4 shows bidimensional representations of the relationship among the three variables in controlling I_w. The values of I_w decrease rapidly with time, especially during the first hours (Fig. 10.4b, c).

The dynamic patterns of soil CO_2 efflux with soil moisture within a wetting-drying cycle are quantified in a water manipulation experiment with

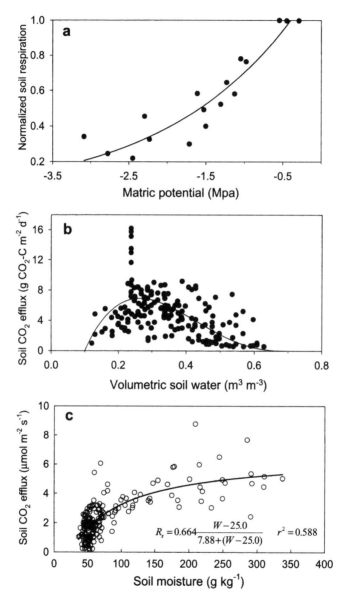

FIGURE 10.3   The responses of soil respiration to moisture dynamics: (a) effect of soil matric water potential on normalized soil respiration during the summer drought (Davidson *et al.* 1998); (b) the parabolic relationship between average daily soil CO_2 efflux and volumetric soil water content in a tallgrass prairie in Texas (Mielnick and Dugas 2000); and (c) relationship between soil moisture (g kg^{-1}W) and soil CO_2 efflux (R_s) in the field at different water treatment (Liu *et al.* 2002a).

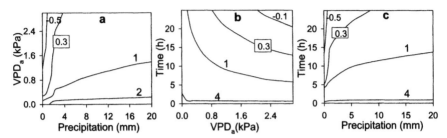

FIGURE 10.4 Contour graphs illustrating the influence of different parameters on the wetting index (Iw). (a) Effect of atmospheric vapor pressure deficit (VPDa) and amount of precipitation during the last rain event 10 h after that rain event. (b) Time since the last rain event and VPDa following 10 mm of precipitation. (c) Time since the last rain event and precipitation during the last rain event at a VPDa of 0.75 kPa. The white box indicates the wetting threshold (0.3) (Redrawn with permission from Tree Physiology: Curiel Yustel *et al.* 2003).

simulated rainfall of 0, 10, 25, 50, 100, 150, 200, and 300 mm in a tallgrass prairie ecosystem (Fig. 5.11, Liu *et al.* 2002a). The time course of soil CO_2 efflux in response to water manipulation is well described by

$$R = R_0 + ate^{-bt} \tag{10.4}$$

where R is soil CO_2 efflux, R_0 is soil CO_2 efflux before water treatment, t is time, and *a* and *b* are coefficients, varying with different water treatments. The equation describes the pattern that soil CO_2 efflux dramatically increases immediately after the water addition, followed by a gradual decline.

SUBSTRATE-RESPIRATION MODELS

Evidence from many experiments supports the idea that substrate supply from canopy photosynthesis significantly regulates respiratory release of CO_2 from soil (see Chapter 5). However, no good relationships have been developed to relate soil respiration directly to substrate supply from canopy photosynthesis. Several surrogate variables have been used to relate soil respiration to substrate supply from photosynthesis or soil carbon pools. Those surrogate variables include LAI (Fig. 5.3, Bremer and Ham 2002, Reichstein *et al.* 2003), annual gross primary productivity (Fig. 5.6, Janssens *et al.* 2001), root biomass (Fig. 3.4, Ryan *et al.* 1996, Thomas *et al.* 2000), litter mass (Fig. 5.4, Maier and Kress 2000), litterfall (Fig. 2.2, Raich and Naderhoffer 1985), mycorrhizal associations (Rygiewicz and Andersen 1994), and the size of soil carbon pool (Fig. 5.5, Franzluebbers *et al.* 2001). In most of the studies, linear equations are used to relate substrate supply to soil respiration. It is not yet clear whether the linear equations truly represent the nature of substrate effects on soil respiration or happen to fit limited data well, since comprehensive data sets are not available.

A Michaelis-Menten kinetics model has been applied to describe responses of soil respiration to O_2 concentration. Oxygen is an essential substrate for aerobic respiration in soil (Sierra and Renault 1995, 1998). The rate of soil respiration asymptotically increases with soil O_2 concentration in organic horizons (0 to 10 cm, 10 to 30 cm) and mineral horizon (Fig. 10.5). The relationship between soil respiration and soil O_2 concentration can be well described:

$$R = R_{max}\left(\frac{C_{O_2}}{k_m + C_{O_2}}\right) \tag{10.5}$$

where R is the rate of O_2 consumption (mol O_2 m^{-3} soil s^{-1}), R_{max} is the maximal rate of O_2 consumption when O_2 does not limit respiration (mol O_2 m^{-3} soil s^{-1}), C_{O_2} is the O_2 concentration (mol O_2 m^{-3} air), and k_m is the Michaelis constant (mol O_2 m^{-3} air).

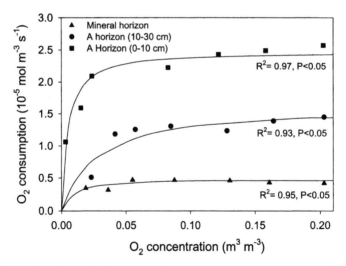

FIGURE 10.5 The relationship between soil respiration (as O_2 consumption) and O_2 concentration for the three upper soil horizons. Fitted lines are by the Michaelis-Menten equation (Redrawn with permission from Soil Science Society of America Journal: Sierra and Renault 1998).

MULTIFACTOR MODELS

As discussed in Chapter 5, soil respiration is affected interactively by many factors. It is highly desirable to develop models that describe interactive effects of multiple factors on soil respiration. Most of the models that have been developed to describe the interactive effects usually use multiplication/division and/or addition/subtraction to combine effects of individual factors (Table 10.3). In the combined temperature-moisture models, the exponential or Arrhenius models or their variants are generally used to describe temperature effects on soil respiration, whereas diverse forms of equations are used to describe effects of soil moisture. For example, three different forms of soil moisture functions are combined with the exponential equation by Gulledge and Schimel (2000) to describe interactive effects of temperature and moisture on soil respiration:

$$R_s = \alpha e^{\beta T}(\chi M) \tag{10.6}$$

$$R_s = \alpha e^{\beta T} - (M - \delta)^2 \tag{10.7}$$

$$R_s = \alpha e^{\beta T}\frac{M}{M + \varepsilon} \tag{10.8}$$

where R_s is soil respiration, T is soil temperature (°C), M is soil moisture (g H_2O g dry soil), α is the flux rate at 0°C, β is a temperature response coefficient, and χ, δ, and ε are different moisture response constants. The quadratic model described in Equation 10.7 assumes an optimum moisture (δ) that allows maximal activity. The asymptotic model of Equation 10.8 assumes that as moisture increases, respiration asymptotically approaches some maximum rates, but moisture does not directly alter the temperature sensitivity. The asymptotic model (Equation 10.8) fits the observed responses of soil respiration to temperature and moisture better than the other two models in taiga forests of interior Alaska (Gulledge and Schimel 2000).

Two exponential equations are combined by Lavigne et al. (2004) to describe responses of soil respiration to changes in temperature and moisture as:

$$R_s = (ce^{d\Psi_s})e^{b(T_s - 10)} \tag{10.9}$$

where c, d, and b are coefficients, Ψ_s is soil water potential, and T_s soil temperature. The equation well describes effects of temperature and moisture on soil respiration in a trenching study in a 40-year-old balsam fir (*Abies balsamea*) forest in New Brunswick, Canada. The estimated temperature sensitivity of soil respiration is not affected by trenching, but basal respiration as

TABLE 10.3 Empirical equations to describe responses of soil respiration (R_s) to multiple factors such as temperature, soil moisture, LAI, and precipitation

Equation	Parameter/Variable	Measurement	Reference
$R_s = \alpha e^{\beta T}(\chi M)$ $R_s = \alpha e^{\beta T} - (M - \delta)^2$ $R_s = \alpha e^{\beta T}\dfrac{M}{M + \varepsilon}$	T = temperature M = soil moisture	Field fluxes in interior Alaska	Gulledge and Schimel (2000)
$R_s = (ce^{d\Psi_s})e^{b(T_s - 10)}$	Ψ_s = soil water potential T_s = soil temperature	Field fluxes in New Brunswick	Lavigne et al. (2004)
$R = R_{ref}f(T_{soil})g(\theta)$	R_{ref} = reference soil respiration $f(T_{soil})$ = temperature function $g(\theta)$ = soil water content function	Field fluxes in central Italy and southern France	Reichstein et al. (2002)
$R_s = 0.88 \pm 0.013W \times T$	W = % gravimetric water content T = temperature	Field fluxes in central Washington	Wildung et al. (1975)
$R_s = 13.6e^{0.087T}(\theta - 0.1)(0.7 - \theta)^{1.46}$	T = soil temperature θ = % volumetric water content	Field fluxes in a Texas grasstawel	Mielnick and Dugas (2000)
$R_s = 0.2439\theta^{0.4199}T^{-0.5581}$	θ = % volumetric water content T = temperature	Field fluxes in California	Qi et al. (2002)
$R_s = ce^{\left(\frac{E}{RT} + \frac{aW_t}{W_t + b}\right)}$	W_t = depth to water table T = temperature R = universal gas constant E = apparent activation energy	Field CO_2 fluxes in Alaskan tundra	Oberbauer et al. (1992)
$R_s = \dfrac{k\theta R_{max}}{k\theta + R_{max}}q^{T/10}\left(1 - \dfrac{C_f}{100}\right)$	θ = % volumetric water content T = temperature R_{max} = maximum flux when θ = 100% C_f = % coarse fraction	Field fluxes in Walker Branch mixed hardwood forest, Tennessee	Hanson et al. (1993)
$R_s = \dfrac{W}{a_1 + W}\dfrac{a_2}{a_2 + W}a_3 a_4^{\frac{T-10}{10}}$	W = % gravimetric water content T = temperature	Field fluxes	Bunnel et al. (1977) Schlentner and Van Cleve (1985) Carlyle and Bathan (1988)

Note: a, a_1, a_2, a_3, a_4, b, c, d, k, α, β, and ε are empirical coefficients to be estimated from regression analysis.

represented by c decreases by 40 to 50% in the trenched plots in comparison with the control plots.

An exponential equation is combined with a parabolic function by Mielnick and Dugas (2000) and Lee et al. (2002) to describe responses of soil respiration to changes in temperature and moisture in a tallgrass prairie in Texas and a cool temperate deciduous broadleaf forest in Japan. The constructed model can explain 96% of the variance of the daily soil CO_2 efflux from daily average temperature and soil water content on sunny days (Fig. 10.6). The modeled soil CO_2 effluxes on rainy days also correlate significantly with the measured ones. However, the measured values from rainy days are, on average, 95% higher than those from sunny days.

The multiplication rule has been applied to quantification of interactive effects of soil temperature (T_{soil}), soil water availability, and vegetation productivity (Reichstein et al. 2003) as:

$$R = R_{ref}(LAI_{max})f(T_{soil}, \theta)g(\theta) \qquad (10.10)$$

where LAI_{max} is the maximum site LAI; R_{ref} (LAI_{max}) is a reference soil respiration at a particular site under constant temperature, no-water-limiting conditions but varies with site-specific productivity as indicated by maximal LAI;

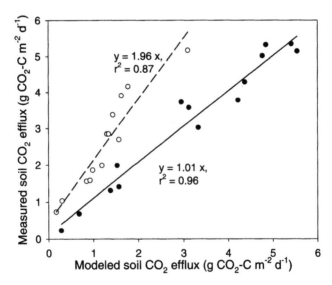

FIGURE 10.6 Relationship between measured and calculated average daily soil carbon fluxes. Dashed line represents the regression line for rainy days. Solid line represents the regression line for sunny days (Redrawn with permission from Ecological Research: Lee et al. 2002).

$f(T_{soil}, \theta)$ represents temperature function as regulated by soil water content relative to that at field capacity (θ); and $g(\theta)$ is the direct effect of relative soil water content on soil respiration. The reference soil respiration is estimated by:

$$R_{ref}(LAI_{max}) = a_{LAI} + b_{LAI} LAI_{max} \qquad (10.11)$$

where a_{LAI} and b_{LAI} are coefficients. The linear relationship occurs between R_{ref} (LAI_{max}) and LAI_{max} in 17 different forest and shrubland sites in Europe and North America (Fig. 5.4). The temperature function is represented by modification of Equation 10.1 as:

$$f(T_{soil}, \theta) = e^{E_0(\theta)\left(\frac{1}{T_{ref}-T_0} - \frac{1}{T_{soil}-T_0}\right)} \qquad (10.12)$$

where $E_o(\theta)$ is the activation-energy-type parameter for soil respiration and varies with θ, T_0 is the lower temperature limit for the soil respiration ($-46°C$), and T_{ref} is reference temperature. The proposed model implies a nonlinear dependency of the apparent Q_{10} on both temperature and soil water content (Fig. 10.7). The direct effect of θ on respiration is:

$$g(\theta) = \frac{\theta}{\theta_{1/2} + \theta} \qquad (10.13)$$

The regression model described by Equation 10.10 well explains soil respiration as dependent on soil temperature, soil water content, and site-specific maximum LAI across 17 forests and shrublands in Europe and North America.

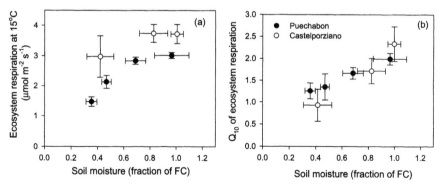

FIGURE 10.7   Relationship between ecosystem respiration at $T_{ref} = 15°C$ and soil moisture (a), and between estimated Q_{10} of ecosystem respiration and soil moisture (b) for the Puéchabon and Castelporziano sites. Horizontal error bars represent standard deviation of soil moisture within moisture classes; vertical bars indicate standard errors of estimate for the parameters (Redrawn with permission from Functional Ecology: Reichstein *et al.* 2002).

The inclusion of LAI as an integrative variable likely accounts for direct influences of canopy photosynthesis on soil respiration and provides a potential link to remote sensing. Since the nature of the interactions is not clear, it is beyond expectation that any of the empirical models developed so far could represent mechanisms undetlying multifactor interactions. To develop mechanistic models for simulation of multifactor interactions in influencing soil respiration, we have to examine how the CO_2 production and transport processes can best be represented in models.

10.2. CO₂ PRODUCTION MODELS

Process modeling is based on our understanding of mechanisms underlying soil respiration. It offers a potential to explain observed temporal and spatial variations in soil CO_2 efflux among different ecosystems and to project soil CO_2 efflux in future climatic conditions. This section describes models that consider only processes of CO_2 production; the production-transport models are explained in the next section.

The CO_2 production models are established on the principle of mass balance of carbon in ecosystems. Most of the biogeochemical models that have been developed in the past decades to simulate terrestrial carbon processes (Parton *et al.* 1987; Rastetter *et al.* 1991, 1997, Comins and McMurtrie 1993, Potter *et al.* 1993, Luo and Reynolds 1999) potentially can be used to examine CO_2 production processes. The biogeochemical models generally share a common structure that partitions carbon input into several pools, from which carbon is released via respiratory processes.

The carbon input into ecosystems is simulated using a variety of methods. A terrestrial carbon sequestration (TCS) model, for example, uses the canopy photosynthetic rates that are estimated by a comprehensive canopy model validated with measured leaf photosynthesis and eddy-covariance measurements of canopy fluxes as the input (Luo *et al.* 2001c). The Carnegie-Ames-Stanford-Approach model uses a function of the absorbed photosynthetically active radiation, the maximum potential light-use efficiency, and temperature and moisture scalars that represent climate stresses on light-use efficiency to estimate NPP as carbon input (Porter *et al.* 1993). The CENTURY model employs a function of live leaf, monthly evapotranspiration, air temperature, available nitrogen supply, and C:N ratio to simulate plant production that drives carbon processes and nutrient cycling in the model (Parton *et al.* 1987).

In the TCS model, the photosynthate is partitioned into four pools in leaves, wood, fine roots, and a labile pool for root exudation (Fig. 10.8). Carbon partitioning into these pools is based on a nitrogen production relationship (Luo and Reynolds 1999). Most of the biogeochemical models do not

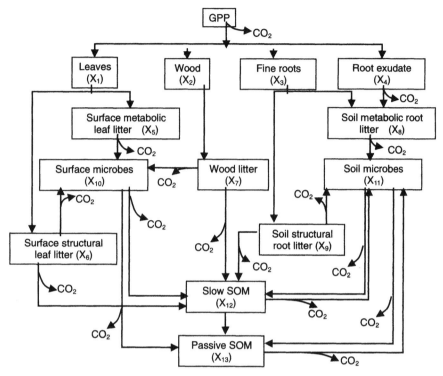

FIGURE 10.8   Carbon pools and pathways of carbon flux in the terrestrial carbon sequestration (TCS) model (Luo *et al.* 2001b).

simulate root exudation, although it potentially can transfer a great amount of carbon from plants to the rhizosphere. The TCS model uses a simple function to allocate a fraction of the total photosynthate to the root exudation pool (Luo *et al.* 2001b). Carbon allocated in leaf, wood, and fine-roots pools is used partly for autotrophic respiration and partly for tissue growth. Dead plant material goes to litter pools. Most of the biogeochemical models simulate dynamics of litter and SOM in soil in a way similar to that used by the Rothamsted (RothC) (Jenkinson and Rayner 1977) or CENTURY models (Parton *et al.* 1987). Leaf and fine-root litters are each divided into metabolic and structural components according to their lignin content and the C:N ratio. Dead wood goes to the structural litter pool. During litter decomposition, carbon substrate in litter is partly released as CO₂ by microbial respiration and partly converted into microbial biomass in the active pool Fig. 10.8. Part of the structural litter is directly transferred to the slow SOM pool. Dead microbes are decomposed to CO₂. Residuals are incorporated into the slow

SOM pool. SOM goes through the formation-decomposition cycle. During each cycle, part of SOM is mineralized into CO_2 by microbial respiration, while part goes back to SOM pools.

In the CO_2 production models, plant respiration is simulated by a variety of methods. The simplest method is to simulate plant respiration by multiplying carbon contents in each of the plant pools (e.g., leaf, wood, and fine root) with their respective specific respiratory rates. The more mechanistic models of plant respiration consider functional components of growth and maintenance (McCree 1970, Thornley 1970). Growth respiration is related to energy required for construction of new plant tissues and its associated CO_2 release. Maintenance respiration is related to energy required to maintain normal functioning of plant tissues. The two components of plant respiration can be expressed in a model as:

$$R_{plant} = \frac{1 - Y_G}{Y_G} \frac{dM}{dt} + mM \qquad (10.14)$$

where R_{plant} is plant respiration; Y_G is growth yield of carbohydrate or biosynthetic efficiency, which is the ratio of mass of carbon incorporated into structure to carbon used for structure plus energy used for synthesis; dM/dt is the growth rate of plant; M is biomass; and m is the maintenance coefficient, as measured by the amount of carbon respired per unit living biomass carbon per unit time. Ion uptake, particularly nitrate, can be costly. Some of the plant respiration model also incorporates ion uptake (Johnson 1983) into a three-component model as:

$$R_{plant} = \left(\frac{1 - Y_G}{Y_G} + af_N \right) \frac{dM}{dt} + mM \qquad (10.15)$$

where a is respiration per unit nitrogen uptake and f_N is fractional nitrogen content of biomass.

Growth respiration is usually estimated according to the amount of plant growth and tissue construction cost. Tissue construction cost varies with chemical composition of plant tissues (McDermit and Loomis 1980, Griffin et al. 1993, Griffin et al. 1996b, Lavigne and Ryan 1997). The maintenance respiration is strongly responsive to environmental change, particularly temperature and tissue nitrogen concentration. Thus, the coefficients of maintenance respiration usually vary with tissue nitrogen concentration and temperature.

Microbial respiration is accompanied by decomposition of litter and SOM in litter and soil pools respectively. The decomposition of litter and SOM is proportional to the amount of carbon in pools and can be described by Equation 3.8. Accordingly, the CO_2 production from each of the litter and SOM pools via microbial decomposition is also proportional to the pool size (X_i) as:

$$R_i = r_i X_i \tag{10.16}$$

where R_i is CO_2 released from pool i and r_i is the coefficient to quantify a fraction of carbon in pool i that is released during decomposition. The modeled soil respiration R_s is the sum of microbial respiration in each of the litter and soil pools plus root respiration as:

$$R_s = \sum R_i + R_{root} \tag{10.17}$$

The root respiration and microbial decomposition are usually regulated by temperature, moisture, O_2 concentration, litter quality, and soil texture. In most of the production models, the effects of those factors on decomposition and respiration are expressed as scalars and combined by multiplication (Table 10.4). For example, in the soil respiration model by Fang and Moncrieff (1999), the specific root and microbial respiration rates are modeled by:

$$r_r = r_{r0}T_sW_sO_s, \tag{10.18}$$

$$r_m = r_{m0}T_sW_sO_s \tag{10.19}$$

where r_r is the specific respiratory rate of the fine root and r_m is the specific microbial respiration rate. r_{r0} and r_{m0} represent the maximum specific respiration rates of roots and microorganisms under optimal conditions at a reference temperature T_0. T_s, W_s, and O_s are scaling factors to represent influences of soil temperature, moisture, and O_2 concentration respectively on root and microbial respiration. Each of the scalars is defined as:

$$T_s = \exp\left(\frac{E}{RT}\frac{T-T_0}{T_0}\right) \tag{10.20}$$

$$W_s = 1 - \exp(-aW + c) \tag{10.21}$$

$$O_s = \frac{1}{1+\dfrac{Km}{[O_2]}} \tag{10.22}$$

where E is the activation energy for respiration, in $kJ\,mol^{-1}$; R is the universal gas constant and T is the absolute temperature in Kelvin (K); a defines the maximal increase in the rate of soil respiration with soil moisture W; c is a constant; and K_m is the Michaelis-Menten constant. W_s and O_s have a value between 0 and 1. Parameter values of E, a, c, and Km can be specified differently for root and microbial respiration.

Burke *et al.* (2003) evaluated eight models of terrestrial biogeochemistry, focusing on model structures governing temperature controls of decomposition rates. The eight models are Rothamsted (RothC) (Jenkinson and Rayner 1977), CENTURY (Parton *et al.* 1983), Terrestrial Ecosystem Model (TEM)

TABLE 10.4 Decomposition processes represented in selected biogeochemical models (Adapted with permission from Princeton University Press: Burke et al. 2003)

Model	Decomposition Equation	Terms	C Pool Structure
Biome-BGC (Hunt et al. 1996)	$k_L = k_q T_s W_s$ $k_s = k_c T_s W_s$	k_L = leaf and root decomposition rate (d⁻¹) k_q = site-specific litter quality constant (d⁻¹) T_s = soil temperature scalar W_s = soil moisture scalar k_s = soil C decomposition rate (d⁻¹) k_c = fixed decomposition rate from CENTURY (d⁻¹)	Leaf- and root-litter carbon Other detrital soil carbon
Forest-BGC (Running and Gower 1991)	$k_L = k_{max} \dfrac{T_s + W_s}{2}$ $k_s = 0.03 k_L$	k_L = leaf and root decomposition rate (yr⁻¹) k_{max} = maximum decomposition rate (yr⁻¹) T_s = soil temperature scalar W_s = soil moisture scalar k_s = soil C decomposition rate (yr⁻¹)	Leaf- and root-litter carbon Other detrital soil carbon
CENTURY (Parton et al. 1994)	$k_1 = k_{max} T_s W_s C_s$ $k_2 = k_{max} T_s W_s Q_s$ $k_3 = k_{max} T_s W$	k_1 = soil microbial decomposition rate (d⁻¹), k_2 = structural plant decomposition rate (d⁻¹), k_1 = all other pools decomposition rate (d⁻¹), k_{max} = fixed maximum decomposition rate (yr⁻¹) T_s = soil temperature scalar W_s = soil moisture scalar C_s = soil texture scalar Q_s = litter quality scalar	Structural litter carbon Metabolic plant carbon Surface microbial carbon Soil microbial carbon Slow soil carbon Passive soil carbon
FAEWE (Van der Peijl and Verhoeven 1999)	$k = k_{max} \dfrac{T_{as}}{T_{ms}}$	k = decomposition rate (wk⁻¹) k_{max} = maximum decomposition rate (wk⁻¹) T_{as} = actual soil temperature scalar T_{ms} = mean annual soil temperature scalar	Detrital soil carbon

Model	Equation	Variables	Carbon pools
Linkages (Pastor and Post 1986)	$k_L = -\ln\{1 - [0.98 + 0.09AET + (0.5 - 0.002AET)\,(L:N)]/100\}$ $k_t = 0.2$ $k_{sw} = 0.1$ $k_{lw} = 0.03$ $k_{dw} = 0.05$ $ks = H\{(-0.0004\,(N:C)/[-0.03 + (N:C)]\}/N$	k_L = root and leaf decomposition rate (yr^{-1}) AET = actual evapotranspiration L:N = litter lignin to nitrogen ratio k_t = twig decomposition rate (yr^{-1}) k_{sw} = small wood decomposition rate (yr^{-1}) k_{lw} = large wood decomposition rate (yr^{-1}) k_{dw} = decayed wood decomposition rate (yr^{-1}) k_s = soil humus decomposition rate (yr^{-1}) H = humus mass (Mg/ha) N = total humus N (Mg/ha) C = total humus C (Mg/ha)	Leaf + root litter carbon Soil humus carbon Twig carbon Small wood carbon Large wood carbon Decayed wood carbon
PnET-II (Aber et al. 1997)	$R = 27.46e^{0.0684T}$	R = soil respiration (g m^{-2} mo^{-1}) T = mean monthly temperature	No detrital carbon pools
RothC (Coleman and Jenkinson 1999)	$k = 1 - e^{-\frac{T_s W_s S_s k_{max}}{12}}$	k = decomposition rate for each pool (mo^{-1}) k_{max} = maximum decomposition rate (yr^{-1}) T_s = air temperature scalar W_s = soil moisture scalar S_s = soil cover scalar	Metabolic litter carbon Structural litter carbon Microbial biomass carbon Humic organic carbon Detrital carbon
TEM (Raich et al. 1991)	$k = k_q W_s e^{0.0693T}$	k = decomposition rate (mo^{-1}) k_q = site-specific litter quality constant (mo^{-1}) W_s = soil moisture/texture scalar T = monthly mean air temperature	

(Raich *et al.* 1991), PnEt-II (Aber *et al.* 1995), Linkages (Pastor and Post 1986), Forest-BGC (Running and Coughlan 1988), Biome-BGC (Hunt *et al.* 1996), and the functional analysis of European wetland ecosystems (FAEWE) model (Van der Peijl and Verhoeven 1999). All the models have multiple pools of organic matter. Decomposition rates of organic matter are modeled by multiplication of pool sizes with specific decomposition rates, which vary according to a number of factors. Variations of the specific decomposition rates with temperature are modeled by temperature scalars, which differ greatly among the eight models (Fig. 10.9). Biome-BGC and PnET-II have exponential tem-

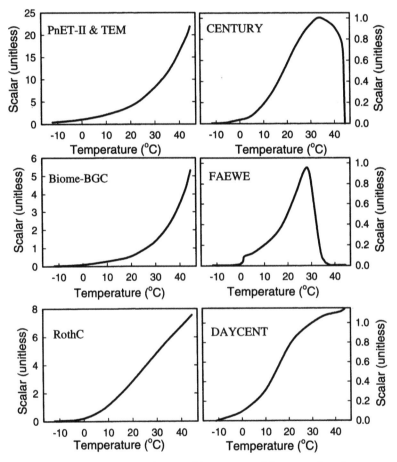

FIGURE 10.9   The temperature scalars used by seven models to simulate the temperature effect on organic matter decompositon (Redrawn with permission from Princeton University Press: Burke *et al.* 2003).

perature scalars that specific decomposition rates increase with temperature in an accelerating fashion. The modeled increase in the specific decomposition rates in PnET-II, however, is five times that in Biome-BGC. RothC has a linear temperature scalar that the specific decomposition rate linearly increases with temperature. CENTURY and FAEWE use optimal functions to model responses of decomposition to changes in temperature. But the two latter models have different optimal temperatures at which decomposition rates reach the maximum. The optimal functions are identical for different organic matter pools and do not vary when the models are applied to ecosystems in different geographical regions. The daily version (DAYCENT) of the CENTURY model uses an arctangent function.

The temperature scalars are modified by other factors to determine temperature sensitivities of decomposition in models. The exponential temperature scalar in TEM is strongly modified by moisture limitation, resulting in nearly zero temperature sensitivity (Fig. 10.10a). Moisture limitation dampens responses of decomposition to changes in temperature with either the linear scalar in the RothC model or the exponential scalar in Biome-BGC. Optimal temperatures for the decomposition shift in CENTURY and FAEWE under water stress. The responses of the decomposition rates in Figure 10.10a can be used to calculate Q_{10} values. PnET, TEM, and Biome-BGC have constant Q_{10} over the entire temperature range (Fig. 10.10b). The Q_{10} values are high at low temperatures and low at high temperatures in the other models.

When all the models were parameterized for the Konza Prairie Long-Term Ecological Research site in Kansas, Burke et al. (2003) found that these models predict different decomposition rates of organic matter and CO_2 releases by an order of magnitude due to differences in model structures (in terms of number of carbon pools), temperature scalars, and moisture interactions. The differences in model structure, temperature scalars, and moisture modifiers among the biogeochemical models reflect the paucity of our knowledge on response functions of decomposition to the environmental variables and interactive effects of multiple factors in the real world. Well-controlled field experiments that permit us to probe fundamental mechanisms of organic matter decomposition and to characterize response functions to temperature and moisture are required to constrain soil respiration models.

The CO_2 production models have been used to address a variety of questions on soil respiration. Luo et al. (2001b) used the TCS model to examine responses of soil respiration to elevated CO_2 in the Duke Forest FACE site. Their analysis shows that fast carbon transfer processes, such as root exudation, may play a minor role in the ecosystem carbon cycling in the forest and are not affected by elevated CO_2. Gu et al. (2004) used the Rothamsted SOC model to demonstrate that the temperature sensitivity of soil respiration is

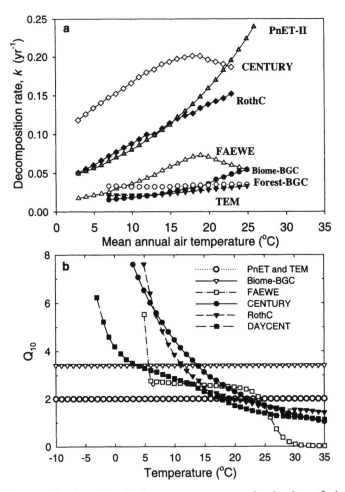

FIGURE 10.10   Panel a: the relationship between temperature and realized specific decomposition rates by seven biogeochemical models at the Konza Prairie Long-Term Ecological Research site. Panel b: The relationship between temperature and the realized Q_{10} in the seven models (Burke *et al.* 2003).

overestimated when seasonal variations of labile carbon pools and temperature are in phase and underestimated when they are out of phase. Wang *et al.* (2002) modified the CENTURY model to simulate short-term soil respiration as observed in a laboratory experiment with different wheat straw types, straw placements, and soil water regimes (continuous moist and alternating moist-dry conditions). The CENTURY model successfully simulates daily CO_2

fluxes except during rewetting periods in comparison with the observation. Frolking *et al.* (1996) developed a daily-step model of the carbon balance that well simulates both the asymmetrical seasonality and short-term variability in soil respiration and other carbon processes for a black spruce/moss boreal forest ecosystem near Thompson, Manitoba.

10.3. CO₂ PRODUCTION-TRANSPORT MODELS

CO_2 efflux from the soil surface is determined by both CO_2 production (see Chapter 3) and transport processes (see Chapter 4). To examine dynamics of soil CO_2 efflux, particularly the short-term fluctuation, we have to use the CO_2 production-transport models. The production-transport models take into account vertical profiles of root biomass, soil carbon pool, temperature, CO_2 concentration, air-filled pore space, moisture, and O_2 concentration. The soil profile is stratified into a number of layers. In each of the soil layers, CO_2 is produced via respiration of live roots and microbes. Usually, the CO_2 production submodel in each layer is relatively simple and based mostly on simple relationships of root respiration to root biomass and microbial CO_2 production to amounts of litter and SOM.

Transport of CO_2 in the soil is usually simulated according to one-dimension CO_2 transport in both gas and liquid phase in the soil on the basis of mass balance (Wood *et al.* 1993, Fang and Moncrieff 1999). The CO_2 mass balance in one soil layer is modeled by Equation 4.7, which considers CO_2 fluxes caused by (1) diffusion in the gaseous phases, (2) diffusion in the liquid phases, (3) gas convection, (4) water vertical movement, and (5) the CO_2 production rate within the layer. The CO_2 diffusions in the gaseous and water phases are calculated by Equations 4.3 and 4.4 respectively. The CO_2 fluxes caused by mass flows of soil gas and water can be estimated by Equations 4.1 and 4.2 respectively. Influences of wind gusts and fluctuating atmospheric pressure on CO_2 releases at the soil surface are generally not considered in any of the production-transport models.

Many production-transport models have been developed to predict soil CO_2 concentration and CO_2 efflux. A one-dimensional mathematical model developed by Ouyang and Boersma (1992a, b) consists of coupled movement of water, heat, and gases through the unsaturated soils for dynamic O_2 and CO_2 exchange between soil and atmosphere. The model simulates rates of CO_2 production by roots and soil microorganisms and incorporates forcing factors of solar radiation, rainfall, water evaporation, and air temperature. The model was modified by Ouyang and Zheng (2000) to examine effects of solar radiation, air temperature, relative humidity, rainfall, soil water movement, heat flux, and CO_2 production on CO_2 diffusive flux into the atmosphere from soil. Solar radiation is identified as one of the most important factors that regulate CO_2 fluxes on the daily time-scale, and rainfall is the factor that controls the monthly CO_2 efflux.

A relatively complex simulation model constructed by Šimůnek and Suarez (1993) is based on relationships of soil CO_2 efflux with soil water potential, temperature, and CO_2/O_2 concentration at different depths along a soil profile. One-dimensional water flow and multiphase transport of CO_2 are simulated by equations of convection, diffusion, and heat flow. Parameter values for the model are obtained independently from the literature. The parameterized model is evaluated by comparing model simulations to published field data from Missouri for three different crops and two growing seasons, as well as a data set from Riverside, California. The model well reproduces observed CO_2 efflux and concentration in the root zone.

A soil CO_2 production-transport model developed by Fang and Moncrieff (1999) also includes one-dimensional water flow and multiphase transport of CO_2, as well as a CO_2 production. The CO_2 production component of the model considers decomposition rates of labile and recalcitrant organic matter and separates roots into three different size classes. The model is validated and applied to a mature slash pine plantation in Florida and modified by Hui and Luo (2004) to evaluate soil CO_2 production and transport in a CO_2 enrichment experiment in the Duke Forest, North Carolina (Fig. 4.2, Table 9.3). Elevated CO_2 increases annual soil CO_2 efflux, but CO_2 transport is not a critical process to regulate soil surface CO_2 efflux on daily or longer time-scales.

A simple and easily parameterized dynamic model developed by Pumpanen *et al.* (2003) describes responses of root and microbial respiration and CO_2 diffusion to soil temperature and moisture (Fig. 10.11). The soil profile is divided into O (humus layer), A (eluvial), B (illuvial), and C (parent material) horizons. All processes and soil properties are described separately for each layer. The CO_2 movement between layers is mediated by diffusion, which is dependent on the total porosity of soil layers, soil water content, layer thickness, and the concentration gradient between the layers. The modeled CO_2 efflux and soil CO_2 concentration are closely related to those observed in the field in southern Finland. A simplified CO_2 production model, coupled with simultaneous transport of soil water, heat, and CO_2, is used to simulate diurnal and seasonal variations in forest floor CO_2 efflux and CO_2 concentration profiles (Jassal *et al.* 2004).

Overall, these production-transport models can well simulate soil CO_2 efflux and CO_2 concentration in different layers and have the potential to explain temporal variations in soil CO_2 efflux. The production-transport models are probably most valuable for quantifying relative contributions of root and microbial activities at different depths to the total soil CO_2 efflux.

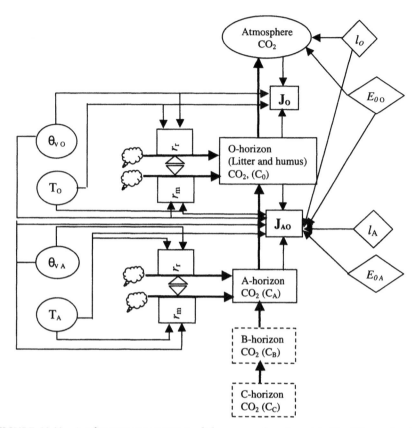

FIGURE 10.11 A schematic presentation of the process-based model. The CO_2 production in each soil layer consists of microbial respiration (r_m) and root respiration (r_r), which are controlled by temperature (T) and by soil water content (θ_v). Carbon dioxide moves between the layers by diffusion; the CO_2 flux (J) depends on the total porosity of soil (E_0), the soil water content, and the thickness of the layers (l), as well as the concentration gradient between the layers. The CO_2 fluxes are denoted by thick arrows, and thin arrows represent information between parameters and processes. C denotes the amount of CO_2 in a soil horizon; O, A, B, and C denote soil layers. In the figure, processes are presented only for O- and A-horizons (Redrawn with permission from Soil Science Society of America Journal: Pumpanen *et al.* 2003).

10.4. MODELING SOIL RESPIRATION AT DIFFERENT SCALES

Models of soil respiration have been applied to plot-scale studies, incorporated into regional and global CO_2 flux models (Raich and Potter 1995, Raich *et al.* 2002), and built into coupled climate-carbon cycle models (Cox *et al.*

2000). The respiration models on different scales share some common features and differ substantially in model structure and parameterization. Almost all the models, for example, consider the effects of temperature and moisture on soil respiration. Most of the process-based models simulate respiration in proportion to pool sizes of carbon in litter and SOM because the fundamental processes that control decomposition are similar across all the spatial scales from plots to continents and the globe. The model structures, however, are different partly because controlling factors are likely to be different between scales (Reichstein *et al.* 2003) and partly because different modelers have a different appreciation of important processes to be incorporated into models. Generally speaking, models oriented to global simulations tend to have a relatively simple structure and fewer parameters than plot-scale models (Ito and Oikawa 2002), although simple models are also widely used to study plot-scale respiration. The large-scale models usually have to make a trade-off among computing power, spatial and temporal resolution, and complexity of model structures. In principle, Occam's razor should dictate model development striving for the simplest model structure that adequately represents important processes and should suffice to address the questions the model is intended to address.

To model a global distribution of soil respiration, for example, Raich and his collaborators considered temperature and moisture regulations only in an early study (Raich and Potter 1995) and effects of temperature, moisture, and production in a later study (Raich *et al.* 2002). At the monthly time step, the soil respiration is approximated by:

$$R_{month} = (R_{LAI=0} + S_{LAI} \times LAI)e^{QT_a} \frac{P + P_0}{K + P + P_0} \qquad (10.23)$$

where R_{month} is the monthly mean soil respiration ($g\,C\,m^{-2}\,mo^{-1}$), $R_{LAI=0}$ is the soil respiration at LAI = 0 and at 0°C without moisture limitation ($g\,C\,m^{-2}\,mo^{-1}$), Q is the temperature sensitivity parameter to determine the exponential relationship between soil respiration and temperature (°C^{-1}), T_a is monthly average soil temperature (°C), K is the half-saturation constant of the hyperbolic relationship of soil respiration with monthly precipitation (cm), and P is monthly precipitation sum (cm). The term ($R_{LAI=0} + S_{LAI} \times LAI$) describes a linear dependency of the basal rates of soil respiration (S_{LAI}) on site peak LAI, and P_0 is related to soil respiration in months without rain. When P = 0, the term of moisture effect becomes $P_0/(K + P_0)$, which mimics the soil water storage effect and results in a strong negative influence on soil respiration when monthly precipitation declines below 20 mm. The response is near saturation when precipitation is above 30 mm. The model does not take into account the soil water storage capacity that keeps accumulated precipitation from previous months. With LAI incorporated into Equation 10.23, the estimates of globally distributed soil respiration can be improved by satellite estimation of LAI.

Global annual CO_2 emissions as predicted by the model are shown in Figure 10.12. Rates of soil respiration are highest in the tropical moist forest regions and lowest in cold tundra and dry desert regions. The model estimates that soil respiration releases $77\,Pg\,C\,yr^{-1}$ from the land ecosystems to the atmosphere (Raich and Potter 1995). The model described by Equation 10.23 simulates CO_2 efflux only and does not consider CO_2 production as functions of pool sizes in roots, litter, and SOM. The flux-based model may reasonably be applied to study spatial distributions of soil respiration but probably not to predict its long-term dynamics. The long-term dynamics of soil respiration are determined largely by changes of carbon pools over time.

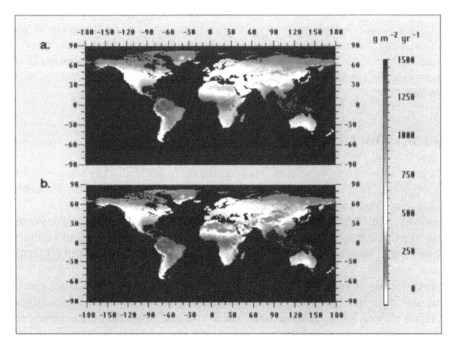

FIGURE 10.12   Global annual soil CO_2 emissions as predicted by the (a) log-transformed model, $\log R_s = F + (QT\dfrac{P}{K+P})$ $R_s = e^{\log R_s} - 1.0$ and (b) untransformed model $R_s = F_e{}^{QT}\dfrac{P}{K+P}$, where R_s $(g\,C\,m^{-2}d^{-1})$ is soil CO_2 efflux, F $(g\,C\,m^{-2}d^{-1})$ is the efflux rate when temperature is zero, $Q(^{\circ}C^{-1})$ is temperature coefficient, $T(^{\circ}C)$ is mean monthly air temperature, $P(cm)$ is mean monthly precipitation, and $K(cm\ month^{-1})$ defines the half-saturation coefficient of the precipitation function (Provided by J. W. Raich with permission form Global Biogeochemical Cycles: Raich and Potter 1995). (see color insert in back of the book)

Most of the global biogeochemical models are CO_2 production models that simulate long-term dynamics of soil respiration according to sizes of carbon pools and corresponding decomposition coefficients (Cramer *et al.* 2001, McGuire *et al.* 2001). For example, the terrestrial carbon cycle model, TRIFFID, simulates net primary production and carbon allocation into leaf, wood, and root pools (Cox 2001). Carbon from dead plant parts is partially released through litter decomposition and partially enters the soil. Soil carbon is eventually broken down by microbes and released back into the atmosphere as soil respiration. The rate of soil respiration is dependent on soil temperature, moisture, and carbon content. TRIFFID is integrated into climate models to predict carbon cycle feedback on global warming (Cox *et al.* 2000) in the context of historical and future climate change. The coupled climate-carbon cycle model predicts 8°C of global terrestrial warming by the year 2100 rather than the 5.5°C predicted without the climate-carbon cycle connection. Similarly, a positive feedback between the carbon cycle and climate occurs in the other coupled modeling study (Friedlingstein *et al.* 2001, 2003; Dufresne *et al.* 2002). The positive feedback effect of the terrestrial carbon cycle on climate warming is based on a model assumption that temperature sensitivity of soil respiration is constant across regions and biomes.

10.5. MODEL DEVELOPMENT AND EVALUATION

Now that we have examined many different models in this chapter, one natural question arises: how can we develop a simple but mechanistic model that can accurately predict soil respiration in different ecosystems? The answer to this question may not be easy. When simple, robust scaling patterns do exist in nature, a simple model, once identified, can be used to connect and extend insights gained from small-scale studies to understand patterns and processes on large scales. For example, Equation 3.8, which describes litter decomposition, is originally derived from empirical analysis of experimental data but has been tested time and again by experimental studies. Although it is still debatable whether we should use different k values to describe three phases of decomposition, the general functional form of first-order linear differential equations, as expressed in Equation 3.8, represents a generic pattern. That is, decomposition is donor-pool-controlled processes for almost all vegetation types in different climatic zones. This equation now becomes a cornerstone of process-based models no matter how complicated a model is. Thus, identifying such a generic pattern from empirical data can greatly improve our ability to predict soil respiration across scales (Harte 2002).

In spite of the fact that Equation 3.8 is generic in representing the nature of decomposition, values of coefficient k vary with plant materials, temperature, moisture, and other environmental and biological factors, leading to different rates of CO_2 releases from microbial decomposition. Consideing just the response function of soil respiration to temperature, an exponential equation can usually effectively summarize data from numerous ecosystems. The relationship provides a way to evaluate new data from different ecosystems and serves as an equation in comprehensive models to simulate responses of soil respiration to temperature. The empirical temperature-respiration relationship itself is a subject of debate in term of its representation of temperature sensitivity. Like any other empirical models, the temperature-respiration relationship derived from regression analysis of experimental data reflects convolution of multiple processes and does not explicitly illustrate mechanisms. Thus, the empirical relationship can hardly be used to predict long-term feedbacks among many processes.

Empirical models for moisture-respiration and substrate-respiration relationships are even less useful than the temperature-respiration models in scaling up plot-level studies to predict large-scale and/or long-term changes in soil respiration. No consistent patterns have yet been revealed on responses of soil respiration to dynamics of soil moisture. We have just started to recognize the complexity of the interactions of soil carbon processes with fluctuation of soil moisture and have not yet fully characterized major processes. Without basic understanding of mechanisms, empirical models derived from observed responses of soil respiration to dynamics of moisture environments cannot be incorporated into large-scale models to predict global and regional soil respiration realistically.

While no explicit equation has yet been developed to link soil respiration directly to substrate supply from aboveground parts of plants, the carbon supply to various processes of soil respiration can be simulated using the principle of mass balance in process-based models. The process-based models are usually developed from systems analysis on processes involved in soil respiration and relationships among the processes. The models are expected to synthesize theory and data from the scientific community into a cohesive representation of the state of knowledge of ecosystem carbon cycles (Burke *et al.* 2003) and thus enable us to evaluate complex interactions between processes and forcing variables and/or among processes themselves. Applications of the models to various scenarios can offer opportunities to test and develop biogeochemical theory and discover their implications for global change. More important, the process-based models are probably the only tool that allows us to extrapolate experimental results quantitatively to longer time-scales and broader spatial extents than the scale on which we can make measurements.

However, accuracy of projections of the process-based models relies on how well the processes are represented in the models. In the CO_2 production models, for example, carbon allocation is still one of the difficult processes to be well represented in models. When some of the fundamental relationships are largely unknown, it is beyond the reach of any process-oriented models to predict soil respiration realistically in future climate scenarios. Moreover, when a comprehensive model incorporates too many processes, the model easily becomes so complicated as to suffer from the difficulty of estimating huge amounts of parameters and to yield substantical uncertainty. Although the models may effectively fit some of the CO_2 efflux data, the quality of fitting may result from tuning or calibration of a large number of parameters. Thus, it is critical to make goal-oriented trade-offs between complexity and trackability for a specific model.

Process-based models are derived from experimental evidence and relevant theory. The latter also emerges from experimental evidence with our process thinking according to human rationality. Thus, our lack of ability to develop a mechanistic model that can realistically predict soil respiration is a consequence of the paucity of appropriate data. Without good data that contain information on the processes we are trying to model, we cannot model the processes well. Ultimately, we have to design innovative experiments to generate types of data that can guide model developments. Models can help evaluate the state of our knowledge and point to next critical experiments we have to conduct.

Commercial Systems and Homemade Chambers of Soil Respiration Measurement

This appendix describes several systems of soil respiration measurements that are either commercially available or developed by researchers.

A—LI-6400 SOIL CO_2 FLUX SYSTEM (LI-COR INC., LINCOLN, NE, USA)

The LI-6400 soil CO_2 flux system consists of the soil CO_2 flux chamber (6400-09) with the infrared gas analyzer and LI-6400 console (Fig. A). This closed dynamic system has been designed to minimize perturbation in the soil-gas concentration gradient and provides maximum operational convenience. The key to the 6400-09 chamber design is that having infrared gas analyzers (CO_2 and H_2O) on the soil chamber makes an ideal system.

The standard procedure for measuring soil respiration follows below. Before starting the measurement, ambient CO_2 concentration at the soil surface is measured. Once the chamber is installed, the CO_2 scrubber is used to draw the CO_2 in the closed system down below the ambient concentration.

FIGURE A   Schematic showing path of air flow between 6400-09 and LI-6400 console (left) and the measurement of soil CO_2 efflux in the field (right). (see color insert in back of the book)

The scrubber is turned off, and soil CO_2 flux causes the CO_2 concentration in the chamber headspace to rise. Data are logged while the CO_2 concentration rises through the ambient level. The software then computes the flux appropriate for the ambient concentration. Major features include:

- No time delays and pressure gradients from an elaborate plumbing system.
- Air is thoroughly mixed inside the chamber while minimizing pressure gradients.
- Water vapor dilution correction results in consistently accurate data.
- Automatic scrub to just below an ambient target maintains the CO_2 gradient to within a few ppm of the natural, undisturbed value.

B—LI-8100 AUTOMATED SOIL CO_2 FLUX SYSTEM (LI-COR INC., LINCOLN, NE, USA)

The system is designed for continuous and unattended long-term measurements to obtain the high temporal resolution of soil CO_2 flux when used with a 20-cm long-term chamber. The long-term chamber moves completely away from the soil measurement area when a measurement is not in progress to ensure that the moisture and temperature of the soil within the measurement collar are similar to the surrounding soil. LI-8100 also supports rapid survey measurements when used either with a 10-cm survey chamber or with a 20-

cm survey chamber. The LI-8100 is a non-steady state, transient system (i.e., closed dynamic system). The flux is estimated using the initial slope of a fitted exponential curve at the ambient CO$_2$ concentration. This is done to minimize the impact of the altered CO$_2$ concentration gradient across the soil surface after the chamber is closed. The LI-8100 has a novel feature to prevent pressure differences between the inside and outside of the chamber. Major features include:

- Continuous, unattended long-term measurements
- Fast, convenient, repeatable survey measurements
- Designed to minimize environmental perturbations
- Fluxes are determined at ambient CO$_2$ concentrations
- Novel vent design allows chamber pressure to track ambient pressure under calm and windy conditions

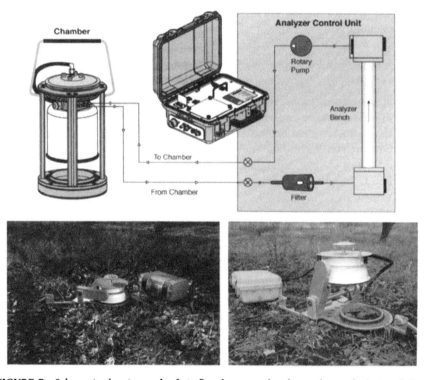

FIGURE B Schematic showing path of air flow between chamber and console (up) and the continuous measurement of soil CO$_2$ efflux with closed (left at bottom) and open (right at bottom) chamber in the field. (see color insert in back of the book)

C—SOIL RESPIRATION SYSTEM (PP SYSTEMS, AMESBURY, MA, USA, AND HITCHIN, UK)

The Soil Respiration System consists of the SRC-1 Soil Respiration Chamber and either the EGM Environmental Gas Monitor or CIRAS Differential CO_2/ H_2O Infrared Gas Analyzers (Fig. C).

Soil respiration is measured when a chamber of known volume is placed on the soil and the rate of increase in CO_2 within the chamber is monitored. Once the SRC-1 chamber has been placed on the soil, the air within the chamber is carefully mixed to ensure representative sampling without generating pressure differences. The CO_2 concentration is measured every 8 seconds with the EGM or CIRAS, and a quadratic equation fitted to the relationship between the increasing CO_2 concentration and elapsed time to determine the rate of increase at time 0. The soil respiration measurement will automatically terminate if the system CO_2 concentration increases more than 60 ppm or if there is an elapsed time of 120 seconds from placement of the chamber on the soil. The measured parameters are recorded either when one of these conditions occurs, or when requested by the operator.

The system can be readily adapted to animal respiration studies or to other measurements of CO_2 gas exchange. The major features include:

- Choice of two systems available (for EGM or CIRAS)
- Simultaneous measurement of soil temperature as an optional extra
- Robust chamber construction
- Ergonomic system design allows for rapid and easy measurements
- On-line statistical data analysis
- Full data storage capability
- RS232 output for transfer to a computer or printer

FIGURE C   Soil respiration system that SRC-1 connected with EGM (left) and CIRAS (right) from PP Systems. (see color insert in back of the book)

D—CFX-2 SOIL CO$_2$ FLUX SYSTEMS (PP SYSTEMS, AMESBURY, MA, USA, AND HITCHIN, UK)

The CFX-2 system is suitable for unattended operation in the field and designed to give accurate measurement of soil net CO$_2$ flux. It is an "Open System" where the measurements are based on concentration differences between air entering and leaving the chamber and the flow rate. The chamber is designed to minimize the internal over-pressure to such a degree that normal soil-atmosphere exchanges are maintained. It also has an integral stainless steel ring to provide a good seal with the soil surface.

The Control Interface Module (CIM) includes an integrated CO$_2$/H$_2$O analyzer, a mass flow controlled air supply providing 100 l/min of ambient air to an opaque chamber, data-logging and storage capability, and keyboard for setting up the system and retrieval of stored data. Major features include:

- Stand-alone operation
- Lightweight and field portable
- Accurate, integral CO$_2$ analyzer and H$_2$O sensor
- Open system measurement
- User friendly operation
- Simple setup
- Full data logging, storage and output of data

FIGURE D CFX-2 from PP Systems. (see color insert in back of the book)

E—SRC SERIES PORTABLE SOIL RESPIRATION SYSTEMS (DYNAMAX INC., HOUSTON, TX, USA)

Two portable soil flux systems, SRS1000 and SRS2000, have been developed for soil respiration measurement. Both the systems consist of a console programming unit and a soil respiration chamber. They work in an "open system" mode.

Both the SRS1000 and SRS2000 systems have a highly accurate CO_2 infrared gas analyzer (IRGA) housed directly adjacent to the soil chamber, ensuring the fastest possible responses to gas exchanges from the soil. The IRGA has an operating range of 0-2000 ppm CO_2, with a resolution of 1 ppm. The IRGA has been designed to have minimal drift and excellent measurement stability. All measurements are automatically compensated for changes in atmospheric pressure and temperature.

Soil respiration chamber is specifically designed for short-term soil flux measurements. The chamber consists of a lower stainless steel collar and an upper measurement compartment. There are sensors for measuring PAR and soil temperature. Major features include:

- Highly portable
- Highly accurate CO_2 IRGA
- Optimized soil chamber with no pressure gradients, being insensitive to wind, and stainless steel soil collar
- Automatic CO_2/H_2O control
- Easy to use

FIGURE E   The measurement of soil respiration in the field by SRS1000 Ultra compact soil flux system (left) and SRS2000 Intelligent portable soil flux system (right). (see color insert in back of the book)

F—SRC-MV5 SOIL RESPIRATION CHAMBER (DYNAMAX INC., HOUSTON, TX, USA)

SRC-MV5 is an automated system for continuously monitoring soil respiration. Traditional point-in-time SRC only allows users to operate over a very short period of time to avoid altering the natural microclimate inside the chamber. Dynamax SRC-MV5 has an automated switching system that is programmed to sequentially open and close the chamber in concert with an infrared gas analysis system. This automated feature permits operation over long time periods without supervision. Major features include:

- The Dynamax SRC-MV5 is constructed using lightweight, durable aluminum for portability and long-term reliability.
- The automated system allows normal wetting and drying of the soil inside the chamber between measurements.
- A flexible neoprene lid, stretched tightly over the chamber, provides an airtight seal.
- Smart design of a deflector over the lid keeps temperatures constant inside the chamber when closed.
- Specially designed inlet and outlet fittings ensure there is no internal pressure gradient, which could affect the evolution of CO_2 from the soil surface.
- Sealed electronics and cover permits the system to operate in almost any environment.
- Adjustable stands allow the SRC to work on uneven surfaces and elevate the major components above flooding.

FIGURE F   SRC-MV5 system. (see color insert in back of the book)

G—AUTOMATED SOIL RESPIRATION SYSTEM FROM WOODS HOLE RESEARCH CENTER (WHRC, USA)

From: http://www.whrc.org/new_england/Methodology/auto_soil_r.htm

This automated soil respiration system is a closed dynamic system and was built based on the designs of Patrick Crill (University of New Hampshire) and Greg Winston (University of California at Irvine). Fig. H1 shows one of the automated chambers in the open position. The chamber top is a schedule 40 PVC pipe cap. The gray structure supporting the chamber top is also made of rectangular PVC bar. A pneumatic piston is attached from the structure to the chamber top.

Pressurizing the piston with compressed air is what lifts or lowers the chamber top onto the collar. The collar is made from schedule 80 PVC pipe cut to 3 inches in length. One end of the pipe is beveled so that it can be inserted into the ground at approximately 3 cm depth. Automobile weather stripping is used as an O-ring on the topside of the collar such that when the chamber top lowers, pressure is applied through the piston forcing a seal with the weather stripping onto the collar. The control system for the pneumatics and the flow to and from the chamber consists of a Licor 6252 Infrared Gas Analyzer (IRGA), and a Campbell CR10X. The chamber tops are raised and lowered by pressurized pistons with an air compressor to supply the pressure. A Campbell relay controller controls the raising and lowering of the chamber tops. A second Campbell relay Controller controls the flow to and from each chamber. The timing for both controllers is controlled by a CR10X datalogger. A pump draws the air (at a rate of $0.7\,L\,min^{-1}$) from the chamber through the flow control solenoids to the IRGA then to the pump and flowmeter, then to the return flow solenoids and back to the chamber. To learn more about building an automated soil respiration system, WHRC has provided links to parts lists and more detailed descriptions and wiring diagrams of the system in website above.

FIGURE G   Automated soil respiration chamber in the open position. (see color insert in back of the book)

H—THE AUTOMATIC CARBON EFFLUX SYSTEM (A.C.E.S, USA)

From: http://www.srs.fs.usda.gov/soils/research/aces.html
ACES was developed by John Butnor, Chris Maier, and Kurt Johnsen at the Forestry Sciences Laboratory, Research Triangle Park, North Carolina, as a multiport, dynamic gas sampling system that utilizes an open flow-through design to measure CO_2 fluxes from the forest floor with a variety of chamber styles. Up to sixteen soil chambers are measured sequentially (fixed or variable time step) using a single infra-red gas analyzer. Air is supplied to each chamber in a push-pull fashion where air flow entering the chamber is greater than exiting to maintain a slight positive chamber pressure. Excess air is vented out the top of the chamber and ensures that the chamber pressure is held near ambient. Chamber pressure can be verified with a digital manometer. Flow rates are measured with mass flow meters. All chambers are continuously evacuated when not being sampled. The soil respiration chambers are constructed of PVC (25 cm diameter, 10 cm height, 4900 cm³) with a lexan lid. Each chamber has an air and soil thermocouple, pressure equilibration with the atmosphere, and reflective insulation that prevents "greenhouse"

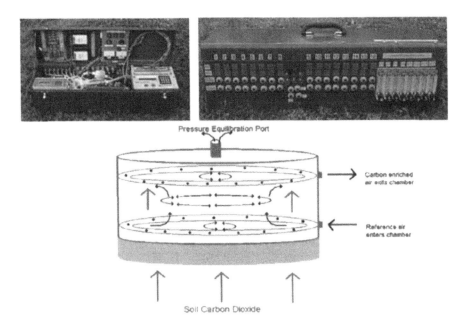

FIGURE H The console of 16-port Automatic Carbon Efflux System (up) and soil respiration chamber showing air flow (bottom). (see color insert in back of the book)

heating in the chamber even in full sunlight. A soil moisture reflectometer is used to take soil moisture readings in each chamber and can be installed in a common location for continuous measurement. The ACES is fully automatic requiring only calibration checks twice per week. Under AC power the system can run continuously, using a DC power supply the ACES can go up to 48 hours without recharging.

References

Aber, J. D., Nadelhoffer, K. J., Stuedler, P., and Melillo, J. M. (1989) Nitrogen saturation in northern forest ecosystems. *Bioscience* 39, 378–386.

Aber, J. D., Ollinger, S. V., Federer, C. A., Reich, P. B., Goulden, M. L., Kicklighter, D. W., Melillo, J. M., and Lathrop, R. G. (1995) Predicting the effects of climate change on water yield and forest production in the northeastern United States. *Climate Research* 5, 207–222.

Aber, J. D., Ollinger, S. V., Federer, C. A., and Driscoll, C. (1997) Modeling nitrogen saturation in forest ecosystems in response to land use and atmospheric deposition. *Ecological Modelling* 101, 61–78.

Abnee, A. C., Thompson, J. A., Kolka, R. K., D'Angelo, E. M., and Coyne, M. S. (2004) Landscape influences on potential soil respiration in a forested watershed of southeastern Kentucky. *Environmental Management* 33 (Supplement 1), S160–S167.

Abril, G., and Borges, A. V. (2005) Carbon dioxide and methane emissions from estuaries. In *Greenhouse gas emissions, fluxes and processes* (A. Tremblay, L. Varfalvy, C. Roehm, and M. Garneau, eds.), pp. 187–207, Springer-Verlag, Berlin.

Achard, F., Eva, H. D., Stibig, H.-J., Mayaux, P., Gallego, J., Richards, T., and Malingreau, J. P. (2002) Determination of deforestation rates of the world's humid tropical forests. *Science* 297, 999–1002.

Acharya, C. L., and Prihar, S. S. (1969) Vapor losses through soil mulch at different wind velocities. *Agronomy Journal* 61, 666–668.

Aguirrezabal, L. A. N., Deleens, E., and Tardieu, F. (1994) Root elongation rate is accounted for by intercepted PPFD and source-sink relations in field and laboratory grown sunflower. *Plant, Cell and Environment* 17, 443–450.

Al-Kaisi, M. M., and Yin, X. (2005) Tillage and crop residue effects on soil carbon and carbon dioxide emission in corn-soybean rotations. *Journal of Environmental Quality* 34, 437–445.

Allard, V., Newton, P. D., Lieffering, M., Soussana, J. F., Grieu, P., and Matthew, C. (2004) Elevated CO_2 effects on decomposition processes in grazed grassland. *Global Change Biology* 10, 1553–1564.

Allen, M. F. (1991) *The ecology of mycorrhizae.* Cambridge University Press, New York.

Allen, A. S., Andrews, J. A., Finzi, A. C., Matamala, R., Richter, D. D., and Schlesinger, W. H. (2000) Effects of free-air CO_2 enrichment (FACE) on belowground processes in a *Pinus taeda* forest. *Ecological Applications* 10, 437–448.

Allison, G. B., Barnes, C. J., and Hughes, M. W. (1983) The distribution of deuterium and [18]O in dry soils: 2, Experimental. *Journal of Hydrology* 64, 377–397.

Almendinger, J. C. (1990) The decline of soil organic matter, total-N, and available water capacity following the late-Holocene establishment of jack pine on sandy mollisols, north-central Minnesota. *Soil Science* 150, 680–694.

Alvarez, R., Santanatoglia, O. J., and Garcia, R. (1995) Soil respiration and carbon inputs from crops in a wheat soybean rotation under different tillage systems. *Soil Use and Management* 11(2), 45–50.

Amos, B., Arkebauer, T. J., and Doran, J. W. (2005) Soil surface fluxes of greenhouse gases in an irrigated maize-based agroecosystem. *Soil Science Society of America Journal* 69(2), 387–395.

Amthor, J. S. (1991) Respiration in a future, higher-CO_2 world. *Plant, Cell and Environment* 14(1), 13–20.

Amthor, J. S. (1994) Respiration and carbon assimilate use. In *Physiology and determination of crop yield* (K. J. Boote, J. M. Bennett, T. R. Sinclair, and G. M. Paulsen, eds.), pp. 221–250, American Society of Agronomy, Madison, WI.

Amthor, J. S. (2000) Plant respiratory responses to the environment and their effects on the carbon balance. In *Plant-environment interactions* (R. E. Wilkinson, ed.), pp. 501–504, Marcel Dekker, New York.

Andersen, C. P., and Scagel, C. F. (1997) Nutrient availability alters belowground respiration of ozone-exposed ponderosa pine. *Tree Physiology* 17, 377–387.

Anderson, J. M. (1973) Carbon dioxide evolution from two temperate deciduous woodland soils. *Journal of Applied Ecology* 10, 361–375.

Anderson, J. P. E. (1982) Soil respiration. In *Methods of soil analysis: Part 2, chemical and microbiological properties* (A. L. Page, R. H. Miller, and D. R. Keeney, eds.), pp. 831–853, 2nd edition, American Society of Agronomy-SSSA Inc., Madison, WI.

Anderson, D. W., and Coleman, D. C. (1985) Dynamics of organic matter in grassland soils. *Journal of Soil and Water Conservation* 40, 211–216.

Andrews, J. A., Harrison, K. G., Matamala, R., and Schlesinger, W. H. (1999) Separation of root respiration from total soil respiration using carbon-13 labeling during Free-Air Carbon Dioxide Enrichment (FACE). *Soil Science Society of America Journal* 63, 1429–1435.

Andriesse, J. P. (1988) Nature and management of tropical peat soils. *FAO Soils Bulletin* 59, 165.

Armentano, T. V., and Menges, E. S. (1986) Patterns of change in the carbon balance of organic soil wetlands of the temperate zone. *Journal of Ecology* 74, 755–774.

Armstrong, R. D., Millar, G., Halpin, N. V., Reid, D. J., and Standley, J. (2003) Using zero tillage, fertilisers, and legume rotations to maintain productivity and soil fertility in opportunity cropping systems on shallow vertisol. *Australian Journal of Experimental Agriculture* 43, 141–153.

Arrhenius, S. (1898) The effect of constant influences upon physiological relationships. *Scandinavian Archives of Physiology* 8, 367–415.

Aselmann, I., and Crutzen, V. (1990) A global inventory of wetland distribution and seasonality, net primary productivity, and estimated methane emissions. In *Soils and the greenhouse effect* (A. F. Bouwman, ed.), pp. 441–449, John Wiley & Sons, New York.

Atjay, G. L., Ketner, P., and Duvigneaud, P. (1987) Terrestrial primary production and phytomass. In *The global carbon cycle* (Scope 13) (B. Bolin, ed.), pp. 129–181, John Wiley & Sons, New York.

Atkin, O. K., Edwards, E. J., and Loveys, B. R. (2000) Response of root respiration to changes in temperature and its relevance to global warming. *New Phytologist* 147, 141–154.

Atkin, O. K., Zhang, Q. S., and Wiskich, J. T. (2002) Effect of temperature on rates of alternative and cytochrome pathway respiration and their relationship with the redox poise of the quinone pool. *Plant Physiology* 128(1), 212–222.

Atkin, O. K., and Tjoelker, M. G. (2003) Thermal acclimation and the dynamic response of plant respiration to temperature. *Trends in Plant Science* 8, 343–351.

Augustine, D. J., and McNaughton, S. J. (1998) Ungulate effects on the functional species composition of plant communities, herbivore selectivity and plant tolerance. *Journal of Wildlife Management* 62, 1165–1183.

Austin, A. T., and Sala, O. E. (2002) Carbon and nitrogen dynamics across a natural precipitation gradient in Patagonia, Argentina. *Journal of Vegetation Science* 13, 351–360.

Baker, J. M. (2000) Conditional sampling revisited. *Agricultural and Forest Meteorology* 104, 59–65.

Baldocchi, D. D., Verma, S. B., and Rosenberg, N. J. (1981) Mass and energy exchanges of a soybean under various environmental regimes. *Agronomy Journal* 73, 706–710.

Baldocchi, D. D., Verma, S. B., Matt, D. R., and Anderson, D. E. (1986) Eddy correlation measurements of carbon dioxide efflux from the floor of a deciduous forest. *Journal of Applied Ecology* 23, 967–975.

Baldocchi, D. D., Meyers, T. P. (1991) Trace gas exchange above the floor of a deciduous forest: 1, Evaporation and CO_2 efflux. *Journal of Geophysical Research* 96(D4), 7171–7285.

Baldocchi, D. D. (1997) Flux footprints within and over forest canopies. *Boundary-Layer Meteorology* 85(2), 273–292.

Baldocchi, D., Falge, E., Gu, L., Olson, R., Hollinger, D., Running, S., Anthoni, P., Bernhofer, C., Davis, K., Evans, R., Fuentes, J., Goldstein, A., Katul, G., Law, B., Lee, X. H., Malhi, Y., Meyers, T., Munger, W., Oechel, W., U, K. T. P., Pilegaard, K., Schmid, H. P., Valentini, R., Verma, S., Vesala, T., Wilson, K., and Wofsy, S. (2001) FLUXNET: A new tool to study the temporal and spatial variability of ecosystem-scale carbon dioxide, water vapor, and energy flux densities. *Bulletin of the American Meteorological Society* 82(11), 2415–2434.

Baldocchi, D. D. (2003) Assessing the eddy covariance technique for evaluating carbon dioxide exchange rates of ecosystems, past, present and future. *Global Change Biology* 9(4), 479–492.

Baldock, J. A. (2002) Interactions of organic materials and microorganisms with minerals in the stabilization of structure. In *Interactions between soil particles and microorganisms: Impact on the terrestrial ecosystem* (IUPAC series on analytical and physical chemistry of environmental systems, Vol. 8) (P. M. Huang, J.-M. Bollag, and N. Senesi, eds.), pp. 85–131, John Wiley & Sons, Chichester, UK.

Bargdett, R. D., Wardle, D. A., and Yeates, G. (1998) Linking above-ground and below-ground interactions: How plant responses to foliar herbivory influence soil organisms. *Soil Biology and Biochemistry* 14, 1867–1878.

Barney, C. W. (1951) Effects of soil temperature and light intensity on root growth of loblolly pine seedlings. *Plant Physiology* 26, 146–163.

Behera, N., Joshi, S. K., and Pati, D. P. (1990) Root contribution to total soil metabolism in a tropical forest soil from Orissa, India. *Forest Ecology and Management* 36, 125–134.

Beier, C. (2004) Climate change and ecosystem function: Full-scale manipulations of CO_2 and temperature. *New Phytologist* 162, 243–245.

Bekku, Y., Koizumi, H., Nakadai, T., and Iwaki, H. (1995) Measurement of soil respiration using closed chamber method, an IRGA technique. *Ecological Research* 10, 369–373.

Bekku, Y., Kimura, M., Ikeda, H., and Koizumi, H. (1997a) Carbon input from plant to soil through root exudation in *Digitaria adscendens* and *Ambrosia artemisiifolia*. *Ecological Research* 12, 305–312.

Bekku, Y., Koizumi, H., Oikawa, T., and Iwaki, H. (1997b) Examination of four methods for measuring soil respiration. *Applied Soil Ecology* 5, 247–254.

Bekku, Y. S., Kume, A., Masuzawa, T., Kanda, H., Nakatsubo, T., and Koizumi, H. (2004a) Soil respiration in a high Arctic glacier foreland in Ny-Ålesund, Svalbard. *Polar Bioscience* 17, 36–46.

Bekku, Y. S., Nakatsubo, T., Kume, A., and Koizumi, H. (2004b) Soil microbial biomass, respiration rate, and temperature dependence on a successional glacier foreland in Ny-Ålesund, Svalbard. *Arctic, Antarctic, and Alpine Research* 36(4), 395–399.

Ben Zioni, A., Vaadia, Y., and Lips, S. H. (1971) Nitrate uptake by roots as regulated by nitrate reduction products of the shoot. *Physiologia Plantarum* 24, 288–290.

Benoit, G. R., and Kirkham, D. (1963) The effect of soil surface conditions on evaporation of soil water. *Soil Science Society of American Processing* 27, 495–498.

Berg, B., Wessen, B., and Ekbohm, G. (1982) Nitrogen level and lignin decomposition in Scots pine needle litter. *Oikos* 38, 291–296.

Berg, B. (1986) Nutrient release from litter and humus in coniferous soils: A mini review. *Journal of Forest Research* 1, 359–369.

Berg, B., and Ekbohm, G. (1991) Litter mass-loss rates and decomposition patterns in some needle and leaf litter types VII Long-term decomposition in a Scots pine forest. *Canadian Journal of Botany* 69, 1449–1456.

Berg, B., and Matzner, E. (1997) Effect of N deposition on decomposition of plant litter and soil organic matter in forest systems. *Environmental Review* 5, 1–25.

Berry, L. J. (1949) The influence of oxygen tension on the respiration rate in different segments of onion roots. *Journal of Cellular and Comparative Physiology* 33(1), 41–66.

Beyer, L. (1991) Intersite characterization and variability of soil respiration in different arable and forest soils. *Biology and Fertility of Soils* 12(2), 122–126.

Bhupinderpal-Singh, A. N., Löfvenius, M. O., Högberg, M. N., Mellander P. E., and Högberg, P. (2003) Tree root and soil heterotrophic respiration as revealed by girdling of boreal Scots pine forest, extending observations beyond the first year. *Plant, Cell and Environment* 26, 1287–1296.

Bijracharya, R. M., Lal, R., and Kimble, J. M. (2000) Diurnal and seasonal CO^2-C flux from soil as related to erosion phases in central Ohio. *Soil Science Society of America Journal* 64, 286–293.

Billings, S. A., Richter, D. D., and Yarie, J. (1998) Soil carbon dioxide fluxes and profile concentrations in two boreal forests. *Canadian Journal of Forest Research* 28, 1773–1783.

Birch, H. F., and Friend, M. T. (1956) Humus decomposition in east Africa soils. *Nature* 178, 500–501.

Birch, H. F. (1958) The effect of soil drying on humus decomposition and nitrogen availability. *Plant and Soil* 10, 9–31.

Birch, H. F. (1959) Further observations on the humus decomposition and nitrification. *Plant and Soil* 11, 262–286.

Bird, M. I. and Pousai, P. (1997) $\delta^{13}C$ variations in the surface soil organic carbon pool. *Global Biogeochemical Cycles* 11, 313–322.

Blackmer, A. M., and Bremner, J. M. (1977) Gas chromatographic analysis of soil atmospheres. *Soil Science Society of America Journal* 41, 908–912.

Bolin, B. (1970) The carbon cycle. *Scientific American* 223(3), 124–130.

Bolin, B., and Sukumar, R. (2000) Global perspective. In *Land use, land-use change and forestry: A special report of the IPCC* (R. T. Watson, I. R. Noble, B. Bolin, N. H. Ravindranath, D. J. Verardo, and D. J. Dokken, eds.), pp. 23–51, Cambridge University Press, UK.

Bonan, G. B., and Shugart, H. H. (1989) Environmental factors and ecological processes in boreal forests. *Annual Review of Ecology and Systematics* 20, 1–28.

Bonan, G. B. (1995) Land-atmosphere CO_2 exchange simulated by a land surface process model coupled to an atmospheric general circulation model. *Journal of Geophysical Research* 100D, 2817–2831.

Bond-Lamberty, B., Wang, C., and Gower, S. (2004) A global relationship between the heterotrophic and autotrophic components of soil respiration? *Global Change Biology* 10(10), 1756–1766.

Boone, R. D., Nadelhoffer, K. J., Canary, J. D., and Kaye, J. P. (1998) Roots exert a strong influence on the temperature sensitivity of soil respiration. *Nature* 396, 570–572.

Bootinga, Y., and Craig, H. (1969) Oxygen isotopic fractionation between CO_2 and water, and the isotopic composition of marine atmospheric CO_2. *Earth and Planetary Science Letters* 5, 285–295.

Borken, W., Xu, Y.-J., Brumme, R., and Lamersdorf, N. (1999) A climate change scenario for carbon dioxide and dissolved organic carbon fluxes from a temperate forest soil: Drought and rewetting effects. *Soil Science Society of America Journal* 63, 1848–1855.

Borken, W., Muhs, A., and Beese, F. (2002) Application of compost in spruce forests: Effects on soil respiration, basal respiration and microbial biomass. *Forest Ecology and Management* 159, 49–58.

Bouma, T. J., De Visser, R., Janssen, J. H. J. A., De Kock, M. J., Van Leeuwen, P. H., and Lambers, H. (1994) Respiratory energy requirements and rate of protein turnover in vivo determined by use of an inhibitor of protein synthesis and a probe to assess its effect. *Physiologia Plantarum* 92, 585–594.

Bouma, T. J., Broekhuysen, A. G. M., and Veen, B. W. (1996) Analysis of root respiration of *Solanum tuberosum* as related to growth, ion uptake and maintenance of biomass. *Plant Physiology and Biochemistry* 34(6), 795–806.

Bouma, T. J., Nielsen, K. L., Eissenstata, D. M., and Lynch, J. P. (1997) Estimating respiration of roots in soil, interactions with soil CO_2, soil temperature and soil water content. *Plant and Soil* 195, 221–232.

Bouma, T. J., and Bryla, D. R. (2000) On the assessment of root and soil respiration for soils of different textures, interactions with soil moisture contents and soil CO_2 concentrations. *Plant and Soil* 227(1–2), 215–221.

Boussingault, J. B., and Levy, B. (1853) Mémoire sur la composition de l'aire confine dans la terre végétal. *Annals of Chemistry and Physics* 37, 5–50.

Bowden, R. D., Nadelhoffer, K. J., Boone, R. D., Melillo, J. M., and Garrison, J. B. (1993) Contributions of aboveground litter, belowground litter, and root respiration to total soil respiration in a temperate mixed hardwood forest. *Canadian Journal of Forest Research* 23, 1402–1407.

Bowden, R. D., Newkirk, K. M., and Rullo, G. M. (1998) Carbon dioxide and methane fluxes by a forest soil under laboratory-controlled moisture and temperature conditions. *Soil Biology and Biochemistry* 30, 1591–1597.

Bowden, R. D., Rullo, G., Stevens, G. R., and Steudler, P. A. (2000) Soil fluxes of carbon dioxide, nitrous oxide, and methane at a productive temperate deciduous forest. *Journal of Environmental Quality* 29(1), 268–276.

Bowdish, S. (2002) The effects of warming and land-use on species composition in the tallgrass prairie, M.S. Thesis, University of Oklahoma, Norman, OK.

Bowling, D. R., McDowell, N. G., Bond, B. J., Law, B. E., Ehleringer, J. R. (2002) ^{13}C content of ecosystem respiration is linked to precipitation and vapor pressure deficit. *Oecologia* 131, 113–124.

Boxman, A. W., van Dam, D., van Dyck, H. F. G., Hogervorst, R. F., and Koopmans, C. J. (1995) Ecosystem responses to reduced nitrogen and sulphur inputs into two coniferous forest stands in the Netherlands. *Forest Ecology and Management* 71, 7–30.

Brambilla, I., Bertani, A., and Reggiani, R. (1986) Effect of inorganic nitrogen nutrition (ammonium and nitrate) on aerobic metabolism in excised rice roots. *Journal of Plant Physiology* 123, 419–428.

Braswell, B. H., Schimel, D. S., Linder, E., and Moore III, B. (1997) The response of global terrestrial ecosystems to interannual temperature variability. *Science* 287, 870–872.

Bremer, J. D., Ham, J. M., Owensby, C. E., and Knapp, A. K. (1998) Responses of soil respiration to clipping and grazing in a tallgrass prairie. *Journal of Environmental Quality* 27, 1539–1548.

Bremer, J. D., and Ham, J. M. (2002) Measurement and modeling of soil CO_2 flux in a temperate grassland under mowed and burned regimes. *Ecological Modeling* 15, 1318–1328.

Bridgham, S. D., Updegraff, K., and Pastor, J. (1998) Carbon, nitrogen and phosphorus mineralization in northern wetlands. *Ecology* 79, 1545–1561.

Broder, W. M., and Wagner, G. H. (1988) Microbial colonization and decomposition of corn, wheat, and soybean residue. *Soil Science Society of America Journal* 52, 112–117.

Brooks, P. D., McKnight, D., and Elder, K. (2005) Carbon limitation of soil respiration under winter snowpacks: Potential feedbacks between growing season and winter carbon fluxes. *Global Change Biology* 11(2), 231–238.

Brown, S. and Lugo, A. E. (1982) The storage and production of organic matter in tropical forests and their role in the global carbon cycle. *Biotropica* 14, 161–187.

Brumme, R. and Beese, F. (1992) Effects of liming and nitrogen fertilization on emissions of CO_2 and N_2O from a temperate forest. *Journal of Geophysical Research* 97, 12851–12858.

Brumme, R. (1995) Mechanisms of carbon and nutrient release and retention in beech forest gaps: III, Environmental regulation of soil respiration and nitrous oxide emissions along a microclimatic gradient. *Plant and Soil* 169, 593–600.

Bryant, J. P., Provenza, F. D., Pastor, J., Reichardt, P. B., Clausen, T. P., and du Toit, J. T. (1991) Interactions between woody plants and browsing mammals mediated by secondary metabolites. *Annual Review of Ecology and Systematics* 22, 431–446.

Bryla, D. R., Bouma, T. J., and Eissenstat, D. M. (1997) Root respiration in citrus acclimates to temperature and slows during drought. *Plant Cell and Environment* 20(11), 1411–1420.

Buchmann, N., Guehl, J. M., Barigah, T., and Ehleringer, J. R. (1997) Interseasonal comparison of CO_2 concentrations, isotopic composition, and carbon cycling in an Amazonian rainforest (French Guiana). *Oecologia* 110, 120–131.

Buchmann, N. (2000) Biotic and abiotic factors controlling soil respiration rates in *Picea abies* stands. *Soil Biology and Biochemistry* 32, 1625–1635.

Bunnell, F. L., Tait, D. E. N., and Flanagan P. W. (1977) Microbial respiration and substrate weight loss. 2, A model of the influences of chemical composition. *Soil Biology and Biochemistry* 9, 41–47.

Bunt, J. S., and Rovira, A. D. (1954) Oxygen uptake and carbon dioxide evolution of heat-sterilized soil. *Nature* 173, 1242.

Burchuladze, A. A., Chudy, M., Eristavi, I. V., Pagava, S. V., Povinec, P., Sivo, A., and Togonidze, G. I. (1989) Anthropogenic ^{14}C variations in atmospheric CO_2 and wines. *Radiocarbon* 31, 771–776.

Burke, I. C., Kaye, J. P., Bird, S. P., Hall, S. A., McCulley, R. L., and Sommerville, G. L. (2003) Evaluating and testing models of terrestrial biogeochemistry, The role of temperature in controlling decomposition. In *Models in ecosystem science* (C. D. Canham, J. J. Cole, and W. K. Lauenroth, eds.), pp. 225–253, Princeton University Press, Princeton, NJ.

Burton, A. J., Pregitzer, K. S., Zogg, G. P., and Zak, D. R. (1996) Latitudinal variation in sugar maple fine root respiration. *Canadian Journal of Forest Research* 26, 1761–1768.

Burton, A. J., Pregitzer, K. S., Zogg, G. P., and Zak, D. R. (1998) Drought reduces root respiration in sugar maple forests. *Ecological Application* 8, 771–778.

Burton, A. J., Pregitzer, K. S., and Hendrick, R. L. (2000) Relationships between fine root dynamics and nitrogen availability in Michigan northern hardwood forests. *Oecologia* 125(3), 389–399.

Buscot F. (2005) What are soils? In *Microorganisms in soils: Roles in genesis and functions.* (F. Buscot, and A. Varma, eds.), pp. 3–17, Soil Biology Series, Vol. 3, Springer-Verlag, Heidelberg, Germany.

Businger, J. A. (1975) Aerodynamics of vegetated surfaces. In *Heat and mass transfer in the biosphere: I, transfer processes in the plant environment* (D. A. de Vries, and N. H. Afgan, eds.), Scripta Book Co., Washington, DC.

Buyanovsky, G. A., and Wagner, G. H. (1983) Annual cycles of carbon dioxide level in soil air. *Soil Science Society of America Journal* 47, 1139–1144.

Buyanovsky, G. A., Kucera, C. L., and Wagner, G. H. (1985) Comparative carbon balance in natural and agricultural ecosystems. *Bulletin of the Ecological Society of America* 66, 149–150.

Buyanovsky, G. A., Kucera, C. L., and Wagner, G. H. (1987) Comparative analyses of carbon dynamics in native and cultivated ecosystems. *Ecology* 68, 2023–2031.

Calfapietra, C., Gielen, B., Galema, A. N. J., Lukac, M., Angelis, P. D., Moscatelli, M. C., Ceulemans, R., and Mugnozza, G. S. (2003) Free-air CO_2 enrichment (FACE) enhances biomass production in a short-rotation poplar plantation. *Tree Physiology* 23, 805–814.

Callaway, R. M., DeLucia, E. H., Thomas, E. M., and Schlesinger, W. H. (1994) Compensatory responses of CO_2 exchange and biomass allocation and their effects on the relative growth rate of ponderosa pine in different CO_2 and temperature regimes? *Oecologia* 98, 159–166.

Cambardella, C. A., Moorman, T. B., Novak, J. M., Parkin, T. B., Karlen, D. L., Turco, F. F., and Konopka, A. E. (1994) Field-scale variability of soil properties in central Iowa soil. *Soil Science Society of America Journal* 58, 1501–1511.

Campbell, J. L., Sun, O. J., and Law, B. E. (2004) Supply-side controls on soil respiration among Oregon forests. *Global Change Biology* 10, 1857–1969.

Cao, G., Tang, Y., Mo, W., Wang, Y., Li, Y., and Zhao, X. (2004) Grazing intensity alters soil respiration in an alpine meadow on the Tibetan plateau. *Soil Biology and Biochemistry* 36, 237–243.

Carlyle, J. C., and Bathan, U. (1988) Abiotic controls of soil respiration beneath an eighteen year old *Pinus radiata* stand in south-east Australia. *Journal of Ecology* 76, 654–662.

Carreiro, M. M., Sinsabaugh, R. L., Repert, D. A., and Parkhurst, D. F. (2000) Microbial enzyme shifts explain litter decay responses to simulated nitrogen deposition. *Ecology* 81, 2359–2365.

Carroll, S. B., and Bliss, L. C. (1982) Jack pine-lichen woodland on sandy soils in northern Saskatchewan and northeastern Alberta. *Canadian Journal of Botany* 60(11), 2270–2282.

Castro, M. S., Melillo, J. M., Steudler, P. A., and Chapman, J. W. (1994) Soil moisture as a predictor of methane uptake by temperate forest soils. *Canadian Journal of Forest Research* 24, 1805–1810.

Ceulemans, R., and Mousseau, M. (1994) Effects of elevated atmospheric CO_2 on woody plants. *New Phytologist* 127, 425–446.

Chapin III, F. S., Matson, P. A., and Mooney, H. A. (2002) *Principles of terrestrial ecosystem ecology.* Springer-Verlag, New York.

Chapman, S. B. (1971) Simple conductimetric soil respirometer for field use. *Oikos* 22(3), 348.

Chapman, S. B. (1979) Some interrelationships between soil and root respiration in lowland calluna heathland in southern England. *Journal of Ecology* 67(1), 1–20.

Chen, H., Harmon, M. E., Griffiths, R. P., and Hicks, W. (2000) Effects of temperature and moisture on carbon respired from decomposing woody roots. *Forest Ecology and Management* 138(1–3), 51–64.

Cheng, W., Coleman, D. C., Corroll, C. R., and Hoffman, C. A. (1994) Investigating short-term carbon flows in the rhizospheres of different plant species, using isotopic trapping. *Agronomy Journal* 86, 782–788.

Cheng, W., and Coleman, D. C. (1990) Effect of living roots on soil organic-matter decomposition. *Soil Biology and Biochemistry* 22(6), 781–787.

Cheng, W. (1996) Measurement of rhizosphere respiration and organic matter decomposition using natural ^{13}C. *Plant and Soil* 183, 263–268.

Cheng, W., Zhang, Q. L., Coleman, D. C., Carroll, C. R., and Hoffman, C. A. (1996) Is available carbon limiting microbial respiration in the rhizosphere? *Soil Biology and Biochemistry* 28, 1283–1288.

Cheng, W., and Johnson, D. W. (1998) Elevated CO_2, rhizosphere processes, and soil organic matter decomposition. *Plant and Soil* 202, 167–174.

Cheng, W. (1999) Rhizosphere feedback in elevated CO_2. *Tree Physiology* 19, 313–320.

Chmielewski, F. M., and Rötzer, T. (2001) Response of tree phenology to climate change across Europe. *Agricultural and Forest Meteorology* 108, 101–112.

Choi, Y., and Wang, Y. (2004) Dynamics of carbon sequestration in a coastal wetland using radiocarbon measurements. *Global Biogeochemical Cycles* 18(4), GB4016, doi: 10.1029/2004GB 002261.

Christensen, T. R., Jonasson, S., Michelsen, A., Callaghan, T. V., and Havstrom, M. (1998) Environmental controls on soil respiration in the Eurasian and Greenlandic Arctic. *Journal of Geophysical Research-Atmospheres* 103(D22), 29015–29021.

Ciais, P., Denning, A. S., Tans, P. P., Berry, J. A., Randall, D. A., Collatz, G. J., Sellers, P. J., White, J. W. C., Trolier, M., Meijer, H. A. J., Francey, R. J., Monfray, P., and Heimann, M. (1997) A three-dimensional synthesis study of \delta^{18}O in atmospheric CO_2: 1, Surface fluxes. *Journal of Geophysical Research-Atmosphere* 102(D5), 5857–5872.

Clifford, S. C., Stronach, I. M., Mohamed, A. D., Azam-Ali, S. N., and Crout, N. M. J. (1993) The effects of elevated atmospheric carbon dioxide and water stress on light interception, dry matter production and yield in stands of groundnut (*Arachis hypogaea* L). *Journal of Experimental Botany* 44, 1763–1770.

Cohen, Y., and Pastor, J. (1991) The responses of a forest model to serial correlations of global warming. *Ecology* 72(3), 1161–1165.

Coleman, D. C. (1973a) Soil carbon balance in a successional grassland. *Oikos* 24, 195–199.

Coleman, D. C. (1973b) Compartmental analysis of "total soil respiration": An exploratory study. *Oikos* 24, 361–366.

Coleman, D. C., Crossley, Jr. D. A., Hendrix, P. F. (2004) *Fundamentals of soil ecology*. Elsevier/ Academic Press, Boston. 386 pp.

Coleman, D. C., and Fry, B. (1991) *Carbon isotope techniques*. Academic Press, Woods Hole, MA.

Coleman, D. C. (1994) Compositional analysis of microbial communities: Is there room in the middle? In *Beyond the biomass* (K. Ritz, J. Dighton, and K. E. Giller, eds.), pp. 201–220, John Wiley & Sons, Chichester, UK.

Coleman, K., and Jenkinson, D. S. (1999) *RothC-26.3: A model for the turnover of carbon in soil: Model description and user's guide*. Lawes Agricultural Trust, Harpenden, UK.

Coleman, M. D., Dickson, R. E., and Isebrands, J. G. (2000) Contrasting fine-root production, survival and soil CO_2 efflux in pine and poplar plantations. *Plant and Soil* 225, 129–139.

Collin, M., and Rasmuson, A. A. (1988) Diffusion of radon through soils: A pore distribution model—Comments. *Soil Science Society of America Journal* 52(3), 897–898.

Comins, H. N., and McMurtrie, R. E. (1993) Long-term response of nutrient-limited forests to CO_2 enrichment; equilibrium behavior of plant-soil models. *Ecological Applications* 3(4), 666–681.

Conant, R. T., Klopatek, J. M., Malin, R. C., and Klopatek, C. C. (1998) Carbon pools and fluxes along a semiarid gradient in northern Arizona. *Biogeochemistry* 43, 43–61.

Conant, R. T., Klopatek, J. M., and Klopatek, C. C. (2000) Environmental factors controlling soil respiration in three semiarid ecosystems. *Soil Science Society of America Journal* 64, 383–390.

Concilio, A., Ma, S. Y., Li, Q. L., LeMoine, J., Chen, J. Q., North, M., Moorhead, D., and Jensen, R. (2005) Soil respiration response to prescribed burning and thinning in mixed-conifer and hardwood forests. *Canadian Journal of Forest Research* 35(7), 1581–1591.

Conen, F., and Smith, K. A. (2000) An explanation of linear increases in gas concentration under closed chambers used to measure gas exchange between soil and the atmosphere. *European Journal of Soil Science* 51, 111–117.

Connin, S. L., Virginia, R. A., and Chamberlain, C. P. (1997) Carbon isotopes reveal soil organic matter dynamics following arid land shrub expansion. *Oecologia* 110, 374–386.

Cotrufo, M. F., Ineson, P., and Scott, A. (1998) Elevated CO_2 reduces the nitrogen concentration of plant tissues? *Global Change Biology* 4, 43–54.

Cox, P. M., Betts, R. A., Jones, C. D., Spall, S. A., and Totterdell, I. J. (2000) Acceleration of global warming due to carbon-cycle feedbacks in a coupled climate model. *Nature* 408, 184–187.

Cox, P. M. (2001) Description of the TRIFFID dynamic global vegetation model. Hadley Centre, Met Office Tech. Note 24, 16 pp.

Craft, C., Reader, J., Sacco, J. N., and Broome, S. W. (1999) Twenty-five years of ecosystem development of constructed *Spartina alterniflora* (Loisel) marshes. *Ecological Applications* 9, 1405–1419.

Craft, C., Broome, S., and Campbell, C. (2002) Fifteen years of vegetation and soil development after brackish-water marsh creation. *Restoration Ecology* 10, 248–258.

Craine, F. M., Wedin, D. A., and Chapin, F. S. (1999) Predominance of ecophysiological controls on soil CO_2 flux in a Minnesota grassland. *Plant and Soil* 207, 77–86.

Craine, J. M., Wedin, D. A., and Reich, P. B. (2001) Grassland species effects on soil CO_2 flux track the effects of elevated CO_2 and nitrogen. *New Phytologist* 150, 425–434.

Craine, F. M., and Wedin, D. A. (2002) Determinants of growing season soil CO_2 flux in a Minnesota grassland. *Biogeochemistry* 59, 303–313.

Craine, J. M., Tilman, D. G., Wedin, D. A., Reich, P. B., Tjoelker, M. J., and Knops, J. M. H. (2002) Functional traits, productivity and effects on nitrogen cycling of 33 grassland species. *Functional Ecology* 16, 563–574.

Cramer, W., Bondeau, A., Woodward, F. I., Prentice, I. C., Betts, R. A., Brovkin, V., Cox, P. M., Fisher, V., Foley, J. A., Friend, A. D., Kucharik, C., Lomas, M. R., Ramankutty, N., Sitch, S., Smith, B., White, A., and Young-Molling, C. (2001) Global response of terrestrial ecosystem structure and function to CO_2 and climate change, results from six dynamic global vegetation models. *Global Change Biology* 7, 357–374.

Crawford, R. M. M. (1992) Oxygen availability as an ecological limit to plant distribution. *Advances in Ecological Research* 23, 93–185.

Crawford, M. C., Grace, P. R., and Oades, J. M. (2000) Allocation of carbon to shoots, roots, soil and rhizosphere respiration by barrel medic (*Medicago truncatula*) before and after defoliation. *Plant and Soil* 227, 67–75.

Crookshanks, M., Taylor, G., and Broadmeadow, M. (1998) Elevated CO_2 and tree root growth: Contrasting responses in *Fraxinus excelsior, Quercus petraea* and *Pinus sylvestris? New Phytologist* 138, 241–250.

Curiel Yuste, J., Janssens, I. A., Carrara, A., Meiresonne, L., and Ceulemans, R. (2003) Interactive effects of temperature and precipitation on soil respiration in a temperate maritime pine forest. *Tree Physiology* 23, 1263–1270.

Curiel Yuste, J., Janssens, I. A., Carrara, A., and Ceulemans, R. (2004) Annual Q_{10} of soil respiration reflects plant phonological patterns as well as temperature sensitivity. *Global Change Biology* 10, 161–169.

Curiel Yuste, J., Janssens, I. A., and Ceulemans, R. (2005) Calibration and validation of an empirical approach to model soil CO_2 efflux in a deciduous forest. *Biogeochemistry* 73(1), 209–230.

Currie, J. A. (1970) Movement of gases in soil respiration: Sorption and transport processes in soils. *SCI Monograph Series* 37, 152–171.

Curtin, D., Selles, F., Wang, H., Campbell, C. A., and Biederbeck, V. O. (1998) Carbon dioxide emissions and transformation of soil carbon and nitrogen during wheat straw decomposition. *Soil Science Society of America Journal* 62, 1035–1041.

Curtin, D., Wang, H., Selles, F., McConkey, B. G., and Campbell, C. A. (2000) Tillage effects on carbon fluxes in continuous wheat and fallow-wheat rotations. *Soil Science Society of America Journal* 64, 2080–2086.

Curtis, P. S., Vogel, C. S., Gough, C. M., Schmid, H. P., Su, H. B., Bovard, B. D. (2005) Respiratory carbon losses and the carbon-use efficiency of a northern hardwood forest, 1999–2003. *New Phytologist* 167(2), 437–456.

Dalal, R. C., and Mayer, R. J. (1986) Long-term trends in fertility of soils under continuous cultivation and cereal cropping in southern Queensland: II, Total organic carbon and its rate of loss from the soil profile. *Australian Journal of Soil Research* 24, 281–292.

Dao, T. H. (1998) Tillage and crop residue effects on carbon dioxide evolution and carbon storage in a Paleustoll. *Soil Science Society of America Journal* 62, 250–256.

Davidson, E. A., Stark, J. M., Firestone, M. K. (1990) Microbial production and consumption of nitrate in an annual grassland. *Ecology* 71, 1968–1975.

Davidson, E. A., Belk, E., and Boone, R. D. (1998) Soil water content and temperature as independent or confound factors controlling soil respiration in a temperature mixed hardwood forest. *Global Change Biology* 4, 217–227.

Davidson, E. A., Verchot, L. V., Cattanio, J. H., Ackerman, I. L., Carvalho, J. E. M. (2000) Effects of soil water content on soil respiration in forests and cattle pastures of eastern Amazonia. *Biogeochemistry* 48, 53–69.

Davidson, E. A., Savage, K., Bolstad, P., Clark, P. A., Curtis, P. S., Ellsworth, D. S., Hanson, P. J., Law, B. E., Luo, Y., Pregitzer, K. S., Randolph, J. C., and Zak, D. (2002a) Belowground carbon allocation in forests estimated from litterfall and IRGA-based soil respiration measurements. *Agricultural and Forest Meteorology* 113, 39–51.

Davidson, E. A., Savage, K., Verchot, L. V., and Navarrro, R. (2002b) Minimizing artifacts and biases in chamber-based measurements of soil respiration. *Agricultural and Forest Meteorology* 113(1–4), 21–37.

Davidson, E. A., Richardson, A. D., Savage, K. E., and Hollinger, D. Y. (2005) A distinct seasonal pattern of the ratio of soil respiration to total ecosystem respiration in a spruce-dominated forest. *Global Change Biology* 11, doi: 10.1111/j.1365–2486.2005.01062.x.

Davidson, E. A., Janssens, I. A., and Luo, Y. (2006) On the variability of respiration in terrestrial ecosystems: moving beyond Q_{10}. *Global Change Biology* 12, 154–164, doi: 10.1111/j.1365–2486.2005.01065.x.

Dawson, T. E., Mambelli, S., Plamboeck, A. H., Templer, P. H., Tu, K. P. (2002) Stable isotopes in plant ecology. *Annual Review of Ecology and Systematics* 33, 507–559.

DeFries, R. S., Townshend, J. R. G., and Hansen, M. (1999) Continuous fields of vegetation characteristics at the global scale at 1-km resolution. *Journal of Geophysical Research-Atmospheres* 104(D14), 16911–16923.

DeFries, R., Houghton, R. A., Hansen, M., Field, C. B., Skole, D., and Townshend, J. (2002) Carbon emissions from tropical deforestation and regrowth based on satellite observations for the 1980s and 1990s. *Proceedings of the National Academy of Sciences* 99, 14256–14261.

Degens, B. P. (1998) Microbial functional diversity can be influenced by the addition of simple organic substrates to soil. *Soil Biology and Biochemistry* 30, 1981–1988.

Dehérain, P. P., and Demoussy, E. (1896) The oxidation of organic matter in soil (French). *Annales Agronomiques* 22, 305–337.

de Jong, E., and Schappert, H. J. V. (1972) Calculation of soil respiration and activity from CO_2 profiles in the soil. *Soil Science* 113, 328–333.

de Jong, E., Schappert, H. J. V., and MacDonald, K. B. (1974) Carbon dioxide evolution from virgin and cultivated soil as affected by management practices and climate. *Canadian Journal of Soil Science* 54, 299–307.

de Jong, E., Redmann, R. E., and Riley, E. A. (1979) A comparison of methods to measure soil respiration. *Soil Science* 127, 300–306.

Delucia, E. H., Callaway, R. M., Thomas, E. M., and Schlesinger, W. H. (1997) Mechanisms of phosphorus acquisition for ponderosa pine seedlings under high CO_2 and temperature. *Annals of Botany* 79(2), 111–120.

Denmead, O. T. (1969) Comparative micrometeorology of a wheat field and a forest of *Pinus radiate. Agricultural Meteorology* 6, 357–371.

Denmead, O. T. (1979) Chamber systems for measuring nitrous oxide emission from soils in the field. *Soil Science Society of America Journal* 43, 89–95.

Denmead, O. T., and Bradley, E. F. (1987) On scalar transport in plant canopies. *Irrigation Science* 8, 131–149

Denmead, O. T. (1995) Novel meteorological methods for measuring trace gas fluxes. *Philosophical Transactions of the Royal Society of London Series A-Mathematical Physical and Engineering Sciences* 351(1696), 383–396.

Denmead, O. T., Raupach, M. R., Dunin, F. X., Cleugh, H. A., and Leuning, R. (1996) Boundary layer budgets for regional estimates of scalar fluxes. *Global Change Biology* 2(3), 255–264.

Denmead, O. T., Harper, L. A., Freney, J. R., Griffith, D. W. T., Leuning, R., and Sharpe, R. R. (1998) A mass balance method for non-intrusive measurements of surface-air trace gas exchange. *Atmospheric Environment* 32(21), 3679–3688.

Denning, A. S., Collatz, G. J., Zhang, C., Randall, D. A., Berry, J. A., Sellers, P. J., Colello, G. D. and Dazlich, D. A. (1996) Simulations of terrestrial carbon metabolism and atmospheric CO_2 in a general circulation model: Part I, Surface carbon fluxes. *Tellus* 48B, 521–542.

de Visser, R. (1985) Efficiency of respiration and energy requirements of N assimilation in roots of *Pisum sativum. Physiologia Plantarum* 65, 209–218.

de Vries, H. L. (1958) Variation in concentration of radiocarbon with time and location on Earth. *Koninklijke Nederlandse Akademie van Wetenschappen* B61, 94–102.

Dilly, O. (2005) Microbial energetic in soil. In *Microorganisms in soils, roles in genesis and functionsm* (F. Buscot, and A. Varma, eds.), pp. 123–138, Soil Biology Series, Vol. 3, Springer-Verlag, Berlin and Heidelberg, Germany.

Dilustro, J. J., Collins, B., Duncan, L., and Crawford, C. (2005) Moisture and soil texture effects on soil CO_2 efflux components in southeastern mixed pine forests. *Forest Ecology and Management* 204(1), 85–95.

Dixon R. K., Brown S., Houghton R. A., Solomon A. M., Trexler M. C., and Wisniewski J, (1994) Carbon pools and flux of global forest ecosystems. *Science* 22, 185–190,

Doran, J. W., Mielke, I. N., and Power, J. F. (1991) Microbial activity as regulated by soil water-filled pore space. In *Ecology of soil microorganisms in the microhabital environments,* pp. 94–99, Transactions of the 14th International Congress of Soil Science Symposim III-3.

Dörr, H., and Münnich, K. O. (1986) Annual variations in the $^{14}CO_2$ content of soil CO_2. *Radiocarbon* 28, 338–345.

Dörr, H., and Münnich, K. O. (1987) Annual variation in soil respiration in selected areas of the temperate zone. *Tellus*, 39B, 114–121.

Dörr, H., and Münnich, K. O. (1990) ^{222}Rn flux and soil air concentration profiles in West Germany: Soil ^{222}Rn as tracer for gas transport in the unsaturated zone. *Tellus* 42B, 20–28.

Douce, R., and Neuburger, M. (1989) The uniqueness of plant mitochondria. *Annual Review of Plant Physiology and Plant Molecular Biology* 40, 371–414.

Drobnik, J. (1962) The effect of temperature on soil respiration. *Folia Microbiology* 7, 132–140.

Dufresne, J. L., Friedlingstein, P., Berthelot, M., Bopp, L., Ciais, P., Fairhead, L., Le Treut, H., and Monfray, P. (2002) On the magnitude of positive feedback between future climate change and the carbon cycle, *Geophysical Research Letter* 29, doi: 10.1029/2001GL013777.

Dunne, J., Harte, J., and Taylor, K. (2003) Response of subalpine meadow plant reproductive phenology to manipulated climate change and natural climate variability. *Ecological Monograph* 73, 69–86.

Dugas, W. A. (1993) Micrometeorological and chamber measurements of CO_2 flux from bare soil. *Agricultural and Forest Meteorology* 67, 115–128.

Dugas, W. A., Reicosky, D. C., and Kiniry, J. R. (1997) Chamber and micrometeorological measurements of CO_2 and H_2O fluxes for three C_4 grasses. *Agricultural and Forest Meteorology* 83, 113–133.

Edmonds, R. L., Marra, J. L., Barg, A. K., and Sparks, G. B. (2000) Influence of forest harvesting on soil organisms and decomposition in western Washington. *USDA Forest Service, General Technical Report* PSW-GTR-178, 53–72.

Edwards, C. A. (1974) Macroarthropods. In *Biology of plant litter decomposition Vol. 2* (C. H. Dickinson, and G. J. F. Pugh, eds.), pp. 533–554, Academic Press, London and New York.

Edwards, N. T., and Sollins, P. (1973) Continuous measurement of carbon dioxide evolution from partitioned forest floor components. *Ecology* 54, 406–412.

Edwards, N. T. (1974) A moving chamber design for measuring soil respiration rates. *Oikos* 25, 97–101.

Edwards, N. T. (1982) The use of soda lime for measuring respiration rates in terrestrial systems. *Pedobiologia* 23, 321–330.

Edwards, N. T., and Ross-Todd, B. M. (1983) Soil carbon dynamics in a mixed deciduous forest following clear-cutting with and without residual removal. *Soil Science Society of America Journal* 47, 1014–1021.

Edwards, N. T. (1991) Root and soil respiration responses to ozone in *Pinus taeda* L. seedlings. *New Phytologist* 118, 315–321.

Edwards, N. T., and Norby, R. J. (1998) Below-ground respiratory responses of sugar maple and red maple saplings to atmospheric CO_2 enrichment and elevated air temperature. *Plant and Soil* 206, 85–97.

Edwards, N. T., and Riggs, J. S. (2003) Automated monitoring of soil respiration: A moving chamber design. *Soil Science Society of America Journal* 67, 1266–1271.

Edwards, N. T., Benham, D. G., Marland, L. A., and Fitter, A. H. (2004) Root production is determined by radiation flux in a temperate grassland community. *Global Change Biology* 10, 209–227.

Ehman, J. L., Schmid, H. P., Grimmond, C. S. B., Randolph, J. C., Wayson, C. A., Hanson, P. J., and Cropley, F. D. (2002) An initial intercomparison of micrometeorological and ecological inventory estimates of carbon exchange in a mid-latitude deciduous forest, *Global Change Biology* 8, 575–589.

Eissenstat, D. M., Wells, C. E., Yanai, R. D., and Whitbeck, J. L. (2000) Building roots in a changing environment, implications for root longevity. *New Phytologist* 147, 33–42.

Ekblad, A., Högberg, P. (2001) Natural abundance of ^{13}C in CO_2 respired from forest soils reveals speed of link between photosynthesis and root respiration. *Oecologica* 127, 305–308.

Ekblad, A., Bostrom, B., Holm, A., and Comstedt, D. (2005) Forest soil respiration rate and $\delta^{13}C$ is regulated by recent above ground weather conditions. *Oecologia* 143, 136–142.

Eliasson, P. E., McMurtrie, R. E., Pepper, D. A., Strömgren, M., Linder, S., and Ågren, G. I. (2005) The response of heterotrophic CO_2 flux to soil warming. *Global Change Biology* 11, 167–181.

Elliott, E. T. (1986) Aggregate structure and carbon, nitrogen, and phosphorus in native and cultivated soils. *Soil Science Society of America Journal* 50, 627–633.

Ellis, R. C. (1969) The respiration of the soil beneath some *Eucalyptus* forest stands as related to the productivity of the stands. *Australian Journal of Soil Research* 7, 349–357.

Emmett, B. A., Beier, C., Estiarte, M., Tietema, A., Kristensen, H. L., Williams, D., Penuelas, J., Schmidt, I., and Sowerby, A. (2004) The response of soil processes to climate change: Results from manipulation studies of shrublands across an environmental gradient. *Ecosystems* 7(6), 625–637.

Epron, D., Le Dantec, V., Dufrene, E., and Granier, A. (2001) Seasonal dynamics of soil carbon dioxide efflux and simulated rhizosphere respiration in a beech forest. *Tree Physiology* 21, 145–152.

Epron, D., Nouvellon, Y., Roupsard, O., Mouvondy, W., Mabiala, A., Saint-André, L., Joffre, R., Jourdan, C., Bonnefond, J. M., Berbigier, P., and Hamel, O. (2004) Spatial and temporal variations of soil respiration in a *Eucalyptus* plantation in Congo. *Forest Ecology and Management* 202, 149–160.

Epstein, H. E., Burke, I. C., and Lauenroth, W. K. (2002) Regional patterns of decomposition and primary production rates in the U.S. Great Plains. *Ecology* 83(2), 320–327.

Eriksson, K-E. L., Blanchette, R. A., and Ander, P. (1990) *Microbial and Enzymatic Degradation of Wood and Wood Components*. Springer-Verlag, Berlin.

Eriksen, J., and Jensen, L. S. (2001) Soil respiration, nitrogen mineralization and uptake in barley following cultivation of grazed grasslands. *Biology and Fertility of Soils* 33, 139–145.

Esser, G., Aselmann, I., and Lieth, H. (1982) Modelling the carbon reservoir in the system compartment "litter." In *Transport of carbon and minerals in major world rivers, Pt. 1* (E. T. Degens, ed.), pp. 39–58, Mitt. Geol.-Paläont. Inst. Univ. Hamburg, SCOPE/UNEP Sonderbd.

Eswaran, H. (2003) *Soil classification, a global desk reference*. CRC Press, Boca Raton, FL.

Eugster, W., and Siegrist, F. (2000) The influence of nocturnal CO_2 advection on CO_2 flux measurements. *Basic Applied Ecology* 1, 177–188.

Euskirchen, E. S., Chen, J. Q., Gustafson, E. J., and Ma, S. Y. (2003) Soil respiration at dominant patch types within a managed northern Wisconsin landscape. *Ecosystems* 6(6), 595–607.

Ewel, K. C., Cropper, W. P., and Gholz, H. L. Jr. (1987a) Soil CO_2 evolution in Florida slash pine plantation: I, Changes through time. *Canadian Journal of Forest Research* 17, 325–329.

Ewel, K. C., Cropper, W. P., and Gholz, H. L. Jr. (1987b) Soil CO_2 evolution in Florida slash pine plantation: 2, Importance of root respiration. *Canadian Journal of Forest Research* 17, 330–333.

Falge, E., Tenhunen, J., Aubinet, M., Bernhofer, C., Clement, R., Granier, A., Kowalski, A., Moors, E., Pilegaard, K., Rannik, U., and Reb, C. (2003) A model-based study of carbon fluxes at ten European forest sites. In *Ecological studies vol. 163: Fluxes of carbon, water and energy of European forests* (R. Valentini, ed.), Springer-Verlag, Heidelburg, Germany.

Fang, C., and Moncrieff, J. B. (1996) An improved dynamic chamber technique for measuring CO_2 efflux from the surface of soil. *Functional Ecology* 10, 297–305.

Fang, C., and Moncrieff, J. B. (1998) An open-top chamber for measuring soil respiration and the influence of pressure difference on CO_2 efflux measurement. *Functional Ecology* 12, 319–325.

Fang, C., Moncrieff, J. B., Gholz, H. L., and Clark, K. L. (1998) Soil CO_2 efflux and its spatial variation in a Florida slash pine plantation. *Plant and Soil* 205, 135–146.

Fang, C., and Moncrieff, J. B. (1999) A model for soil CO_2 production and transport: 1, Model development. *Agricultural and Forest Meteorology* 95, 225–236.

Fang, C., and Moncrieff, J. B. (2001) The dependence of soil CO_2 efflux on temperature. *Soil Biology and Biochemistry* 33, 155–165.

Fang, J., Piao, S., Field, C. B., Pan, Y., Gao, Q., Zhou, L., Peng, C., and Tao, S. (2003) Increasing net primary production in China from 1982 to 1999. *Frontiers in Ecology and the Environment* 1, 293–297.

Fang, C., Smith, P., Moncrieff, J. B., and Smith, J. U. (2005) Similar response of labile and resistant soil organic matter pools to changes in temperature. *Nature* 433, 57–59.

Fernandez, I. J., Son Y, Kraske, C. R., Rustad, L. E., and David, M. B. (1993) Soil carbon dioxide characteristics under different forest types and after harvest. *Soil Science Society of America Journal* 57, 1115–1121.

Field, C. B., Ball, J. T., and Berry, J. A. (1989) Photosynthesis, Principles and field techniques. In *Plant physiological ecology, field methods and instrumentation* (R. W. Pearcy, J. Ehleringer, H. A. Mooney, and P. W. Rundel, eds.), pp. 209–253, Chapman and Hall, New York.

Field, C. B., Jackson, R. B., and Mooney, H. A. (1995) Stomatal responses to increased CO_2: Implications from the plant to the global scale. *Plant, Cell and Environmens* 18, 1214–1225.

Fierer, N., and Schimel, J. P. (2003) A proposed mechanism for the pulse in carbon dioxide production commonly observed following the rapid rewetting of a dry soil. *Soil Science Society of America Journal* 67, 798–805.

Fierer, N., Allen, A. S., Schimel, J. P., and Holden, P. A. (2003) Controls on microbial CO_2 production, a comparison of surface and subsurface soil horizons. *Global Change Biology* 9, 1322–1332.

Finzi, A. C., Allen, A. S., DeLucia, E. H., Ellsworth, D. S., and Schlesinger, W. H. (2001) Forest litter production, chemistry, and decomposition following two years of free-air CO_2 enrichment. *Ecology* 82(2), 470–484.

Fitter, A. H., Fitter, R. S. R., Harris, I. T. B., and Williamson, M. H. (1995) Relationships between first flowering date and temperature in the flora of a locality in central England. *Functional Ecology* 9, 55–60.

Fitter, A. H., Graves, J. D., Self, G. K., Brown, T. K., Bogie, D. S., and Taylor, K. (1998) Root production, turnover and respiration under two grassland types along an altitudinal gradient: Influence of temperature and solar radiation. *Oecologia* 114, 20–30.

Fitter, A. H., Self, G. K., Brown, T. K., Bogie, D. S., Graves, J. D., Benham, D., and Ineson, P. (1999) Root production and turnover in an upland grassland subjected to artificial soil warming respond to radiation flux and nutrients, not temperature. *Oecologia* 120, 575–581.

Flanagan, P. W., and Veum, A. K. (1974) Relationships between respiration, weight loss, temperature and moisture in organic residues on tundra. In *Soil organisms and decomposition in tundra* (A. J. Holding, O. W. Heal, S. F. MacClean, Jr., and P. W. Flanagan, eds.), pp. 249–277, IBP Tundra Biome Steering Committee, Stockholm, Sweden.

Flanagan, L. B., Ehleringer, J. R., and Pataki, D. E. (2005) *Stable isotopes and biosphere-atmosphere interactions: processes and biological controls.* Elsevier Academic Press, San Diego, CA.

Fog, K. (1988) The effect of added nitrogen on the rate of decomposition of organic matter. *Biological Reviews* 63, 433–462.

Fogel, R., and Cromack, K. (1977) Effect of habitat and substrate quality of Douglas fir litter decomposition in western Oregon. *Canadian Journal of Botany* 55, 1632–1640.

Foley, J. A. (1994) Net primary productivity in the terrestrial biosphere: the application of a global model. *Journal of Geophysical Research-Atmosphere* 99(20), 773–783.

Fonte, S., and Schowalter, T. (2004) Decomposition of greenfall vs. senescent foliage in a tropical forest ecosystem in Puerto Rico. *Biotropica* 36, 474–482.

Frank, A. B. (2002) Carbon dioxide fluxes over a grazed prairie and seeded pasture in the northern Great Plains. *Environmental Pollution* 116, 397–403.

Frank, A. B., Liebig, M. A., and Hanson, J. D. (2002) Soil carbon dioxide fluxes in northern semiarid grasslands. *Soil Biology and Biochemistry* 34, 1235–1241.

Franklin, O., Högberg, P., Ekblad, A., and Ågren, J. I. (2003) Pine forest floor carbon accumulation in response to N and PK additions, bomb ^{14}C modelling and respiration studies. *Ecosystems* 6, 644–658.

Franzluebbers, A. J., Haney, R. L., Honeycutt, C. W., Arshad, M. A., Schomberg, H. H., and Hons, F. M. (2001) Climatic influences on active fractions of soil organic matter. *Soil Biology and Biochemistry* 33(7–8), 1103–1111.

Friedlingstein, P., Fung, I., Holland, E., John, J. G., Brasseur, G. P., Erikson, D., and Schimel, D. (1995) On the contribution of CO_2 fertilization to the missing sink. *Global Biogeochemical Cycles* 9(4), 541–556.

Friedlingstein, P., Bopp, L., Ciais, P., Dufresne, J. L., Fairhead, L., LeTreut, H., Monfray, P., and Orr, J. (2001) Positive feedback between future climate change and the carbon cycle. *Geophysical Research Letter* 28, 1543–1546.

Friedlingstein, P., Dufresne, J. L., Cox, P. M., and Rayner, P. (2003) How positive is the feedback between climate change and the carbon cycle? *Tellus* 55B, 692–700.

Friese, C. F., and Allen, M. F. (1991) Tracking the fates of exotic and local VA mycorrhizal fungi: Methods and patterns. *Agriculture Ecosystems and Environment* 34(1–4), 87–96.

Frolking, S., Goulden, M. L., Wofsy, S. C., Fan, S.-M., Sutton, D. J., Munger, J. W., Bazzaz, A. M., Daube, B. C., Crill, P. M., Aber, J. D., Band, L. E., Wang, X., Savage, K., Moore, T., and Harriss, R. C. (1996) Modelling temporal variability in the carbon balance of a spruce/moss boreal forest, *Global Change Biology* 2, 343–366.

Fu, S., Cheng, W. X., and Susfalk, R. (2002) Rhizosphere respiration varies with plant species and phenology, a greenhouse pot experiment. *Plant and Soil* 239, 133–140.

Gallardo, A., and Schlesinger, W. H. (1994) Factors limiting microbial biomass in the mineral soil and forest floor of a warm-temperate forest. *Soil Biology and Biochemistry* 26, 1409–1415.

Gambrell, R. P., and Patrick, W. H. Jr. (1978) Chemical and microbiological properties of anaerobic soils and sediments. In *Plant life in anaerobic environments* (D. D. Hook, and R. M. M. Crawford, eds.), pp. 375–423, Ann Arbor Science Publishers, Ann Arbor, MI.

Gärdenäs, A. I. (2000) Soil respiration fluxes measured along a hydrological gradient in a Norway spruce stand in south Sweden (Skogaby). *Plant and Soil* 221, 273–280.

Garrett, H. E., and Cox, G. S. (1973) Carbon dioxide from the floor of an oak-hickory forest. *Soil Science Society of America Proceedings* 37, 641–611.

Gaudinski, J. B., Trumbore S. E., Davidson E. A., and Zheng S. (2000) Soil carbon cycling in a temperate forest, radiocarbon-based estimates of residence times, sequestration rates and partitioning of fluxes. *Biogeochemistry* 51, 33–69.

George, K., Norby, R. J., Hamilton, J. G., and DeLucia, E. H. (2003) Fine-root respiration in a loblolly pine and sweetgum forest growing in elevated CO_2. *New Phytologist* 160, 511–522.

Ghani, A., Dexter, M., and Perrott, K. W. (2003) Hot-water extractable carbon in soils: A sensitive measurement for determining impacts of fertilization, grazing and cultivation. *Soil Biology and Biochemistry* 35, 1231–1243.

Gholz, H. L., Perry, C. S., Cropper, W. P., and Hendry, L. C. Jr. (1985) Litterfall, decomposition and nitrogen and phosphorus dynamics in a chronosequence of slash pine (*Pinus elliottii*) plantations. *Forest Science* 31, 463–478.

Gholz, H. L., Wedin, D. A., Smitherman, S. M., Harmon, M. E., and Parton, W. J. (2000) Long-term dynamics of pine and hardwood litter in contrasting environments: Toward a global model of decomposition. *Global Change Biology* 6(7), 751–765.

Giardina, C. P., and Ryan, M. G. (2000) Evidence that decomposition rates of organic carbon in mineral soil do not vary with temperature. *Nature* 404, 858–861.

Giardina, C. P., Ryan, M. G., Binkley, D., and Fownes, J. H. (2003) Primary production and carbon allocation in relation to nutrient supply in a tropical experimental forest. *Global Change Biology* 9, 1438–1450.

Giardina, C. P., Binkley, D., Ryan, M. G., Fownes, J. H., and Senock, R. S. (2004) Belowground carbon cycling in a humid tropical forest decreases with fertilization. *Oecologia* 139, 545–550.

Gill, R. A., and Jackson, R. B. (2000) Global patterns of root turnover for terrestrial ecosystems. *New Phytologist* 147(1), 13–31.

Gilmore, J. T., Clark, M. D., and Sigua, G. C. (1985) Estimating net nitrogen mineralization from carbon dioxide evolution. *Soil Science Society of America Journal* 49, 1398–1402.

Gilmanov, T. G., Tieszen, L. L., Wylie, B. K., Flanagan, L. B., Frank, A. B., Haferkamp, M. R., Meyers, T. P., and Morgan, J. A. (2004) Integration of CO_2 flux and remotely-sensed data for primary production and ecosystem respiration analyses in the northern Great Plains: Potential for quantitative spatial extrapolation. *Global Ecology and Biogeography* 14(3), 271–292.

Gliński, J., Stepniewski, W. (1985) *Soil aeration and its role for plants*. CRC Press, Boca Raton, FL.

Gliński, J., and Lipiec, J. (1990). *Soil physical conditions and plant roots*, CRC Press, Boca Raton, FL.

Golley, F. B., Odum, H. T., Wilson, R. F. (1962) The structure and metabolism of a Puerto Rican red mangrove forest in May. *Ecology* 43, 9–19.

Gordon, A. M., Schlenter, R. E., and Van Cleave, K. (1987) Seasonal patterns of soil respiration and CO_2 evolution following harvesting in the white spruce forests of interior Alaska. *Canadian Journal of Forest Research* 17, 304–310.

Gorham, E. (1995) The biogeochemistry of northern peatlands and its possible response to global warming. In *Biotic Processes and Potential Feedbacks* (G. M. Woodwell, and F. T. McKenzie, eds.), pp. 169–187, Oxford Unversity Press, Oxford, UK.

Goulden, M. L., Munger, J. W., Fan, S. M., Daube, B. C., and Wofsy, S. C. (1996) Exchange of carbon dioxide by a deciduous forest: Response to interannual climate variability. *Science* 271, 1576–1578.

Goulden, M. L., and Crill, P. M. (1997) Automated measurements of CO_2 exchange at the moss surface of a black spruce forest. *Tree Physiology* 17, 537–542.

Goulden, M. L., Wofsy, S. C., Harden, J. W., Trumbore, S. E., Crill, P. M., Gower, S. T., Fires, T., Daube, B., Fan, S. M., Sutton, D. J., Bazzaz, A., and Munger, J. W. (1998) Sensitivity of boreal forest carbon balance to soil traw. *Science* 279, 214–217.

Grace, J., and Rayment, M. (2000) Respiration in balance. *Nature* 404(6780), 819–820.

Gray, T. R. G. (1990) Methods for studying the microbial ecology of soil. In *Methods in microbiology: Vol. 22, Techniques in microbial ecology* (R. Grigorova, and J. R. Norris, eds.), pp. 309–342, Elsevier Academic Press, London.

Grayston, S. J., Vaughan, D., and Jones, D. (1996) Rhizosphere carbon flow in trees, in comparison with annual plants: The importance of root exudation and its impact on microbial activity and nutrient availability. *Applied Soil Ecology* 5, 29–56.

Greaves, J. R., and Carter, E. G. (1920) Influence of moisture on the bacterial activities of the soil. *Soil Science* 10, 361–387.

Greaves, M. P., and Darbyshire, J. F. (1972) The ultrastructure of the mucilaginous layer on plant roots. *Soil Biology and Biochemistry* 4, 443–449.

Gregory, P. J., and Atwell, B. J. (1991) The fate of carbon in pulse-labelled crops of barley and wheat. *Plant and Soil* 136, 205–213.

Griffin, D. M. (1981) Water potential as a selective factor in the microbial ecology of soils. In *Water potential relations in soil microbiology* (J. F. Parr, W. R. Gardner, and L. F. Elliott, eds.), pp. 141–151, SSSA Spec. Publ. 9, Madison, WI.

Griffin, K. L., Thomas, R. B., and Strain, B. R. (1993) Effects of nitrogen supply and elevated carbon dioxide on construction cost in leaves of *Pinus taeda* (L.) seedlings. *Oecologia* 95, 575–580.

Griffin, K. L., Ross, P. D., Sims, D. A., Luo, V., Seemann, J. R., Fox, C. A., and Ball, J. T. (1996a) EcoCELLs, tools for mesocosm scale measurements of gas exchange. *Plant, Cell and Environment* 18, 1210–1221.

Griffin, K, L, Winner, W. E., and Strain, B. R. (1996b) Construction cost of loblolly and ponderosa pine leaves grown with varying carbon and nitrogen availability. *Plant, Cell and Environment* 19, 729–738.

Griffin, K. L., Bashkin, M. A., Thomas, R. B., and Strain, B. R. (1997) Interactive effects of soil nitrogen and atmospheric carbon dioxide on root/rhizosphere carbon dioxide efflux from loblolly and ponderosa pine seedlings. *Plant Soil* 190, 11–18.

Griffis, T. J., Rouse, W. R., and Waddington, J. M. (2000) Interannual variability of net ecosystem CO_2 exchange at a subarctic fen. *Global Biogeochemical Cycles* 14, 1109–1121.

Grogan, P. (1998) CO_2 flux measurement using soda lime: Correction for water formed during CO_2 adsorption. *Ecology* 79(4), 1467–1468.

Grogan, P., and Chapin, F. S. (1999) Arctic soil respiration: Effects of climate and vegetation depend on season. *Ecosystems* 2(5), 451–459.

Groleau-Renaud, V., Plantureux, S., and Guckert, A. (1998) Influence of plant morphology on root exudation of maize subjected to mechanical impedance in hydroponic conditions. *Plant and Soil* 201(2), 231–239.

Gu, L., Fuentes, J. D., Shugart, H. H., Staebler, R. M., and Black, T. A. (1999) Responses of net ecosystem exchanges of carbon dioxide to changes in cloudiness: Results from two North American deciduous forests. *Journal of Geophysical Research* 104, 31421–31434.

Gu, L., Post, W. M., and King, A. W. (2004) Fast labile carbon turnover obscures sensitivity of heterotrophic respiration from soil to temperature: A model analysis. *Global Biogeochemical Cycles* 18, 1–11.

Gulledge, J. M., Doyle, A. P., and Schimel, J. P. (1997) Different NH_4^+-inhibition patterns of soil CH_4 consumption: A result of distinct CH_4 oxidizer populations across sites? *Soil Biology and Biochemistry* 29, 13–21.

Gulledge, J., and Schimel, J. P. (2000) Controls on soil carbon dioxide and methane fluxes in a variety of taiga forest stands in interior Alaska. *Ecosystems* 3, 269–282.

Gunadi, B. (1994) Litterfall, litter turnover and soil respiration in two pine forest plantations in central Java, Indonesia. *Journal of Tropical Forest Science* 6, 310–322.

Gundersen, P. (1998) Effects of enhanced nitrogen deposition in a spruce forest at Klosterhede, Denmark, examined by moderate NH_4NO_3 addition. *Forest Ecology and Management* 101, 251–268.

Gupta, S. R., and Singh, J. S. (1977) Effect of alkali concentration, volume and absorption area on measurement of soil respiration in a tropical sward. *Pedobiologia* 17(4), 233–239.

Haber, W. (1958) Ökologische Untersuchungen der Bodenatmung. *Flora* 146, 109–157.

Haider, K., Martin, J. P., and Filip, Z. (1975) *Humus biochemistry, Vol. 4*. Marcel Dekker, New York.

Hamdi, Y. A. (1971) Soil-water tension and the movement of rhizobia. *Soil Biology and Biochemistry* 3, 121–126.

Hanks, R. J., and Woodruff, N. P. (1958) Influence of wind on water vapor transfer through soil, gravel, and straw mulches. *Soil Science* 86, 160–164.

Hanson, P. J., Wullschleger, S. D., Bohlman, S. A., and Todd, D. E. (1993) Seasonal and topographic patterns of forest floor CO_2 efflux from an upland oak forest. *Tree Physiology* 13, 1–15.

Hanson, P. J., Edwards, N. T., Garten, C. T., and Andrews, J. A. (2000) Separating root and soil microbial contributions to soil respiration: A review of methods and observations. *Biogeochemistry* 48, 115–146.

Hanson, P. J., O'Neill, E. G., Chambers, M. L. S., Riggs, J. S., Joslin, J. D., and Wolfe, M. H. (2003). Soil respiration and litter decomposition. In *North American temperate deciduous forest responses to changing precipitation regimes* (P. J. Hanson, and S. D. Wullschleger, eds.), pp. 163–189, Springer, New York.

Harden, J. W., Sundquist, E. T., Stallard, R. F., and Mark, R. K. (1992) Dynamics of soil carbon during deglaciation of the Laurentide ice sheet. *Science* 258, 1921–1924.

Harmon, M. E., Franklin, J. F., Swanson, F. J., Sollins, P., Gregory, S. V., Lattin, J. D., Anderson, N. H., Cline, S. P., Aumen, N. G., Sedell, J. R., Lienkaemper, G. W., Cromack, K., and Cummins, K. W. (1986) Ecology of coarse woody debris in temperate ecosystems. In *Advances in ecological research* (A. MacFayden, and E. D. Ford, eds.), pp. 133–302, Academic Press, London.

Harper, C. W., Blair, J. M., Fay, P. A., Knap, A. K., and Carlisle, J. D. (2005) Increased rainfall variability and reduced rainfall amount decreases soil CO_2 flux in a grassland ecosystem. *Global Change Biology* 11, 322–334.

Harris, R. F. (1981) Effect of water potential on microbial growth and activity. In *Water potential relations in soil microbiology* (J. F. Parr, W. R. Gardner, and L. F. Elliott, eds.), pp. 23–95, Soil Science Society of America, Madison, WI.

Hart, S. C., and Sollins, P. (1998) Soil carbon and nitrogen pools and processes in an old-growth conifer forest 13 years after trenching. *Canadian Journal of Forest Research* 28, 1261–1265.

Harte, J., and Shaw, R. (1995) Shifting dominance within a montane vegetation community: Results of a climate-warming experiment. *Science* 267, 878–880.

Harte, J., Torn, M. S., Chang, F. R., Feifarek, B., Kinzig, A. P., Shaw, R., and Shen, K. (1995) Global warming and soil microclimate: Results from a meadow-warming experiment, *Ecological Applications* 5, 132–150.

Harte, J. (2002) Toward a synthesis of the Newtonian and Darwinian worldviews. *Physics Today* 55(10), 29–34.

Haynes, R. J., and Gower, S. T. (1995) Belowground carbon allocation in unfertilized and fertilized red pine plantations in northern Wisconsin. *Tree Physiology* 15, 317–325.

Hearly, R. W., Striegl, R. G., Russell, T. F., Hutchinson, G. L., and Livingston, G. P. (1996) Numerical evaluation of static chamber measurements of soil-atmosphere gas exchange, Identification of physical processes. *Soil Science Society of America Journal* 60, 740–747.

Heilmeier, H., Erhard, M., and Schulze, E. D. (1997) Biomass allocation and water use under arid conditions. In *Plant resource allocation* (F. A. Bazzaz, and J. Grace, eds.), pp. 93–112, Academic Press, San Diego, CA.

Hendrickson, O. Q., Chatarpaul, L., and Burgess, D. (1989) Nutrient cycling following whole-tree and conventional harvest in northern mixed forest. *Canadian Journal of Forest Research* 19, 725–735.

Hénin, S., Monnier, G., and Turc, L. (1959) Un aspect de la dynamique des matières organiques du sol. *Comptes rendus de l'Académie des Sciences (Paris)* 248, 138–141.

Hesterberg, R., and Siegenthaler, U. (1991) Production and stable isotopic composition of CO_2 in a soil near Bern, Switzerland. *Tellus* 43B, 197–205.

Hibbard, K. A., Law, B. E., Reichstein, M., and Sulzman, J. (2005) An analysis of soil respiration across northern hemisphere temperate ecosystems. *Biogeochemistry* 73(1), 29–70.

Higgins, P. A. T., Jackson, R. B., des Rosiers, J. M., and Field, C. B. (2002) Root production and demography in a California annual grassland under elevated atmospheric carbon dioxide. *Global Change Biology* 8, 841–850.

Hillel, D. (1998) *Environmental Soil Physics.* Academic Press/Elsevier, San Diego, CA.

Hirsch, A. I., Trumbore, S. E., and Goulden, M. L. (2002) Direct measurement of the deep soil respiration accompanying seasonal thawing of a boreal forest soil. *Journal of Geophysical Research-Atmosphere* 108(D23), 8221, doi: 10.1029/2001JD000921.

Hirsch, A. I., Trumbore, S. E., and Goulden, M. L. (2004) The surface CO_2 gradient and pore-space storage flux in a high-porosity litter layer. *Tellus* 56B, 312–321.

Hobbie, S. E. (1996) Temperature and plant species control over litter decomposition in Alaskan tundra. *Ecological Monographs* 66, 503–522.

Hobbie, S. E. (2000) Interactions between litter lignin and soil nitrogen availability during leaf litter decomposition in a Hawaiian montane forest. *Ecosystems* 3, 484–494.

Högberg, P., Nordgren, A., Buchmann, N., Taylor, A. F. S., Ekblad, A., Högberg, M. N., Nyberg, G., Ottosson-Löfvenius, M., and Read, D. J. (2001) Large-scale forest girdling shows that current photosynthesis drives soil respiration. *Nature* 411, 789–792.

Högberg, P., Nordgren, A., and Ågren, G. I. (2002) Carbon allocation between tree root growth and root respiration in boreal pine forest. *Oecologia* 132, 579–581.

Holland, E. A., and Coleman, D. C. (1987) Litter placement effects on microbial and organic-matter dynamics in an agroecosystem. *Ecology* 68, 425–433.

Holthausen, R. S., and Caldwell, M. M. (1980) Seasonal dynamics of root system respiration in *Atriplex confertifolia. Plant and Soil* 55, 307–317.

Hook, P. B., and Burke, I. C. (2000) Biogeochemistry in a shortgrass landscape: Control by topography, soil texture, and microclimate. *Ecology* 81, 2686–2703.

Hoosbeek, M. R., van Breemen, N., Vasander, H., Buttler, A., and Berendse, F. (2002) Potassium limits potential growth of bog vegetation under elevated atmospheric CO_2 and N deposition. *Global Change Biology* 8, 1130–1138.

Houghton J. T., Meira Filho, L. G., Callander, B. A., Harris, N., Kattenberg, A., and Maskell, K. (1996) *Climate Change 1995: The science of climate change.* Cambridge University Press, Cambridge, UK.

Houghton, R. A. (2002) Magnitude, distribution and causes of terrestrial carbon sinks and some implications for policy. *Climate Policy* 2, 71–88.

Houghton, R. A. (2003) Why are estimates of the terrestrial carbon balance so different? *Global Change Biology* 9, 500–509.

Howard, P. J. A. (1966) A method for estimation of carbon dioxide evolved from surface of soil in the field. *Oikos* 17(2), 267–271.

Howard, D. M., and Howard, P. J. A. (1993) Relationships between CO_2 evolution, moisture content, and temperature for a range of soil types. *Soil Biology and Biochemistry* 25, 1537–1546.

Huang, B., and Nobel, P. S. (1993) Hydraulic conductivity and anatomy along lateral roots of cacti: Changes with soil water status. *New Phytologist* 123, 499–507.

Hudson, R. J., Gherini, S. A., and Goldstein, R. A. (1994) Modelling the global carbon cycle: Nitrogen fertilization of the terrestrial biosphere and the "missing" CO_2 sink. *Global Biogeochemical Cycles* 8, 307–333.

Hui, D., Luo, Y., Johnson, D. W., Cheng, W., Coleman, J. S., and Sims, D. A. (2001) Canopy radiation and water use efficiency as affected by elevated CO_2. *Global Change Biology* 7, 75–91.

Hui, D., Luo, Y., and Katul, G. (2003) Partitioning interannual variability in net ecosystem exchange into climatic variability and functional change. *Tree Physiology* 23, 433–442.

Hui, D., and Luo, Y. (2004) Evaluation of soil CO_2 production and transport in Duke Forest using a process-based modeling approach. *Global Biogeochemical Cycles* 18, GB4029, doi: 10.1029/2004GB002297.

Humfeld, H. (1930) A method for measuring carbon dioxide evolution from soil. *Soil Science* 30(1), 1–11.

Hungate, B. A., Chapin, F. S., Zhong, H., Holland, E. A., and Field, C. B. (1997) Stimulation of grassland nitrogen cycling under carbon dioxide enrichment. *Oecologia* 109(1), 149–153.

Hunt, E. R. Jr., Piper, S. C., Nemani, R., Keeling, C. D., Otto, R. D., and Running, S. W. (1996) Global net carbon exchange and intra-annual atmospheric CO_2 concentrations predicted by an ecosystem process model and three-dimensional atmospheric transport model. *Global Biogeochemical Cycles* 10, 431–456.

Hutchinson, G. L., Livingston, G. P., Healy, R. W., and Striegl, R. G. (2000) Chamber measurement of surface-atmosphere trace gas exchange: Numerical evaluation of dependence on soil, interfacial layer, and source/sink properties. *Journal of Geophysical Research-Atmosphere* 105(D7), 8865–8875.

Hutchinson, G. L., and Rochette, P. (2003) Non-flow-through steady-state chambers for measuring soil respiration: numerical evaluation of their performance. *Soil Science Society of American Journal* 67, 166–180.

Hutsch, B. W., Augustin, J., and Merbach, W. (2002) Plant rhizodeposition—an important source for carbon turnover in soils. *Journal of Plant Nutrition and Soil Science* 165(4), 397–407.

Ibrahim, L., Roe, M. F., and Cameron, A. D. (1997) Main effects of nitrogen supply and drought stress upon whole plant carbon allocation in poplar. *Canadian Journal of Forest Research* 27, 1413–1419.

Ilstedt, U., Nordgren, A., and Malmer, A. (2000) Optimum soil water for soil respiration before and after amendment with glucose in humid tropical acrisols and a boreal mor layer. *Soil Biology and Biochemistry* 32, 1591–1599.

Ineson, P., Cotrufo, M. F., Bol, R., Harkness, D. D., and Blum, H. (1996) Quantification of soil carbon inputs under elevated CO_2: C_3 plants in a C_4 soil. *Plant and Soil* 187, 345–350.

Ino, Y., and Monsi, M. (1969) An experimental approach to the calculation of CO_2 amount evolved from several soils. *Japanese Journal of Botany* 20, 153–188.

IPCC (1990) *Climate change: The IPCC scientific assessment* (J. T. Houghton, G. J. Jenkins, and J. J. Ephraums, eds.), Cambridge University Press, Cambridge, UK.

IPCC (1992) *Climate change 1992: The supplementary report to the IPCC scientific assessment* (J. T. Houghton, B. A. Callander, and S. K. Varney, eds.), Cambridge University Press, Cambridge, UK.

IPCC (1995) *Climate change 1994: Radiative forcing of climate change and an evaluation of the IPCC IS92 emission scenarios* (J. T. Houghton, L. G. Meira Filho, J. Bruce, H. Lee, B. A. Callander, E. Haites, N. Harris, and K. Maskell, eds.), Cambridge University Press, Cambridge, UK.

IPCC (2001) Climate change 2001: The scientific basis. Contribution of Working Group I to the *Third assessment report of the intergovernmental panel on climate change* (J. T. Houghton, Y. Ding, D. J. Griggs, M. Noguer, P. J. van der Linden, X. Dai, K. Maskell, and C. A. Johnson, eds.), Cambridge University Press, Cambridge, UK and New York.

Iritz, Z., Lindroth, A., and Gärdenäs, A. (1997) Open ventilated chamber system for measurements of H_2O and CO_2 fluxes from the soil surface. *Soil Techniques* 10, 169–184.

Ivarson, K. C., and Sowden, F. J. (1970) Effect of frost action and storage of soil at freezing temperatures on the free amino acids, free sugars, and respiratory activity of soil. *Canadian Journal of Soil Science* 50, 191–198.

Irvine, J., and Law, B. E. (2002) Contrasting soil respiration in young and old-growth ponderosa pine forests, *Global Change Biology* 8, 1183–1194.

Ito, A., and Oikawa, T. (2002) A simulation model of the carbon cycle in land ecosystems (Sim-CYCLE): A description based on dry-matter production theory and plot-scale validation. *Ecological Modelling* 151, 143–176.

Jackson, R. B., Canadell, J., Ehleringer, J. R., Mooney, H. A., Sala, O. E., and Schulze, E. D. (1996) A global analysis of root distributions for terrestrial biomes. *Oecologia* 108, 389–411.

Janssens, I. A., and Ceulemans, R. (1998) Spatial variability in forest soil CO_2 efflux assessed with a calibrated soda lime technique. *Ecology Letters* 1, 95–98.

Janssens, I. A., Kowalski, A. S., Longdoz, B., and Ceulemans, R. (2000) Assessing forest soil CO_2 efflux: An in situ comparison of four techniques. *Tree Physiology* 20, 23–32.

Janssens I. A., Lankreijer, H., Matteucci, G., Kowalski, A. S., Buchmann, N., Epron, D., Pilegaard, K., Kutsch, W., Longdoz, B., Grünwald, T., Montagnani, L., Dore, S., Rebmann, C., Moors, E. J., Grelle, A., Rannik, Ü., Morgenstern, K., Oltchev, S., Clement, R., Guomundsson, J., Minerbi, S., Berbigier, P., Ibrom, A., Moncrieff, J., Aubinet, M., Bernhofer, C., Jensen, N. O., Vesala, T., Granier, A., Schulze, E.-D., Lindroth, A., Dolman, A. J., Jarvis, P. G., Ceulemans, R., and Valentini, R (2001) Productivity overshadows temperature in determining soil and ecosystem respiration across European forests. *Global Change Biology* 7(3), 269–278.

Janssens, I. A., Dore, S., Epron, D., Lankreijer, H., Buchmann, N., Longdoz, B., Brossaud, J., and Montagnani, L. (2003) Climatic influences on seasonal and spatial differences in soil CO_2 efflux. In *Ecological studies 163: Canopy fluxes of energy, water and carbon dioxide of European forests* (R. Valentini, ed.), pp. 235–256, Springer-Verlag, Berlin and Heidelberg, Germany.

Janzen, H. H., Campbell, C. A., Brandt, S. A., Lafond, G. P., and Townley-Smith, L. (1992) Light-fraction organic matter in soils from long-term crop rotations. *Soil Science Society of America Journal* 56, 1799–1806.

Jassal, R. S., Black, T. A., Drewitt, G. B., Novak, M. D., Gaumont-Guay, D., and Nesic, Z. (2004) A model of the production and transport of CO_2 in soil, predicting soil CO_2 concentrations and CO_2 efflux from a forest floor. *Agricultural and Forest Meterology* 124, 219–236.

Jastrow, J. D., and Miller, R. M. (1997) Soil aggregate stabilization and carbon sequestration, feedbacks through organomineral associations. In *Soil processes and the carbon cycle* (R. Lal, J. M. Kimble, R. F. Follett, and B. A. Stewart, eds.), pp. 207–223, CRC Press, Boca Raton, FL.

Jauhiainen, J., Takahashi, H., Heikkinen, J. E. P., Martikainen, P. J., and Vasander, H. (2005) Carbon fluxes from a tropical peat swamp forest floor. *Global Change Biology* 11(10), 1788–1797.

Jenkinson, D. S., and Rayner, J. H. (1977) The turnover of soil organic matter in some of the Rothamsted classical experiments. *Soil Science* 123, 298–305.

Jenkinson, D. S. (1990) The turnover of organic carbon and nitrogen in soil. *Philosophical Transactions of the Royal Society of London, Series B* 329, 361–368.

Jenny, H., Gessel, S. P., and Bingham, F. T. (1949) Comparative study of decomposition rates of organic matter in temperate and tropical regions. *Soil Science* 68(6), 419–432.

Jenny, H. (1980) *The Soil Source.* Springer-Verlag, Berlin and Heidelberg, Germany, and New York.

Jensen, B. (1993) Rhizodeposition by (CO_2)-C-14-pulse-labeled spring barley grown in small-field plots on sandy loam. *Soil Biology and Biochemistry* 25(11): 1553–1559.

Jensen, L. S., Mueller, T., Tate, K. R., Ross, D. J., Magid, J., and Nielsen, N. E. (1996) Soil surface CO_2 flux as an index of soil respiration *in situ*: A comparison of two chamber methods. *Soil Biology and Biochemistry* 28(10–11), 1297–1306.

Jiang, L., Shi, F., Li, B., Luo, Y., Chen, J., and Chen, J. (2005) Separating rhizosphere respiration from total soil respiration in two larch plantations in northeastern China. *Tree Physiology* 25, 1187–1195.

Jobbágy, E. G., and Jackson, R. B. (2000) The vertical distribution of soil organic carbon and its relation to climate and vegetation. *Ecological Applications* 10, 423–436.

Johnson, E., and Heinen, R. (2004) Carbon trading: Time for industry involvement. *Environment International* 30(2), 279–288.

Johnson, I. R. (1983) Nitrate uptake and respiration in roots and shoots: A model. *Physiologia Plantarum* 58, 145–147.

Johnson, D. W., and Lindberg, S. E. (1992) Atmospheric deposition and forest nutrient cycling: A synthesis of the integrated forest study. *Ecological Studies Series 91*, Springer-Verlag, New York.

Johnson, D. W., Geisinger, D. R., Walker, R. F., Newman, J., Vose, J. M., Elliot, K. J., and Ball, J. T. (1994) Soil pCO_2, soil respiration, and root activity in CO_2-fumigated and nitrogen-fertilized ponderosa pine. *Plant and Soil* 165, 129–138.

Johnson, L. C., Shaver, G. R., Cades, D. H., Rastetter, E., Nadelhoffer, K., Ginlin, A., Laundre, J., and Stanley, A. (2000) Plant carbon-nutrient interactions control CO_2 exchange in Alaskan wet sedge tundra ecosystems. *Ecology* 81, 453–469.

Johnson, L. C., and Matchett, J. R. (2001) Fire and grazing regulate belowground processes in tallgrass prairie. *Ecology* 82, 3377–3389.

Johnson, D. W., Hungate, B. A., Dijkstra, P., Hymus, G., Hinkle, C. R., Stiling, P., and Drake, B. G. (2003) The effects of elevated CO_2 on nutrient distribution in a fire-adapted scrub oak forest. *Ecological Applications* 13, 1388–1399.

Johnson, D. W., Cheng, W., Joslin, J. D., Norby, R. J., Edwards, N. T., and Todd, D. E., Jr. (2004) Effects of elevated CO_2 on nutrient cycling in a sweetgum plantation. *Biogeochemistry* 69, 379–403.

Jonasson, S., Castro, J., and Michelsen, A. (2004) Litter, warming and plant affect respiration and allocation of soil microbial and plant C, N and P in Arctic mesocosms. *Soil Biology and Biochemistry* 36, 1129–1139.

Jones H. G. (1992) *Plants and microclimate: A quantitative approach to environmental plant physiology.* 2nd edition, Cambridge University Press, New York.

Kaal, E. E. J., De Jong, E., and Field, J. A. (1993) Stimulation of ligninolytic peroxidase activity by nitrogen nutrients in the white rot fungus *Bjerkandera* sp. strain BOS55. *Applied and Environmental Microbiology* 59, 4031–4036.

Kabwe, L. K., Hendry, M. J., Wilson, G. W., and Lawrence, J. R. (2002) Quantifying CO_2 fluxes from soil surfaces to the atmosphere. *Journal of Hydrology* 260, 1–14.

Kane, E. S., Pregitzer, K. S., and Burton, A. J. (2003) Soil respiration along environmental gradients in Olympic national park. *Ecosystems* 6, 326–336.

Kanemasu, E. T., Powers, W. L., and Sij, J. W. (1974) Field chamber measurements of CO_2 flux from soil surface. *Soil Science* 118, 233–237.

Kang, S., Kim, S., Oh, S., and Lee, D. (2000) Predicting spatial and temporal patterns of soil temperature based on topography, surface cover and air temperature. *Forest Ecology and Management* 136, 173–184.

Kang, S., Kim, S., and Lee, D. (2002) Spatial and temporal patterns of solar radiation based on topography and air temperature. *Canadian Journal of Forest Research* 32, 487–497.

Kang, S., Doh, S., Lee, D., Jin, V. L., Kimball, J. S. (2003) Topographic and climatic control on soil respiration in six temperate mixed-hardwood forest slopes, Korea. *Global Change Biology* 9, 1427–1437.

Kasper, T. C., and Bland, W. L. (1992) Soil temperature and root growth. *Soil Science* 154, 290–299.

Kasurinen, A., Kokko-Gonzales, P., Riikonen, J., Vapaavuori, E., Holopainen, T. (2004) Soil CO_2 efflux of two silver birch clones exposed to elevated CO_2 and O_3 levels during three growing seasons. *Global Change Biology* 10, 1654–1665.

Katagiri, S. (1988) Estimation of proportion of root respiration in total soil respiration in deciduous broadleaved stands. *Journal of Japan Forest Society* 70, 151–158.

Katul, G. G., Finkelstein, P. L., Clarke, J. F., Ellestad, T. G. (1996) An investigation of the conditional sampling method used to estimate fluxes of active, reactive, and passive scalars. *Journal of Applied Meteorology* 35, 1835–1845.

Katul, G., Oren, R., Ellsworth, D., Hsieh, C., Phillipps, N., and Lewin, K. (1997) A langrangian dispersion model for predicting CO_2 sources and sinks, and fluxes in a uniform loblolly pine (*pinusd taeda* L.) stand. *Journal of Geophysical Research* 102, 9309–9321.

Kaye, J., Barrett, J., and Burke, I. (2002) Stable nitrogen and carbon pools in grassland soils of variable texture and carbon content. *Ecosystems* 5, 461–471.

Keith, H., Jacobsen, K. L., and Raison, R. J. (1997) Effects of soil phosphorus availability, temperature and moisture on soil respiration in Eucalyptus pauciflora forest. *Plant and Soil* 190, 127–141.

Kern, J. S., and Johnson, M.G. (1993) Conservation tillage impacts on national soil and atmospheric carbon levels. *Soil Science Society of America Journal* 57, 200–210.

Keyser, P., Kirk, T. K., and Zeikus, J. G. (1978) Ligninolytic enzyme systems of *Phanerochaete chrysosporium* synthesized in the absence of lignin in response to nitrogen starvation. *Journal of Bacteriology* 135, 790–797.

Killham, K. (1994) *Soil Ecology*. Cambridge University Press, Cambridge, UK.

Kimball, B. A., and Lemon, E. R. (1971) Air turbulence effects upon soil gas exchange. *Soil Science Society of America Proceeding* 35, 16–21.

Kimball, B. A. (1983) Canopy gas exchange, gas exchange with soil. In *Limitations to efficient water use in crop production*, ASA-CSSA-SSSA, Madison, WI.

King, J. A., and Harrison, R. (2002) Measuring soil respiration in the field: An automated closed chamber system compared with portable IRGA and alkali absorption methods. *Communications in Soil Science and Plant Analysis* 33(3–4), 403–423.

King, J. S., Pregitzer, K. S., and Zak, D. R. (1999) Clonal variation in above and belowground growth responses of *Populus tremuloides* Michaux: Influence of soil warming and nutrient availability. *Plant and Soil* 217, 119–130.

King, J. S., Pregitzer, K. S., Zak, D. R., Sober, J., Isebrands, J. G., Dickson, R. E., Hendrey, G. R., and Karnosky, D. F. (2001) Fine-root biomass and fluxes of soil carbon in young stands of paper birch and trembling aspen as affected by elevated atmospheric CO_2 and tropospheric O_3. *Oecologia* 128, 237–250.

King, J. S., Hanson, P. J., Bernhardt, E. Y., Deangelis, P., Norby, R. J., and Pregitzer, K. S. (2004) A multiyear synthesis of soil respiration responses to elevated atmospheric CO_2 from four forest FACE experiments, *Global Change Biology* 10, 1027–1042.

Kirita, H., and Hozumi, K. (1966) Re-examination of the absorption method of measuring soil respiration under field conditions. *Physiological Ecology* 14, 23–31.

Kirita, H. (1971) Re-examination of the absorption method of measuring soil respiration under field conditions: II, Effect of the size of the apparatus on CO_2-absorption rates. *Japanese Journal of Ecology* 21, 27–42.

Kirschbaum, M. U. F. (1995) The temperature dependence of soil organic matter decomposition, and the effect of global warming on soil organic C storage. *Soil Biology and Biochemistry* 27, 753–760.

Kirschbaum, M. U. F. (2004) Soil respiration under prolonged soil warming: Are rate reductions caused by acclimation or substrate loss? *Global Change Biology* 10, 1870–1877.

Klinge, H., and Herrera, R. (1978) Biomass studies in Amazon caatinga forest in southern Venezuela. *Tropical Ecology* 19, 93–110.

Knapp, A. K., Conard, S. L., and Blair, J. M. (1998) Determinants of soil CO_2 flux from a sub-humid grassland, effect of fir and fire history. *Ecological Applications* 8, 760–770.

Knoepp, J. D., and Vose, J. M. (2002) Quantitative comparison of *in situ* soil CO_2 flux measurement methods. Research Paper SRS-28. Asheville, NC. Department of Agriculture, Forest Service, Southern Research Station.

Knörr, W., Prentice, I. C., House, J. I., and Holland, E. A. (2005) Long-term sensitivity of soil carbon turnover to warming. *Nature* 433, 298–301.

Kolari, P., Pumpanen, J., Rannik, Ü., Ilvesniemi, H., Hari, P., and Berninger, F. (2004) Carbon balance of different aged Scots pine forests in southern Finland. *Global Change Biology* 10, 1106–1119.

Koopmans, C. J., van Dam, D., Tietema, A., and Verstraten, J. M. (1997) Natural ^{15}N abundance in two nitrogen saturated forest ecosystems. *Oecologia* 111, 470–480.

Kowalenko, C. G., Ivarson, K. C., and Cameron, D. R. (1978) Effect of moisture content, temperature and nitrogen fertilization on carbon dioxide evolution from field soils. *Soil Biology and Biochemistry* 10, 417–423.

Kucera, C. L., and Kirkham, D. R. (1971) Soil respiration studies in tallgrass prairie in Missouri. *Ecology* 52, 912–915.

Kuikman, P. J., Jansen, A. G., and van Veen, J. A. (1991) ^{15}N-nitrogen mineralization from bacteria by protozoan grazing at different soil moisture regimes. *Soil Biology and Biochemistry* 23, 193–200.

Kummerow, J., and Mangan, R. (1981) Root systems in quercus-dumosa nutt dominated chaparral in southern California. *Acta Oecologica* 2(2), 177–188.

Kushida, K., Kim, Y., and Tanaka, N. (2004) Remote sensing of net ecosystem productivity based on component spectrum and soil respiration observation in a boreal forest, interior Alaska. *Journal of Geophysical Research*, 109, D06108, doi: 10.1029/2003JD003858.

Kutsch, L. W., Staack, A., Wötzel, J., Middelhoff, U., and Kappen, L. (2001) Field measurements of root respiration and total soil respiration in an alder forest. *New Phytologist* 150, 157–168.

Kuzyakov, Y., and Cheng, W. (2001) Photosynthesis controls of rhizosphere respiration and organic matter decomposition. *Soil Biology and Biochemistry* 33 (14), 1915–1925.

Kuzyakov, Y. (2002) Review: Factors affecting rhizosphere priming effects. *Journal of Plant Nutrition and Soil Science* 165 (4), 382–396.

Laidler, K. J. (1972) Unconventional applications of Arrhenius law. *Journal of Chemical Education* 49(5), 343–344.

Lal, R. (2003) Global potential of soil carbon sequestration to mitigate the greenhouse effect. *Critical Reviews in Plant Sciences* 22 (2), 151–184.

Lal, R. (2004) Soil carbon sequestration impacts on global climate change and food security. *Science* 304, 1623–1627.

Lamade, E., Djegui, N., and Leterme, P. (1996) Estimation of carbon allocation to the roots from soil respiration measurements of oil palm. *Plant and Soil* 181, 329–339.

Lambers, H., and Steingrover, E. (1978) Efficiency of root respiration of a flood-tolerant and a flood-intolerant *senecio* species as affected by low oxygen tension. *Physiologia Plantarum* 42, 163.

Lambers, H. (1980) The physiological significance of cyanide-resistant respiration in higher plants. *Plant, Cell and Environment* 3, 293–302.

Lambers, H., Scheurwater, I., and Atkin, O. K. (1996) Respiratory patterns in roots in relation to their functioning. In *Plant roots: The hidden half* (Y. Waisel, A. Eshel, and U. Kafkafi, eds.), pp. 323–362, Marcel Dekker, New York.

Lambers, H., Chapin III, F. S., and Pons, T. (1998) *Plant physiological ecology*. Springer-Verlag, New York.

Lambers, H., Atkin, O. K., and Millenaar, F. F. (2002) Respiratory patterns in roots in relation to their functioning. In *Plant roots: The hidden half* (Y. Waisel, A. Eshel, and U. Kafkafi, eds.), pp. 521–552, 3rd edition, Marcel Dekker, New York.

Lankreijer, H., Janssens, I. A., Buchmann, N., Longdoz, B., Epron, D., and Dore, S. (2003) Measurement of soil respiration. In *Ecological studies Vol. 163: Fluxes of carbon, water and energy of European forests* (R. Valentini, ed.), Springer-Verlag, Berlin and Heidelberg, Germany.

Laporte, M. F., Duchesne, L. C., and Morrison, I. K. (2003) Effect of clearcutting, selection cutting, shelterwood cutting and microsites on soil surface CO_2 efflux in a tolerant hardwood ecosystem of northern Ontario, *Forest ecology and mangagement* 174, 565–575.

Larionova, A. A., Rozanova, L. N., and Samoilov, T. I. (1989) Dynamics of gas exchange in the profile of a gray forest soil. *Soviet Soil Science* 3, 104–110.

Larson, M. M. (1970) Root regeneration and early growth of red oak seedlings: Influence of soil temperature. *Forest Science* 16, 442–446.

Lavigne, M. B., and Ryan, M. G. (1997) Growth and maintenance respiration rates of aspen, black spruce and jack pine stems at northern and southern BOREAS sites. *Tree Physiology* 17, 543–551.

Lavigne, M. B., Foster, R. J., and Goodine, G. (2004) Seasonal and annual changes in soil respiration in relation to soil temperature, water potential and trenching. *Tree Physiology* 24, 415–424.

Law, B. E., Ryan, M. G., and Anthoni, P. M. (1999) Seasonal and annual respiration of a ponderosa pine ecosystem. *Global Change Biology* 5, 169–182.

Law, B. E., Kelliher, F. M., Baldocchi, D. D., Anthoni, P. M., Irvine, J., Moore, D., and Van Tuyl, S. (2001) Spatial and temporal variation in respiration in a young ponderosa pine forest during a summer drought. *Agricultural and Forest Meteorology* 110, 27–43.

Leadley, P. W., Niklaus, P. A., Stocker, R., and Körner, C. (1999) A field study of the effects of elevated CO_2 on plant biomass and community structure in a calcareous grassland. *Oecologia* 118, 39–49.

Leavitt, S. W., Paul, E. A., Kimball, B. A., Hendrey, G. R., Mauney, J. R., Rauschkolb, R., Rogers, H., Lewin, K. F., Nagy, J., Pinter, P. J., and Johnson, H. B. (1994) Carbon isotope dynamics of free-air CO_2 enriched cotton and soils. *Agricultural Forest and Meteorology* 70, 87–101.

Leavitt, S. W., Paul, E. A., Galadima, A., Nakayama, F. S., Danzer, S. R., Johnson, H., and Kimball, B. (1996) Carbon isotopes and carbon turnover in cotton and wheat FACE experiments. *Plant and Soil* 187, 147–155.

Leavitt, S. W., Pendall, E., Paul, E. A., Brooks, T., Kimball, B. A., Pinter, P. J., Johnson, H. B., Matthias, A., Wall, G. W., and La Morte, R. L. (2001) Stable-carbon isotopes and soil organic carbon in wheat under CO_2 enrichment. *New Phytologist* 150(2), 305–314.

Lebedjantzev, A. N. (1924) Drying of soil as one of the natural factors in maintaining soil fertility. *Soil Science* 18, 419–447.

Le Cain, D. R., Morgan, J. A., Schuman, G. E., Reeder, J. D., and Hart, R. H. (2000) Carbon exchange rates in grazed and ungrazed pastures of Wyoming. *Journal of Range Management* 53, 199–206.

Le Dantec, V., Epron, D., and Dufrêne, E. (1999) Soil CO_2 efflux in beech forest: Comparison of two closed dynamic systems. *Plant and Soil* 214, 125–132.

Lee, M. S., Nakane, K., Nakatsubo, T., Mo, W. H., and Koizumi, H. (2002) Effects of rainfall events on soil CO_2 flux in a cool temperate deciduous broad-leaved forest. *Ecological Research* 17, 401–409.

Lemon, E. (1969) Aerodynamic studies of CO_2 exchange between the atmosphere and the plant. In *Harvesting the sun: Photosynthesis in plant life.* (A. San Pietro, F. A. Greer, and T. S. Army, eds.), 3rd edition, pp. 263–290, Academic Press, New York.

Levin, I., and Kromer, B. (1997) Twenty years of atmospheric $^{14}CO_2$ observations at Schauinsland Station, Germany. *Radiocarbon* 39, 205–218.

Levy, P. E., Meir, P., Allen, S. J., and Jarvis, P. G. (1999) The effect of aqueous transport of CO_2 in xylem sap on gas exchange in woody plants. *Tree Physiology* 19(1), 53–58.

Lewicki, J. L., Evans, W. C., Hilley, G. E., Sorey, M. L., Rogie, J. D., and Brantley, S. L. (2003) Shallow soil CO_2 flow along the San Andreas and Calaveras Faults, California, *Journal of Geophysical Research-Solid Earth* 108(B4), 2187, doi: 10.1029/2002JB002141.

Liang, N., Nakadai, T., Hirano, T., Qu, L., Koike, T., Fujinuma, Y., and Inoue, G. (2004) *In situ* comparision of four approaches to estimating soil CO_2 efflux in a northern larch (*Larix kaempferi* Sarg.) forest. *Agricultural and Forest Meteorology* 123, 97–117.

Liang, N., Inoue, G., and Fujinuma, Y. (2005a) A multichannel automated chamber system for continuous measurement of forest soil CO_2 efflux. *Tree Physiology* 23, 825–832.

Liang, N., Hirano, T., Tang, J., Irvines, J., Black, T. A., Takagi, K., Baldocchi, D., Law, B. E., Fujinuma, Y., and Inoue, G. (2006) Long-term continuous measurement of soil CO_2 efflux using automated chamber and soil CO_2 concentration gradient techniques. *Global Change Biology* (in press).

Lichter, J., Barron, S. H., Finzi, A. C., Irving, K. F., Roberts, M. T., Stemmler, E. A., and Schlesinger, W. H. (2005) Soil carbon sequestration and turnover in a pine forest after six years of atmospheric CO_2 enrichment: Soil carbon sequestration under elevated CO_2. *Ecology* 86, 1835–1847.

Lidstrom, M. E. (1992) The genetics and molecular biology of methanol-utilizing bacteria. In *Methane and methanol utilizers* (J. C. Murrell, and H. Dalton, eds.), pp. 183–206, Plenum Press, New York.

Lieffers, V. J., and Rothwell, R. L. (1986) Effects of depth of water table and substrate temperature on root and top growth of *Picea mariana* and *Larix laricina* seedlings. *Canadian Journal of Forest Research* 16, 1201–1206.

Lieth, H., and Ouellette, R. (1962) Studies on the vegetation of the Gaspé Peninsula: 2, The soil respiration of soil plant communities. *Canadian Journal of Botany* 40, 127–140.

Lieth, H., and Werger, M. J. A. (1989) *Tropical rain forest ecosystems: Biogeographical and ecological studies.* Elsevier Academic Press, Amsterdam, the Netherlands.

Liljeroth, E., Kuikman, P., and Vanveen, J. A. (1994) Carbon translocation to the rhizosphere of maize and wheat and influence on the turnover of native soil organic-matter at different soil-nitrogen levels. *Plant and Soil* 161 (2), 233–240.

Lin, G., and Ehleringer, J. R. (1997) Carbon isotopic fractionation does not occur during dark respiration in C_3 and C_4 plants. *Plant Physiology* 114, 391–394.

Lin, G., Ehleringer, J. R., Rygiewicz, P. T., Johnson, M. G., and Tinge, D. T. (1999) Elevated CO_2 and temperature impacts on different components of soil CO_2 efflux in Douglas-fir terracosm. *Global Change Biology* 5, 157–168.

Lin, G., Rygiewicz, P. T., Ehleringer, J. R., Johnson, M. G., and Tingey, D. T. (2001) Time-dependent responses of soil CO_2 efflux components to elevated atmospheric $[CO_2]$ and temperature in experimental forest mesocosms. *Plant and Soil* 229, 259–270.

Linn, D. M., and Doran, J. W. (1984) Effects of water-filled pore space on carbon dioxide and nitrous oxide production in tilled and nontilled soils. *Soil Science Society of America Journal* 48, 1267–1272.

Lipp, C. C., and Andersen, C. P. (2003) Role of carbohydrate supply in white and brown root respiration of ponderosa pine. *New Phytologist* 160, 523–531.

Liski, J., Ilvesniemi, H., Mäkelä, A., and Westman, C. J. (1999) CO_2 emissions from soil in response to climatic warming are overestimated: The decomposition of old soil organic matter is tolerant of temperature. *Ambro* 28, 171–174.

Litton, C. M., Ryan, M. G., Knight, D. H., and Stahl, P. D. (2003) Soil-surface carbon dioxide efflux and microbial biomass in relation to tree density 13 years after a stand replacing fire in a lodgepole pine ecosystem. *Global Change Biology* 9(5), 680–696.

Liu, X., Wan, S., Su, B., Hui, D., and Luo, Y. (2002a) Response of soil CO_2 efflux to water manipulation in a tallgrass prairie ecosystem. *Plant and Soil* 240, 213–223.

Liu, W., Fox, J. E. D., and Xu, Z. (2002b) Litterfall and nutrient dynamics in a montane moist evergreen broad-leaved forest in Ailao Mountains, SW China. *Plant Ecology* 164 (2), 157–170.

Lloyd, J., and Taylor, J. A. (1994) On the temperature dependence of soil respiration. *Functional Ecology*, 8, 315–323.

Loiseau, P., and Soussana, J. F. (1999) Elevated [CO_2], temperature increase and N supply effects on the turnover of below-ground carbon in a temperature grassland ecosystem. *Plant and Soil* 210, 233–247.

Lomander, A. L., Kätterer, T., and Andrén, O. (1998) Modeling the effects of temperature and moisture on CO_2 evolution from top- and subsoil using a multi-compartment approach. *Soil Biology and Biochemistry* 30, 2023–2030.

Londo, A. J., Messina, M. G., and Schoenholtz, S. H. (1999) Forest harvesting effects on soil temperature, moisture, and respiration in a bottomland hardwood forest. *Soil Science Society of America Journal* 63, 637–644.

Longdoz, B., Yernaux, M., and Aubinet, M. (2000) Soil CO_2 efflux measurements in a mixed forest: Impact of chamber disturbances, spatial variability and seasonal evolution. *Global Change Biology* 6, 907–917.

Lucht, W., Prentice, I. C., Myneni, R. B., Sitch, S., Friedlingstein, P., Cramer, W., Bousquet, P., Buermann, W., and Smith, B. (2002) Climatic control of the high-latitude vegetation greening trend and Pinatubo effect. *Science* 296, 1687–1689.

Luken, J. O., and Billings, W. D. (1985) The influence of microtopographic heterogeneity on carbon dioxide efflux from a subarctic bog. *Holarctic Ecology* 8, 306–312.

Lundegårdh, H. (1921) Ecological studies in the assimilation of certain forest plants and shore plants. *Svensk Botaniska Tidskrift* 15, 46–94.

Lundegårdh, H. (1927) Carbon dioxide evolution of soil and crop growth. *Soil Science* 23, 417–453.

Lund, C. P., Riley, W. J., Pierce, L. L., and Field, C. B. (1999) The effects of chamber pressurization on soil-surface CO_2 flux and the implications for NEE measurements under elevated CO_2. *Global Change Biology* 5, 269–281.

Luo, Y., Jackson, R. B., Field, C. B., and Mooney, H. A. (1996) Elevated CO_2 increases belowground respiration in California grasslands. *Oecologia* 108, 130–137.

Luo, Y., and Reynolds, J. F. (1999) Validity of extrapolating field CO_2 experiments to predict carbon sequestration in natural ecosystems, *Ecology* 80, 1568–1583.

Luo, Y., Wan, S., Hui, D., and Wallace, L. (2001a) Acclimatization of soil respiration to warming in a tall grass prairie. *Nature* 413, 622–625.

Luo, Y., Wu, L. H., Andrews, J. A., White, L., Matamala, R., Schafer, K. V. R., and Schlesinger, W. H. (2001b) Elevated CO_2 differentiates ecosystem carbon processes: Deconvolution analysis of Duke Forest FACE data. *Ecological Monographs* 71, 357–376.

Luo, Y., Hui, D., and Zhang, D. (2006) Elevated carbon dioxide stimulates net accumulations of carbon and nitrogen in terrestrial ecosystems: Results of meta-analysis. *Ecology* (in press).

Luo, Y., Medlyn, B., Hui, D., Ellsworth, D., Reynolds, J. F., and Katul, G. (2001c) Gross primary productivity in the Duke Forest: Modeling synthesis of the free-air CO_2 enrichment experiment and eddy-covariance measurements. *Ecological Applications* 11, 239–252.

Lutze, J. L., Gifford, R. M., and Adams, H. N. (2000) Litter quality and decomposition in *Danthonia richardsonii* swards in response to CO_2 and nitrogen supply over four years of growth. *Global Change Biology* 6, 13–24.

Lynch, J. M., and Whipps, J. M. (1990) Substrate flow in the rhizosphere. *Plant and Soil* 129, 1–10.

Lyr, H., and Hoffmann, G. (1967) Growth rates and growth periodicity of roots. *International Review of Forestry Research* 2, 181–236.

Lytle, D. E, and Cronan, C. S. (1998) Comparative soil CO_2 evolution, litter decay, and root dynamics in clearcut and uncut spruce-fir forest. *Forest Ecology and Management* 103, 121–128.

Ma, S. (2003) Interactions between microclimate, soil respiration, and disturbances in a forest ecosystem: Lessons from the Teakettle experimental forest in California's Sierra Nevada. Ph. D. Thesis, University of Toledo, Toledo, OH.

Maggs, J., and Hewett, B. (1990) Soil and litter respiration in rain forests of contrasting nutrient status and physiognomic structure near Lake Eacham, northeast Queensland. *Australian Journal of Ecology* 15, 329–336.

Magill, A. H., and Aber, J. D. (1998) Long-term effects of experimental nitrogen additions on foliar litter decay and humus formation in forest ecosystems. *Plant and Soil* 203, 301–311.

Magill, A. H., Aber, J. D., Berntson, G. M., McDowell, W. H., Nadelhoffer, K. J., Melillo, J. M., and Steudler, P. (2000) Long-term nitrogen additions and nitrogen saturation in two temperate forests. *Ecosystems* 3, 238–253.

Maier, C. A., and Kress, L. W. (2000) Soil CO_2 evolution and root respiration in 11 year-old loblolly pine (*Pinus taeda*) plantations as affected by moisture and nutrient availability. *Canadian Journal of Forest Research* 30(3), 347–359.

Makarov, B. N. (1958) Diurnal variation in soil respiration and in the carbon dioxide content of the layer of air next to the soil. *Soils and Fertility* 21, NO. 978 (Abstract).

Malhi, Y., Baldocchi, D. D., and Jarvis, P. G. (1999) The carbon balance of tropical, temperate and boreal forests. *Plant, Cell and Environment* 22(6), 715–740.

Maltby, E., and Immirzi, P. (1993) Carbon dynamics in peatlands and other wetland soils, regional and global perspectives. *Chemosphere* 27, 999–1023.

Martinez-Yrizar, A., Burquez, A., Nuñez, S., and Miranda, H. (1999) Temporal and spatial variation of litter production in Sonoran Desert communities. *Plant Ecology* 145(1), 37–48.

Matthews, E. (1997) Global litter production, pools, and turnover times: Estimates from measurement data and regression models. *Journal of Geophysical Research* 102, 18771–18800.

Mattson, K. G., and Smith, H. C. (1993) Detrital organic matter and soil CO_2 efflux in forests regeneration from cutting in West Virginia. *Soil Biology and Biochemistry* 25(9), 1241–1248.

Mattson, K. G. (1995) CO_2 efflux from coniferous forest soils: Comparison of measurement methods and effects of added nitrogen. In *Advances in soil science: Soils and global change* (R. Lal, J. Kimble, E. Levigne, and B. A. Stewart, B.A., eds.), pp. 329–341, Lewis Publishers/CRC Press, Boca Raton, FL.

McCree, K. J. (1970) An equation for the respiration of white clover plants grown under controlled conditions. In *Prediction and measurement of photosynthetic productivity* (I. Stelik, ed.), pp. 221–229, Proc. IBP/PP Technical Meeting, Trebon, Czechoslovakia: PUDOC, Wageningen, the Netherlands.

McCulley, R. L., Archer, S. R., Boutton, T. W., Hons, F. M., and Zuberer, D. A. (2004) Soil respiration and nutrient cycling in wooded communities developing in grassland. *Ecology* 85(10), 2804–2817.

McCulley, R. L., Burke, I. C., Nelson, J. A., Lauenroth, W. K., Knapp, A. K., and Kelly, E. F. (2005) Regional patterns in carbon cycling across the Great Plains of North America. *Ecosystems* 8, 106–121.

McDermit, D. K., and Loomis, R. S. (1981) Elemental composition of biomass and its relation to energy content, growth efficiency and growth yield. *Annals of Botany* 48, 275–290.

McFadden, J. P., Eugster, W., and Chapin III, F. S. (2003) A regional study of the controls on water vapor and CO_2 fluxes in Arctic tundra. *Ecology* 84, 2762–2776.

McGuire, A. D., Melillo, J. M., Kicklighter, D. W., and Joyce, L. A. (1995) Equilibrium responses of soil carbon to climate change: Empirical and process-based estimates. *Journal of Biogeography* 22, 785–796.

McGuire, A. D., Sitch, S., Clein, J. S., Dargaville, R., Esser, G., Foley, J., Heimann, M., Joos, F., Kaplan, J., Kicklighter, D. W., Meier, R. A., Melillo, J. M., B., III, Moore Prentice, I. C., Ramankutty, N., Reichenau, T., Schloss, A., Tian, H., Williams, L. J., and Wittenberg, U. (2001) Carbon balance of the terrestrial biosphere in the twentieth century: Analyses of CO_2, climate and land use effects with four process-based ecosystem models. *Global Biogeochemical Cycles* 15(1), 183–206.

McHale, P. J., Mitchell, M. J., and Bowles, F. P. (1998) Soil warming in a northern hardwood forest, trace gas fluxes and leaf litter decomposition. *Canadian Journal of Forest Research* 28, 1365–1372.

McInerney, M., and Bolger, T. (2000) Temperature, wetting cycles and soil texture effects on carbon and nitrogen dynamics in stabilized earthworm casts. *Soil Biology and Biochemistry* 32, 335–349.

McNeill, J. R., and Winiwarter, A. (2004) Breaking the sod: Humankind, history, and soil. *Science* 304(5677), 1627–1629.

McMichael, B. L., and Burke, J. J. (1998) Soil temperature and root growth. *Hort Science* 33, 947–951.

McNaughton, S. J., Banyikwa, F. F., and McNaughton, M. M. (1998) Root biomass and productivity in a grazing ecosystem: The Serengeti. *Ecology* 79(2), 587–592.

Medina, E., and Zelwer, M. (1972) Soil respiration in tropical plant communities. In *Papers from a symposium on tropical ecology with an emphasis on organic productivity* (P. M. Golley, and F. B. Golley, eds.), pp. 245–269, University of Georgia, Athens, GA.

Medlyn, B. E., Badeck, F. W., De Pury, D. G. G., Barton, C. V. M., Broadmeadow, M., Ceulemans, R., De Angelis, P., Forstreuter, M., Jach, M. E., Kellomäki, S., Laitat, E., Marek, S., Philippot, A., Rey, J., Strassemeyer, K., Laitinen, R., Liozon, B., Portier, P., Roberntz, M., Wang, K., and Jstbid, P. G. (1999) Effects of elevated [CO_2] on photosynthesis in European forest species: A meta-analysis of model parameters. *Plant, Cell and Environment* 22(12), 1475–1495.

Meentemeyer, V., Box, E., and Thompson, R. (1982) World patterns and amounts of terrestrial plant litter production. *Bioscience* 32, 125–128.

Meharg, A. A. (1994) A critical review of labelling techniques used to quantify rhizosphere carbon-flow. *Plant and Soil* 166, 55–62.

Melillo, J. M., Abet, J. D., and Muratore, J. F. (1982) Nitrogen and lignin control of hardwood leaf litter decomposition dynamics. *Ecology* 63, 621–626.

Melillo, J. M., McGuire, D. A., Kicklighter, D. W., Moore III, B., Vorosmarty, C. J., and Schloss, A. L. (1993) Global climate change and terrestrial net primary production. *Nature* 363, 234–240.

Melillo, J. M., Steudler, P. A., Aber, J. D., Newkirk, K., Leu, H., Bowles, F. P., Catricala, C., Magill, A., Ahrens, T., and Morriseau, S. (2002) Soil warming and carbon-cycle feedbacks to the climate system. *Science* 298, 2173–2176.

Melling, L., Hatano, R., and Goh, K. J. (2005) Soil CO_2 flux from three ecosystems in tropical peatland of Sarawak, Malaysia. *Tellus* 57B, 1–11.

Mendham, D. S., O'Connell, A. M., and Grove, T. S. (2002) Organic matter characteristics under native forest, long-term pasture, and recent conversion to Eucalyptus plantations in Western Australia: Microbial biomass, soil respiration, and permanganate oxidation. *Australian Journal of Soil Research* 40(5): 859–872.

Merckx, R., Den Hartog, A., and Van Veen, J. A. (1985) Turnover of root-derived material and related microbial biomass formation in soils of different texture. *Soil Biology and Biochemistry* 17, 565–569.

Merritt, C. (1968) Effect of environment and heredity on the root growth pattern of red pine. *Ecology* 49, 34–40.

Metting, F. B. (1993) Structure and physiological ecology of soil microbial communities. In *Soil microbial ecology: Applications in agricultural and environmental management* (F. B. Metting, ed.), pp. 3–25, Marcel Dekker, New York.

Michelsen, A., Graglia, E., Schmidt, I. K., Jonasson, S., Quarmby, C., and Sleep, D. (1999) Differential responses of grass and dwarf shrubs to long term changes in soil microbial biomass C, N and P, following factorial addition of NPK fertilizer, fungicide and labile carbon to a heath. *New Phytologist* 143, 523–538.

Middelburg, J. J., Klaver, G., Nieuwenhuize, J., Wielemaker, A., deHaas, W., Vlug, T., and van der Nat, J. (1996) Organic matter mineralization in intertidal sediments along an estuarine gradient. *Inter Research-Marine Ecology Progress Series* 132, 157–168.

Mielnick, P. C., and Dugas, W. A. (2000) Soil CO_2 flux in a tallgrass prairie. *Soil Biology and Biochemistry* 32, 221–228.

Mikan, C. J., Schimel, J. P., and Doyle, A. P. (2002) Temperature controls of microbial respiration in Arctic tundra soils above and below freezing. *Soil Biology and Biochemistry* 34, 1785–1795.

Milchunas, D. G., Sala, O. E., and Lauenroth, W. K. (1988) A generalized model of effects of grazing by large herbivores on grassland community structure. *American Naturalist* 132, 87–106.

Minderman, G. (1968) Addition, decomposition, and accumulation of organic matter in forests. *Journal of Ecology* 56, 355–362.

Minderman, G., and Vulto, J. C (1973) Comparison of techniques for the measurement of carbon dioxide evolution from soil. *Pedobiologia* 13, 73–80.

Misson, L., Tang, J., Xu, M., McKay M, and Goldstein, A. (2005) Influences of recovery from clear-cut, climate variability, and thinning on the carbon balance of a young ponderosa pine plantation. *Agricultural and Forest Meteorology* 130, 207–222.

Mitchell, R. J., Runion, G. B., Prior, S. A., Rogers, H. H., Amthor, J. S., and Henning, F. P. (1995) Effects of nitrogen on *Pinus palustris* foliar respiratory responses to elevated atmospheric CO_2 concentration. *Journal of Experimental Botany* 46, 1561–1567.

Moldrup, P., Kruse, C. W., Rolston, D. E., and Yamaguchi, T. (1996) Modeling diffusion and reaction in soils: III, Predicting gas diffusivity from the Campbell soil-water model. *Soil Science* 161, 366–375.

Moldrup, P., Olesen, T., Gamst, J., Schjønning, P., Yamaguchi, T., and Rolston, D. E. (2000a) Predicting the gas diffusion coefficient in undisturbed soil from soil water characteristics. *Soil Science Society of America Journal* 64, 94–100.

Moldrup, P., Olesen, T., Gamst, J., Schjønning, P., Yamaguchi, T., and Rolston, D. E. (2000b) Predicting the gas diffusion coefficient in repacked soil, water-induced linear reduction model. *Soil Science Society of America Journal* 64, 1588–1594.

Möller, J. (1879) Über die freie Kohlensäure im Boden. *Forschende Gebiete-Agricultural Physiology* 2, 329–338.

Monteith, J. L. (1962) Gas exchange in plant communities. In *Environmental control of plant growth* (L. T. Evans, ed.), pp. 95–112, Academic Press, New York.

Monteith, J. L., Sceicz, G., and Yabuky, K. (1964) Crop photosynthesis and the flux of carbon dioxide below the canopy. *Journal of Applied Ecology* 1, 321–337.

Monteith, J. L., and Unsworth, M. H. (1990) *Principles of environmental physics*. 2nd edition. Academic Press/Elsevier, New York.

Mooney, H. A., Drake, B. G., Luxmoore, R. J., Oechel, W. C., and Pitelka, L. F. (1991) Predicting ecosystem responses to elevated CO_2 concentrations. *Bioscience* 41(2), 96–104

Moore, T. R. (1986) Carbon dioxide evolution from subarctic peatlands in eastern Canada. *Arctic Alpine Research* 18, 189–193.

Moore, T. R., and Knowles, R. (1989) The influence of water-table levels on methane and carbon-dioxide emissions from peatland soils. *Canadian Journal of Soil Science* 69(1), 33–38.

Moorhead, K. K., Graetza, D. A., and Reddy, R. K. (1987) Decomposition of fresh and anaerobically digested plant biomass in soil. *Journal of Environmental Quality* 16, 25–28.

Morley, C. R., Trofymow, J. A., Coleman, D. C., and Cambardella, C. (1983) Effects of freeze-thaw stress on bacterial population in soil microcosms. *Microbial Ecology* 9, 329–340.

Murphy, K. L., Burke, I. C., Vinton, M. A., Lauenroth, W. K., Aguiar, M. R., Wedin, D. A., Virginia, R. A., and Lowe, P. N. (2002) Regional analysis of litter quality in the central grassland region of North America. *Journal of Vegetation Science* 13, 395–402.

Nadelhoffer, K. J. (2000) The potential effects of nitrogen deposition on fine-root production in forest ecosystems. *New Phytologist* 147, 131–139.

Nakane, K. (1975) Dynamics of soil organic matter in different parts on a slope under evergreen oak forest. *Japan Journal of Ecology* 25, 204–216

Nakane, K., Yamamoto, M., and Tsubota, H. (1983) Estimation of root respiration rate in a mature forest ecosystem. *Japanese Journal of Ecology* 33, 397–408.

Nakane, K., Tsubota, H., and Yamamoto, M. (1986) Cycling of soil carbon in a Japanese red pine forest: II, Changes occurring in the first year after clearfelling. *Ecological Research* 1, 47–58.

Nakane, K. (1995) Soil carbon cycling in a Japanese cedar (*Cryptomeria japonica*) plantation. *Forest Ecology and Management* 72, 185–197.

Nakane, K., Kohno, T., and Horikoshi, T. (1996) Root respiration rate before and just after clear-felling in a mature, deciduous, broad-leaved forest. *Ecological Research* 11, 111–119.

National Research Council (1986) *Global change in the geosphere-biosphere: Initial priorities for an IGBP*. U.S. Committee for an International Geosphere-Biosphere Program, National Academy Press, Washington, DC.

Nay, S. K., Mattson, K. G., and Bormann, B. T. (1994) Biases of chamber methods for measuring soil CO_2 efflux demonstrated with a laboratory apparatus. *Ecology* 75(8), 2460–2463.

Nay, S. M., and Bormann, B. (2000) Soil carbon changes: Comparing flux monitoring and mass balance in a box lysimeter experiment. *Soil Science Society of America Journal* 64, 943–948.

Neilson, J. W., and Pepper, I. L. (1990) Soil respiration as an index of soil aeration. *Soil Science Society of America Journal* 54, 428–432.

Nielsen, D. R., van Genuchten, M. Th., and Biggar, J. W. (1986) Water flow and solute transport processes in the unsaturated zone. *Water Resource Research* 22, 89–108.

Niinistö, S. M., Silvola, J., and Kellomäki, S. (2004) Soil CO_2 efflux in a boreal pine forest under atmospheric CO_2 enrichment and air warming. *Global Change Biology* 10, 1–14.

Niklaus, P. A., Wohlfender, M., Siegwolf, R., and Körner, C. (2001) Effects of six years atmospheric CO_2 enrichment on plant, soil, and soil microbial C of a calcareous grassland. *Plant and Soil* 233, 189–202.

Nitschelm, J. J., Lüscher, A., Hartwig, U. A., and van Kessel, C. (1997) Using stable isotopes to determine soil carbon input differences under ambient and elevated atmospheric CO_2 conditions. *Global Change Biology* 3, 411–416.

Nobel, P. S., and Palta, J. A. (1989) Soil O_2 and CO_2 effects on root respiration of cacti. *Plant and Soil* 120, 263–271.

Nobel, P. S. (2005) Physiochemical and environmental plant physiology. Elsevier Academic Press, Amsterdam, the Netherlands.

Norby, R. J., O'Neill, E. G., Hood, W. G., and Luxmoore, R. G. (1987) Carbon allocation, root exudation and mycorrhizal colonization of *Pinus echinata* seedlings grown under CO_2 enrichment. *Tree Physiology* 3, 203–210.

Norby, R. J., Cotrufo, M. F., Ineson, P., O'Neill, E. G., and Canadell, J. G. (2001) Elevated CO_2, litter chemistry, and decomposition: A synthesis. *Oecologia* 127, 153–165.

Norby, R. J., Hartz-Rubin, J. S., and Verbrugge, M. J. (2003) Phenological responses in maple to experimental atmospheric warming and CO_2 enrichment. *Global Change Biology* 9, 1792–1801.

Norby, R. J., and Luo, Y. (2004) Evaluating ecosystem responses to rising atmospheric CO_2 and global warming in a multi-factor world. *New Phytologist* 162, 281–293.

Nordt, L. C., Wilding, L. P., and Drees, L. R. (2001) Pedogenic carbonate transformations in leaching soil systems: Implications for the global C cycle. In *Global climate change and pedogenic carbonates* (R. Lal, J. M. Kimble, H. Eswaran, and B. A. Stewart, eds.), pp. 43–63, CRC/Lewis Publishers, Boca Raton, FL.

Norman, J. M., Garcia, R., and Verma, S. B. (1992) Soil surface CO_2 fluxes and the carbon budget of a grassland. *Journal of Geophysical Research-Atmosphere* 97, 18845–18853.

Norman, J. M., Kucharik, C. J., Gower, S. T., Baldocchi, D. D., Crill, P. M., Rayment, M., Savage, K., and Striegl, R. G. (1997) A comparison of six methods for measuring soil-surface carbon dioxide fluxes. *Journal of Geophysical Research*, 102, 28771–28777.

Oads, J. M. (1989) An introduction to organic matter in mineral soils. In *Minerals in soil environmentas* (J. B. Dixon, and S. B. Weed, eds.), pp. 89–159, Soil Science Society of America Inc., Madison, WI.

Oberbauer, S. F., Gillespie, C. T., Cheng, W., Gebauer, R., Sala Serra, A., and Tenhunen, J. D. (1992) Environmental effects of CO_2 efflux from riparian tundra in the northern foothills of the Brooks Range, Alaska. *Oecologia* 92, 568–577.

O'Connell, A. M. (1987) Litter decomposition, soil respiration and soil chemical and biochemical properties at three contrasting sites in karri (*Eucalyptus diversicolor* F. Muell.) forests of south-western Australia. *Australian Journal of Ecology* 12, 31–40.

O'Connell, A. M. (1990) Microbial decomposition (respiration) of litter in eucalypt forests of south-western Australia: An empirical model based on laboratory incubations. *Soil Biology and Biochemistry* 22, 153–160.

Oechel, W. C., Vourlitis, G. L., Hasting, S. J., and Bochkarev, S. A. (1995) Change in Arctic CO_2 flux over two decades, Effects of climate change at Barrow, Alaska, *Ecological Applications* 5(3), 846–855.

Oechel, W. C., Vourlitis, G. L., and Hastings, S. J. (1997) Cold season CO_2 emission from Arctic soil. *Global Biogeochemical Cycles* 11, 163–172.

Ohashi, M., and Satio, A. (1998) Problems of plant roots: Methods and ecological significance of root respiration measurement. *Agriculture and Horticulture* 73, 67–71.

Ohashi, M., Gyokusen, K., and Saito, A. (1999) Measurement of carbon dioxide evolution from a Japanese cedar (*Cryptomeria japonica* D. Don) forest floor using an open-flow chamber method. *Forest Ecology and Management* 123, 105–114.

Ohashi, M., Gyokusen, K., and Saito, A. (2000) Contribution of root respiration to total soil respiration in a Japanese cedar (*Cryptomeria japonica* D. Don) artificial forest. *Ecological Research* 15, 323–333.

Ohtonen, R., and Vare, H. (1998) Vegetation composition determines microbial activities in a boreal forest soil. *Microbial Ecology* 36(3), 328–335.

O'Leary, M. H. (1988) Carbon isotopes in photosynthesis. *Bioscience* 38, 325–336.

Olson, J. S. (1963) Energy storage and the balance of producers and decomposers in ecological systems. *Ecology* 44, 322–331.

Olsson, P., Linder, S., Giesler, R., and Högberg, P. (2005) Fertilization of boreal forest reduces both autotrophic and heterotrophic soil respiration. *Global Change Biology* 11, 1745–1753.

O'Neill, K. P., Kasischke, E. S., and Richter, D. D. (2002) Environmental controls on soil CO_2 flux following fire in black spruce, white spruce, and aspen stands of interior Alaska. *Canadian Journal of Forest Research* 32, 1525–1541.

O'Neill, K. P., Kasischke, E. S., and Richter, D. D. (2003) Seasonal and decadal patterns of soil carbon uptake and emission along an age sequence of burned black spruce stands in interior Alaska. *Journal of Geophysical Research-Atmospheres* 108 (D1), 8155, doi: 10.1029/2001 JD000443.

Orchard, V. A., and Cook, F. (1983) Relationship between soil respiration and and soil moisture. *Soil Biology and Biochemistry* 15, 447–453.

Oren, R., Ellsworth, D. S., Johnson, K. H., Phillips, N., Ewers, B. E., Maier, C., Schäfer, K. V. R., McCarthy, H., Hendrey, G., McNulty, S. G., and Katul, G. G. (2001) Soil fertility limits carbon sequestration by a forest ecosystem in a CO_2-enriched atmosphere. *Nature* 411, 469–472.

Ouyang, Y., and Boersma, L. (1992a) Dynamic oxygen and carbon dioxide exchange between soil and atmosphere: I, Model development. *Soil Science Society of America Journal* 56, 1695–1702.

Ouyang, Y., and Boersma, L. (1992b) Dynamic oxygen and carbon dioxide exchange between soil and atmosphere: II, Model simulation. *Soil Science Society of America Journal* 56, 1702–1710.

Ouyang, Y., and Zheng, C. (2000) Surficial processes and CO_2 flux in soil ecosystem. *Journal of Hydrology* 234, 54–70.

Ovenden, L. (1990) Peat accumulation in northern wetlands. *Quarterly Research* 33, 377–386.

Pajari, B. (1995) Soil respiration in a poor upland site of Scots pine stand subjected to elevated temperatures and atmospheric carbon concentration. *Plant and Soil* 169, 563–570

Palta, J. A., and Nobel, P. S. (1989) Root respiration for *Agave deserti*: Influence of temperature, water status and root age on daily patterns. *Journal of Experimental Botany* 40: 181–186.

Papendick, R. I., and Campbell, G. S. (1981) Theory and measurement of water potential. In *Water potential relations in soil microbiology* (J. F. Parr, W. R. Gardner, and L. F. Elliott, eds.), pp. 1–22, Soil Science Society of America, Special Publication No. 9, Madison, WI.

Parkin, T. B., and Kaspar, T. C. (2004) Temporal variability of soil carbon dioxide flux: Effect of sampling frequency on cumulative carbon loss estimation. *Soil Science Society of America Journal* 68, 1234–1241.

Parsons, A. N., Barrett, J. E., Wall, D. H., and Virginia, R. A. (2004) Soil carbon dioxide flux in Antarctic dry valley ecosystems. *Ecosystems*, 7, 286–295.

Parton, W. J., Anderson, D. W., Cole, C. V., and Stewart, J. W. B. (1983) Simulation of soil organic matter formations and mineralization in semiarid agroecosystems. In *Nutrient Cycling in Agricultural Ecosystems* (R. R. Lowrance, R. L. Todd, L. E. Asmussen, and R. A. Leonard, eds.), pp. 533–550, Special Publication No. 23, University of Georgia, College of Agricultural Experiment Stations, Athens, GA.

Parton, W. J., Schimel, D. S., Cole, C. V., and Ojima, D. S. (1987) Analysis of factors controlling soil organic matter levels in Great Plains grasslands. *Soil Science Society of America Journal* 51, 1173–1179.

Parton, W. J., Schimel, D., Ojima, D., and Cole, C. (1994) A general study model for soil organic model dynamics, sensitivity to litter chemistry, texture, and management. *Soil Science Society of America Special Publication* 39, 147–167.

Parton, W. J., Scurlock, J. M. O., Ojima, D. S., Schimel, D. S., Hall, D. O., and SCOPE GRAM group members (1995) Impact of climate change on grassland production and soil carbon worldwide. *Global Change Biology* 1, 13–22.

Pastor, J., and Post, W. M. (1986) Influence of climate, soil moisture, and succession on forest carbon and nitrogen cycles. *Biogeochemistry* 2, 3–27.

Pastor, J., and Post, W. M. (1988) Response of northern forests to CO_2-induced climate change. *Nature* 334 (6177), 55–58.

Pataki, D. E., Ellsworth, D. S., Evans, R. D., Gonzalez-Meler, M., King, J., Leavitt, S. W., Lin, G., Matamala, R., Pendall, E., Siegwolf, R., Kessel, C. V., and Ehleringer, J. R. (2003) Tracing changes in ecosystem function under elevated carbon dioxide conditions. *Bioscience* 53(9), 805–818.

Pate, J. S., and Layzell, D. B. (1990) Energetics and biological costs of nitrogen assimilation. In *The biochemistry of plants, Vol. 16: Intermediary nitrogen metabolism* (B. J. Miflin, ed.), pp. 1–42, Academic Press, San Diego, CA.

Paterson, E., Hall, J. M., Rattray, E. A. S., Grifiths, B. S., Ritz, K., and Killham, K. (1997) Effect of elevated CO_2 on rhizosphere carbon flow and soil microbial processes. *Global Change Biology* 3, 363–377.

Pati, D. P., Behera, N., and Dash, M. C. (1983) Microbial and root contribution to total soil metabolism in a tropical grassland soil from Orissa, India. *Revue d' Ecologie et de Biologie du Sol* 20, 183–190.

Pattey, E., Desjardins, R. L., Boudreau, F., and Rochette, P. (1992) Impact of density fluctuations on flux measurements of trace gases: Implications for the relaxed eddy accumulation technique. *Boundary Layer Meteorology* 59, 195–203.

Pattey, E., Desjardins, R. L., and Rochette, P. (1993) Accuracy of the relaxed eddy-accumulation technique, evaluated using CO_2 flux measurements. *Boundary Layer Meteorology* 66, 341–355.

Patwardhan, A. S., Nieber, J. L., and Moore, I. D. (1988) Oxygen, carbon-dioxide, and water transfer in soils: Mechanisms and crop response. *Transactions of the ASAE* 31(5), 1383–1395.

Paul, E. A., and Clark, F. E. (1996) *Soil Microbiology and Biochemistry.* 2nd edition, Academic Press/Elsevier, San Diego, CA.

Paustian, K., Andrén, O., Clarholm, M., Hansson, A. C., Johansson, G., Lagerlöf, J., Lindberg, T., Pettersson, R., and Sohlenius, B. (1990) Carbon and nitrogen budgets of four agroecosystems with annual and perennial crops, with and without N fertilization. *Journal of Applied Ecology* 27, 60–84.

Paustian, K., Ågren, G. I., and Bosatta, E. (1997) Modelling litter quality effects on decomposition and soil organic matter dynamics. In *Driven by nature: Plant litter quality and decomposition* (G. Cadisch, and K. E. Giller, eds.), pp. 313–335, CAB International, Wallingford, UK.

Paw, U. K. T., Qiu, J., Su, H., Watanabe, T., and Brunet, Y. (1995) Surface renewal analysis: A new method to obtain scalar fluxes. *Agricultural and Forest Meteorology* 74, 119–137.

Payne, D., and Gregory, P. J. (1988) The temperature of the soil. In *Russell's soil conditions and plant growth* (A. Wild, ed.), pp. 282–297, 11th edition, Longman Scientific and Technical, Harlow, UK.

Pendall, E., Grosso, S. D., King, J. Y., Le Cain, D. R., Milchunas, D. G., Morgan, J. A., Mosier, A. R., Ojima, D., Parton, W. A., Tans, P. P., and White, J. W. C. (2003) Elevated atmospheric CO_2 effects and soil water feedbacks on soil respiration components in a Colorado grassland. *Global Biogeochemical Cycles* 17(2), doi: 10.1029/2001GB001821.

Penning de Vries, F. W. T. (1975) The cost of maintenance processes in plant cells. *Annals of Botany* 39, 77–92.

Pera, A., Vallini, G., Sireno, I., Bianchin, M. L., and Debertoldi, M. (1983) Effect of organic-matter on rhizosphere microorganisms and root development of sorghum plants in 2 different soils. *Plant and Soil* 74(1): 3–18.

Persson, T., Lundkvist, H., Wiren, A., Hyvönen, R., and Wessen, B. (1989) Effects of acidification and liming on carbon and nitrogen mineralisation and soil organisms in mor humus. *Water, Air and Soil Pollution* 45, 77–96.

Peterjohn, W. T., Melillo, J. M., and Bowles, S. T. (1993) Soil warming and trace gas fluxes: Experimental design and preliminary flux results. *Oecologia* 93, 18–24.

Peterjohn, W. T., Melillo, J. M., Steudler, P. A., Newkirk, K. M., Bowles, F. P., and Aber, J. D. (1994) Responses of trace gas fluxes and N availability to experimentally elevated soil temperatures. *Ecological Applications* 4, 617–625.

Pillers, M. D., and Stuart, J. D. (1993) Leaf litter accretion and decomposition in interior and coastal redwood stands. *Canadian Journal of Forest Research* 7, 680–699.

Pinck, L. A., Allison, F. E., and Sherman, M. S. (1950) Maintenance of soil organic matter: II, Losses of carbon and nitrogen from young and mature plant materials during decomposition in soil. *Soil Science* 69, 391–401.

Pinto, A. S., Bustamante, M. M. C., Kisselle, K., Burke, R., Zepp, R., Viana, L. T., Varella, R. F., and Molina, M. (2002) Soil emissions of N_2O, NO, and CO_2 in Brazilian savannas: Effects of vegetation type, seasonality, and prescribed fires. *Journal of Geophysical Research-Atmospheres* 107(D20), 8089, doi: 10.1029/2001JD000342.

Poff, R. J. (1996) Effects of silvicultural practices and wildfire on productivity of forest soils. In *Sierra Nevada ecosystem project, final report to Congress, Volume II: Assessments and scientific basis for management options*, pp. 477–495, University of California-Davis, CA.

Pol-van Dasselaar, A., Corre, W, J., Prieme, A., Klemedtsson, A. K., Weslien, P., Stein, A., Klemedtsson, L., and Oenema, O. (1998) Spatial variability of methane, nitrous oxide, and carbon dioxide emissions from drained grasslands. *Soil Science Society of America Journal* 62, 810–817.

Post, W. M., Emanuel, W. R., Zinke, P. J., and Stangenberger, A. G. (1982) Soil carbon pools and world life zones. *Nature* 298, 156–159.

Potter, C. S., Randerson, J. T., Field, C. B., Matson, P. A., Vitousek, P. M., Mooney, H. A., and Klooster, S. A. (1993) Terrestrial ecosystem production: A process model based on global satellite and surface data. *Global Biogeochemical Cycles* 7, 811–841.

Pregitzer, K. S., Laskowski, M. J., Burton, A. J., Lessard, V. C., and Zak, D. R. (1998) Variation in sugar maple root respiration with root diameter and soil depth. *Tree Physiology* 18, 665–670.

Pregitzer, K. S., King, J. S., Burton, A. J., and Brown, S. E. (2000) Responses of tree fine roots to temperature. *New Phytologist* 147, 105–115.

Pregitzer, K. S. (2003) Woody plants, carbon allocation and fine roots. *New Phytologist* 158, 421–423.

Pregitzer, K. S., and Euskirchen, E. S. (2004) Carbon cycling and storage in world forests: Biome patterns related to forest age. *Global Change Biology* 10(12), 2052–2077.

Prescott, C. E., Zabek, L. M., Staley, C. L., and Kabzems, R. (2000) Decomposition of broadleaf and needle litter in forests of British Columbia: Influences of litter type, forest type, and litter mixtures. *Canadian Journal of Forest Research* 30(11), 1742–1750.

Price, M. V., and Waser, N. M. (1998) Effects of experimental warming on plant reproductive phenology in a subalpine meadow. *Ecology* 79, 1261–1271.

Pritchett, W. L., and Fisher, R. F. (1987) *Properties and management of forest soil.* 2nd edition, John Wiley & Sons, New York.

Pumpanen, J., Ilvesniemi, H., Keronen, P., Nissinen, A., Pohja, T., Vesala, T., and Hari, P. (2001) An open chamber system for measuring soil surface CO_2 efflux: Analysis of error sources related to the chamber system. *Journal of Geophysical Research-Atmosphere* 106(D8), 7985–7992.

Pumpanen, J., Ilvesniemi, H., and Hari, P. (2003) A process-based model for predicting soil carbon dioxide efflux and concentration. *Soil Science Society of America Journal* 67(2), 402–413.

Pumpanen, J., Kolari, P., Ilvesniemi, H., Minkkinen, K., Vesala, T., Niinisto, S., Lohila, A., Larmola, T., Morero, M., Pihlatie, M., Janssens, I., Yuste, J. C., Grunzweig, J. M., Reth, S., Subke, J. A., Savage, K., Kutsch, W., Ostreng, G., Ziegler, W., Anthoni, P., Lindroth, A., Hari, P. (2004) Comparison of different chamber techniques for measuring soil CO_2 efflux. *Agricultural and Forest Meteorology* 123(3–4), 159–176.

Qi, J., Marshall, J. D., and Mattson, K. G. (1994) High soil carbon dioxide concentrations inhibit root respiration of Douglas fir. *New Phytologist* 128, 435–442.

Qi, Y., Xu, M., and Wu, J. (2002) Temperature sensitivity of soil respiration and its effects on ecosystem carbon budget: nonlinearity begets surprises. *Ecological Modeling* 153, 131–142.

Raguotis, A. D. (1967) Biological activity of sod-podzolic forest soils of Lithuanian SSR. *Soviet Soil Science*-USSR (6), 751.

Raich, J. W., and Nadelhoffer, K. J. (1989) Belowground carbon allocation in forest ecosystems: Global trends. *Ecology* 70(5), 1346–1354.

Raich, J. W., Rastetter, E. B., Melillo, J. M., Kicklighter, D. W., Steudler, P. A., Peterson, B. J., Grace, A. L., Moore, B. I., and Vörösmarty, C. J. (1991) Potential net primary productivity in South America: Application of a global model. *Ecological Applications* 1, 399–429.

Raich, J. W., and Schlesinger, W. H. (1992) The global carbon dioxide flux in soil respiration and its relationship to vegetation and climate. *Tellus* 44B, 81–99.

Raich, J. W., and Potter, C. S. (1995) Global patterns of carbon dioxide emissions from soils. *Global Biogeochemical Cycles* 9, 23–36.

Raich, J. W. (1998) Aboveground productivity and soil respiration in three Hawaiian rain forests. *Forest Ecology and Management* 107, 309–318.

Raich, J. W., and Tufekcioglu, A. (2000) Vegetation and soil respiration: Correlations and controls. *Biogeochemistry* 48, 71–90.

Raich, J. W., Potter, C. S., and Bhagawati, D. (2002) Interannual variability in global soil respiration, 1980–94. *Global Change Biology* 8, 800–812.

Rao, D. L. N., and Pathak, H. (1996) Ameliorative influence of organic matter on biological activity of salt affected soils. *Arid Soil Research and Rehabilitation* 10, 311–319.

Rastetter, E. B., Ryan, M. G., Shaver, G. R., Melillo, J. M., Nadelhoffer, K. J., Hobbie, J. E., and Aber, J. D. (1991) A general biogeochemical model describing the responses of the carbon and nitrogen cycles in terrestrial ecosystems to changes in carbon dioxide, climate and nitrogen deposition. *Tree Physiology* 9, 101–126.

Rastetter, E. B., Ågren, G. I., and Shaver, G. R. (1997) Responses of N-limited ecosystems to increased CO_2: A balanced-nutrition, coupled-element-cycles model. *Ecological Applications* 7, 444–460.

Ratkowsky, D. A., Olley, J., McMeekin, T. A., and Ball, A. (1982) Relationship between temperature and growth rate of bacterial cultures. *Journal of Bacteriology* 149, 1–5.

Raupach, M. R. (1987) A Lagrangian analysis of scalar transfer in vegetation canopies. *Quarterly Journal of the Royal Meteorological Society* 113, 107–120.

Raupach, M. R. (1989a) A practical Lagrangian method for relating scalar concentrations to source distributions in vegetation canopies. *Quarterly Journal of the Royal Meteorological Society* 115, 609–632.

Raupach, M. R. (1989b) Applying Lagrangian fluid mechanics to infer scalar source distributions from concentration profiles in plant canopies. *Agricultural Forest and Meteorology* 47, 85–108.

Rayment, M. B., and Jarvis, P. G. (1997) An improved open chamber system for measuring soil CO_2 effluxes of a boreal black spruce forest. *Journal of Geophysiological System-Atmosphere* 102, 28779–28784.

Rayment, M. B. (2000) Investigating the role of soils in terrestrial carbon balance: Harmonizing methods for measuring soil CO_2 efflux. LESC exploratory workshop, Edinburgh, 6–8 April, 2000, European Science Foundation. http//www.esf.org/generic/163/2073aappitem5.1a.pdf.

Rayment, M. B., and Jarvis, P. G. (2000) Temporal and spatial variation of soil CO_2 efflux in a Canadian boreal forest. *Soil Biology and Biochemistry* 32, 35–45.

Reich, P. B., Walters, M. B., Tjoelker, M. G., Vanderklein, D., and Buschena, C. (1998) Photosynthesis and respiration rates depend on leaf and root morphology and nitrogen concentration in nine boreal tree species different in relative growth rate. *Functional Ecology* 12, 395–405.

Reich, P. B., Knops, J., Tilman, D., Craine, J., Ellsworth, D., Tjoelker, M., Lee, T., Wedin, D., Naeem, S., Bahauddin, D., Hendrey, G., Jose, S., Wrage, K., Goth, J., and Bengston, W. (2001) Plant diversity enhances ecosystem responses to elevated CO_2 and nitrogen deposition. *Nature* 411, 809–824.

Reichstein, M., Tenhunen, J. D., Roupsard, O. *et al.* (2002) Ecosystem respiration in two Mediterranean evergreen Holm oak forests: Drought effects and decomposition dynamics. *Functional Ecology* 16, 27–39.

Reichstein, M., Rey, A., Freibauer, A., Tenhunen, J., Valentini, R., Banza, J., Casals, P., Cheng, Y., Grünzweig, J. M., Irvine, J., Joffre, R., Law, B. E., Loustau, D., Miglietta, F., Oechel, W., Ourcival, J. M., Pereira, J. S., Peressotti, A., Ponti, F., Qi, Y., Rambal, S., Rayment, M., Romanya, J., Rossi, F., Tedeschi, V., Tirone, G., Xu, M., and Yakir, D. (2003) Modeling temporal and large-scale spatial variability of soil respiration from soil water availability, temperature and vegetation productivity indices. *Global Biogeochemical Cycles* 17(4), 1104 doi: 10.1029/2003GB002035.

Reichstein, M., Subke, J. A., Angeli, A. C., and Tenhunen, J. D. (2005) Does the temperature sensitivity of decomposition of soil organic matter depend upon water content, soil horizon, or incubation time? *Global Change Biology* 11, 1754–1767.

Reiners, W. A. (1968) Carbon dioxide evolution from the floor of three Minnesota forests. *Ecology* 49, 471–483.

Reinke, J. J., Adriano, D. C., and Mcleod, K. W. (1981) Effects of litter alteration on carbon dioxide evolution from a South Carolina pine forest floor. *Soil Science Society of America Journal* 45, 620–623.

Repnevskaya, M. A. (1967) Liberation of CO_2 from soil in the pine stands of the Kola Peninsula. *Soviet Soil Science* 68, 1067–1072.

Resh, S. C., Binkley, D., and Parrotta, J. A. (2002) Greater soil carbon sequestration under nitrogen-fixing trees compared with *Eucalyptus* species. *Ecosystems* 5, 217–231.

Rey, A., Pegoraro, E., Tedeschi, V., Parri, I. D., Jarvis, P. G., and Valentini, R. (2002) Annual variation in soil respiration and its components in a coppice oak forest in central Italy. *Global Change Biology* 8, 851–866.

Rice, C. W., and Smith, M. S. (1982) Denitrification in no-till and plowed soil. *Soil Science Society of America Journal* 17, 11–16.

Richards, B. N. (1974) *Introduction to the soil microbiology.* John Wiley & Sons, New York.

Richards, B. N. (1987) *The microbiology of terrestrial ecosystems.* John Wiley & Sons, New York.

Richardson, A. D., and Hollinger, D. Y. (2005) Statistical modeling of ecosystem respiration using eddy covariance data: Maximum likelihood parameter estimation, and Monte Carlo simulation of model and parameter uncertainty, applied to three simple models. *Agricultural Forest and Meteorology* 131, 191–208.

Richey, J. E., Melack, J. M., Aufdenkampe, A. K., Ballester, V. M., and Hess, L. L. (2002) Outgassing from Amazonian rivers and wetlands as a large tropical source of atmospheric CO_2. *Nature* 416(6881), 617–620.

Richter, D. D., and Markewitz, D. (1995) How deep is soil? *Bioscience* 45(9), 600–609.

Rillig, M. C. (2004) Arbuscular mycorrhizae and terrestrial ecosystem processes. *Ecology Letters* 7, 740–754.

Risk, D., Kellman, L., and Beltrami, H. (2002a) Soil CO_2 production and surface flux at four climate observations in eastern Canada. *Global Biogeochemical Cycles* 16(4), 1122, doi: 10.1029/2001GB001831.

Risk, D., Kellman, L., and Beltrami, H. (2002b) Carbon dioxide in soil profiles, production and temperature dependence. *Geophysical Research Letters* 29(6), doi: 10.1029/2001GL014002.

Robinson, D., and Scrimgeour, C. M. (1995) The contribution of plant C to soil CO_2 measured using $\delta^{13}C$. *Soil Biology and Biochemistry* 27, 1653–1656.

Rochette, P., Desjardins, R. L., and Pattey, E. (1991) Spatial and temporal variability of soil respiration in agricultural fields. *Canadian Journal of Soil Science* 71, 189–196.

Rochette, P., Gregorich, E. G., and Desjardins, R. L. (1992) Comparison of static and dynamic closed chambers for measurement of soil respiration under field conditions. *Canadian Journal of Soil Science* 72, 605–609.

Rochette, P., Flanagan, L. B., and Gregorich, E. G. (1999) Separating soil respiration into plant and soil components using analyses of the natural abundance of carbon-13. *Soil Science Society of America Journal* 63, 1207–1213.

Rochette, P., and Hutchinson, G. L. (2003) Measurement of soil respiration *in situ*: Chamber techniques. In *Micrometeorological measurements in agricultural systems* (J. L. Hatfield, ed.), ASA, CSSA, and SSSA, Madison WI.

Rodeghiero, M., and Cescatti, A. (2005) Main determinants of forest soil respiration along an elevation/temperature gradient in the Italian Alps. *Global Change Biology* 11(7), 1024–1041.

Roehm, C. L. (2005) Respiration in wetland ecosystems. In *Respiration in aquatic ecosystem* (P. A. Giorgio, and P. J. I. B. Williams, eds.), pp. 93–102, Oxford University Press, New York.

Rogers, H. H., Prior, S. A., Runion, G. B., and Mitchell, R. J. (1996) Root to shoot ratio of crops as influenced by CO_2. *Plant and Soil* 187, 229–248.

Rolston, D. E. (1986) Methods of soil analysis part 1. In *Physical and mineralogical methods* (A. Klute, ed.), pp. 1103–1119, 2nd edition, American Society of Agronomy and Soil Science Society of America, Madison, WI.

Rouhier, H., Billès, G. A., El Kohen, A., Mousseau, M., and Bottner, P. (1994) Effect of elevated CO_2 on carbon and nitrogen distribution within a tree (*Castanea sativa* mill.) soil system. *Plant and Soil* 162, 281–292.

Rouhier, H., Billès, G., Billès, L., and Bottner, P. (1996) Carbon fluxes in the rhizosphere of sweet chestnut seedlings (*Castanea sativa*) grown under two atmospheric CO_2 concentrations, ^{14}C partitioning after pulse labelling. *Plant and Soil* 180, 101–111.

Ruddiman, W. F. (2003) The anthropogenic greenhouse era began thousands of years ago. *Climatic Change* 61(3), 261–293.

Rudolph, T. D., and Laidly, P. R. (1990) *Pinus banksiana* Lamb. (Jack Pine). In *Silvics of North America, Vol. 1: Conifers* (R. M. Burns, and B. H. Honkala, eds.), pp. 280–293, USDA Forest Service Agricultural Handbook, Washington, DC.

Ruess, R. W., and Seagle, S. W. (1994) Landscape patterns in soil microbial processes in the Serengeti National Park, Tanzania. *Ecology* 75(4), 892–904.

Ruess, R. W., Hendrick, R. L., and Bryant, J. P. (1998) Regulation of fine root dynamics by mammalian browsers in early successional Alaskan taiga forests. *Ecology* 79(8), 2706–2720.

Ruess, L., Michelsen, A., Schmidt, I. K., and Jonasson, S. (1999) Simulated climate change affecting microorganisms, nematode density and biodiversity in subarctic soils. *Plant and Soil* 212, 63–73.

Rühlmann, J. (1999) A new approach to estimating the pool of stable organic matter in soil using data from long-term field experiments. *Plant and Soil* 213(1–2), 149–160.

Running, S. W., Nemani, R. R., and Hungerford, R. D. (1987) Extrapolation of synoptic meteorological data in mountainous terrain and its use for simulating forest evapotranspiration and photosynthesis. *Canadian Journal of Forest Research* 17, 472–483.

Running, S. W., and Coughlan, J. C. (1988) A general model of forest ecosystem process for regional applications: I, Hydrologic balance, canopy gas exchange and primary production processes. *Ecological Applications* 42, 125–154.

Running, S. W., and Gower, S. T. (1991) Forest BGC: A general model of forest ecosystem processes for regional applications: II, Dynamic carbon allocation and nitrogen budgets. *Tree Physiology* 9, 147–160.

Russell, E. J., and Appleyard, A. (1915) The atmosphere of the soil, its composition and the causes of variation. *Journal of Agricultural Science* 7, 1–48.

Russell, C. A., and Voroney, R. P. (1998) Carbon dioxide efflux from the floor of a boreal aspen forest: I, Relationship to environmental variables and estimates of C respired. *Canadian Journal of Soil Science* 78, 301–310.

Rustad, L. E., and Fernandez, I. J. (1998) Experimental soil warming effects on CO_2 and CH_4 flux from a low elevation spruce-fir forest soil in Maine, U.S.A. *Global Change Biology* 4, 597–605.

Rustad, L. E., Melillo, J. M., Mitchell, M. J., Fernandez, I. J., Steudler, P. A., and McHale, P. J. (2000) Effects of soil warming on C and N cycling in northern U.S. forest soils. In *Responses of northern U.S. forests to environmental change* (R. Mickler, R. Birdsey, and J. Hom, eds.), Springer-Verlag, New York.

Rustad, L. E., Campbell, J. L., Marion, G. M., Norby, R. J., Mitchell, M. J., Hartley, A. E., and Gurevitch, J. (2001) A meta-analysis of the response of soil respiration, net nitrogen mineralization, and aboveground plant growth to experimental ecosystem warming. *Oecologia* 126, 543–562.

Ryan, M. G. (1991) The effects of climate change of plant respiration. *Ecological Applications* 1, 157–167.

Ryan, M. G. (1995) Foliar maintenance respiration of subalpine and boreal trees and shrubs in relation to nitrogen content. *Plant, Cell and Environment* 18, 765–772.

Ryan, M. G., Hubbard, R. M., Pongracic, S., Raison, R. J., and McMurtrie, R. E. (1996) Foliage, fine-root, woody-tissue and stand respiration in *Pinus radiata* in relation to nitrogen status. *Tree Physiology* 16(3), 333–343.

Rygiewicz, P. T., and Andersen, C. P. (1994) Mycorrhizae alter quality and quantity of carbon allocated belowground. *Nature* 369, 58–60.

Sabine, C. S., Hemann, M., Artaxo, P., Bakker, D., Chen, C. T. A., Field, C. B., Gruber, N., Le Quere, C., Prinn, R. G., Richey, J. E., Romero-Lankao, P., Sathaye, J., and Valentini, R. (2003) Current status and past trends of the carbon cycle. In *Toward CO_2 stabilization: Issues, strategies, and consequences* (C. B. Field, and M. R. Raupac, eds.), Island Press, Washington, DC.

Saggar, S., and Hedley, C.B. (2001) Estimating seasonal and annual carbon inputs and root decomposition rates in a temperate pasture following field ^{14}C pulse-labeling. *Plant and Soil* 236, 91–103.

Saiya-Cork, K. R., Sinsabaugh, R. L., and Zak, D. R. (2002) The effects of long term nitrogen deposition on extracellular enzyme activity in an *Acer saccharum* forest soil. *Soil Biology Biochemistry* 34, 1309–1315.

Saleska, S. R., Shaw, M. R., Fischer, M. L., Dunne, J. A., Still, C. J., Holman, M. L., and Harte, J. (2002) Plant community composition mediates both large transient decline and predicted long-term recovery of soil carbon under climate warming. *Global Biogeochemical Cycles*, 16 (4), 1055, doi: 10.1029/2001GB001573.

Salimon, C. I., Davison, E. A., Victoria, R. L., and Melo, A. W. F. (2004) CO_2 flux from soil in pastures and forests in southwestern Amazonia. *Global Change Biology* 10, 1–11.

Saglio, P. H., and Pradet, A. (1980) Soluble sugars, respiration, and energy charge during aging of excised maize root tips. *Plant Physiology* 66, 516–519.

Sanderman, J., Amundson, R. G., Baldocchi, D. D. (2003) Application of eddy covariance measurements to the temperature dependence of soil organic matter mean residence time. *Global Biogeochemical Cycles* 17, 1061, doi:10.1029/2001GB001833.

Šantrůčková, H., Bird, M. I., Kalaschnikov, Y. N., Grund, M., Elhottova, D., Simek, M., Grigoryev, S., Gleixner, G., Arneth, A., Schulze, E. D., and Lloyd, J. (2003) Microbial characteristics of soils on a latitudinal transect in Siberia. *Global Change Biology* 9(7), 1106–1117.

Sanz, M. J., Schulze, E.-D., and Valentini, R. (2004) International policy framework on climate change, Sinks in recent international agreements. In *The global carbon cycle: Integrating humans, climate, and the natural world* (SCOPE series 62) (C. B. Field, and M. R. Raupach, eds.), Island Press, Washington, DC.

Sauerbeck, D. R. (2001) CO_2 emissions and C sequestration by agriculture: Perspectives and limitations. *Nutrient Cycling in Agroecosystems* 60(1–3), 253–266.

Saunders, S. C., Chen, J., Drummer, T. D., Crow, T. R., Brosofske, K. D., and Gustafson, E. J. (2002) The patch mosaic and ecological decomposition across spatial scales in a managed landscape of northern Wisconsin, USA. *Basic and Applied Ecology* 3, 49–64.

Savage, K. E., and Davidson, E. A. (2001) Interannual variation of soil respiration in two New England forests. *Global Biogeochemical Cycles* 15(2), 337–350.

Saviozzi, A., Levi-Minzi, R., Cardelli, R., and Riffaldi, T. (2001) A comparison of soil quality in adjacent, cultivated forest and native grassland soils. *Plant and Soil* 233, 251–259.

Sawamoto, T., Hatano, R., Yajima, R., Takahashi, K., and Isaev, A. P. (2000) Soil respiration in Siberian taiga ecosystems with different histories of forest fire. *Soil Science and Plant Nutrition* 46(1), 31–42.

Schery, S. D., Gaeddert, D. H., and Wilkening, M. H. (1984) Factors affecting exhalation of radon from a gravelly sandy loam. *Journal of Geophysical Research-Atmosphere* 89, 7299–7309.

Schilling, E. B., Lockaby, B. G., and Rummer, R. (1999) Belowground nutrient dynamics following three harvest intensities on the pearl river floodplain, Mississippi. *Soil Science Society of America Journal* 63, 1856–1868.

Schimel, D. S., Braswell, B. H., Holland, A. B., McKeown, R., Ojima, D. S., Painter, T. H., Parton, W. J., and Townsend, A. R. (1994) Climatic, edaphic, and biotic controls over the storage and turnover of carbon in soils. *Global Biogeochemistry Cycles* 8, 279–293.

Schimel, D. S. (1995) Terrestrial ecosystems and the carbon cycle, *Global Change Biology* 1, 77–91.

Schimel, J. P., and Clein, J. S. (1996) Microbial response to freeze-thaw cycles in tundra and taiga soils. *Soil Biology and Biochemistry* 28, 1061–1066.

Schimel, D. S., House, J. I., Hibbard, K. A., Bousquet, P., Ciais, P., Peylin, P., Braswell, B. H., Apps, M. J., Baker, D., Bondeau, A., Canadell, J., Churkina, G., Cramer, W., Denning, A. S., Field, C. B., Friedlingstein, P., Goodale, C., Heimann, M., Houghton, R. A., Melillo, J. M., Moore, B., Murdiyarso, D., Noble, I., Pacala, S. W., Prentice, I. C., Raupach, M. R., Rayner, P. J., Scholes, R. J., Steffen, W. L., and Wirth, C. (2001) Recent patterns and mechanisms of carbon exchange by terrestrial ecosystems. *Nature* 414(6860), 169–172.

Schjønning, P., Thomsen, I. K., Moldrup, P., and Christensen, B. T. (2003) Linking soil microbial activity to water- and air-phase contents and diffusivities. *Soil Science Society of America Journal* 67, 156–165.

Schlentner, R. E., and Van Cleve, K. (1985) Relationships between CO_2 evolution from soil, substrate temperature, and substrate moisture in four mature forest types in interior Alaska. *Canadian Journal of Forest Research* 15, 97–106.

Schlesinger, W. H. (1977) Carbon balance in terrestrial detritus. *Annual Review of Ecology Systematics* 8, 51–81.

Schlesinger, W. H. (1997) *Biogeochemistry: An analysis of global change*. Academic Press/Elsevier, San Diego, CA.

Schlesinger, W. H., and Lichter, J. (2001) Limited carbon storage in soil and litter of experimental forest plots under increased atmospheric CO_2. *Nature* 411, 466–469.

Schonwitz, R., Stichler, W., and Ziegler, H. (1986) $\delta^{13}C$ values of CO_2 from soil respiration on sites with crops of C_3 and C_4 type of photosynthesis. *Oecologia* 69, 305–308.

Schulze, E. (1967) Soil respiration of tropical vegetation types. *Ecology* 48, 652–653.

Schulze, E. D. (1982) Plant life forms as related to plant carbon, water and nutrient relations. In *Encyclopedia of plant physiology, physiological plant ecology, Vol. 12B: Water relations and photosynthetic productivity* (O. L. Lange, P. S. Nobel, C. B. Osmond, and H. Ziegler, eds.), pp. 615–676, Berlin and Heidelberg, Germany.

Schwartz, M. D. (1998) Green-wave phenology. *Nature* 394, 839–840.

Schwendenmann, L., Veldkkamp, E., Brenes, T., O'Brien, J. J., and Mackensen, J. (2003) Spatial and temporal variation in soil CO_2 efflux in an old-growth neotropical rain forest, La Selva, Costa Rica. *Biogeochemistry* 64, 111–128.

Scott-Denton, L. E., Sparks, K. L., and Monson, R. K. (2003) Spatial and temporal control of soil respiration rate in a high-elevation, subalpine forest. *Soil biology and biochemistry* 35, 525–534.

Seneviratne, R., and Wild, A. (1985) Effect of mild drying on the mineralization of soil nitrogen. *Plant and Soil* 84, 175–179.

Šesták, Z., Catsky, J., and Jarvis, P. G. (1971) Plant photosynthetic production manual of methods. Dr. W. Junk, Norwell, MA.

Shaver, G. R., Johnson, L. C., Cades, D. H., Murray, G., Laundre, J. A., Rastetter, E. B., Nadelhoffer, K. J., and Giblin, A. E. (1998) Biomass and CO_2 flux in wet sedge tundras: Responses to nutrients, temperature, and light. *Ecological Monographs* 68, 75–97.

Shaver, G. R., Canadell, J., Chapin III, F. S., Gurevitch, J., Henry, G., Ineson, P., Jonasson, S., Melillo, J., Pitelka, L., and Rustad, L. (2000) Global warming and terrestrial ecosystems: A conceptual framework for analysis. *Bioscience* 50, 871–882.

Shaw, M. R., and Harte, J. (2001) Responses of nitrogen cycling to simulated climate change: Differential responses along a subalpine ecotone. *Global Change Biology* 7, 193–210.

Shibistova, O., Lloyd, J., Evgrafova, S., Savushkina, N., Zrazhevskaya, G., Arneth, A., Knohl, A., and Kolle, O. (2002) Seasonal and spatial variability in soil CO_2 efflux rates for a central Siberian *Pinus sylvestris* forest. *Tellus* 54B, 552–567.

Sierra, J., and Renault, P. (1995) Oxygen consumption by soil microorganisms as affected by oxygen and carbon dioxide. *Applied Soil Ecology* 2, 175–184.

Sierra, J., and Renault, P. (1998) Temporal pattern of oxygen concentration in the interaggregate pore space of a hydromorphic soils. *Soil Science Society of America Journal* 62, 1398–1405.

Silver, W. L. (1998) The potential effects of elevated CO_2 and climate change on tropical forest soils and biogeochemical cycling. *Climatic Change* 39(2–3), 337–361.

Silver, W. L., Lugo, A. E., and Keller, M. (1999) Soil oxygen availability and biogeochemistry along rainfall and topographic gradients in upland wet tropical forest soils. *Biogeochemistry* 44(3), 301–328.

Silver, W. L., Ostertage, R., and Lugo, A. E. (2000) The potential for carbon sequestration through reforestation of abandoned tropical agricultural and pasture lands. *Restoration Ecology* 8(4), 394–407.

Silver, W. L., and Miya, R. K. (2001) Global patterns in root decomposition: Comparisons of climate and litter quality effects. *Oecologia* 129(3), 407–419.

Silver, W. L., Thompson, A. W., McGroddy, M. E., Varner, R. K., Dias, J. D., Silva, H., Crill, P. M., and Keller, M. (2005) Fine root dynamics and trace gas fluxes in two lowland tropical forest soils. *Global Change Biology* 11(2), 290–306.

Silvola, J., Välijoki, J., and Aaltonen, H. (1985) Effect of draining and fertilization on soil respiration at three ameliorated peatland sites. *Acta Forestalia Fennica* 191, 1–32.

Silvola, J. (1986) Carbon-dioxide dynamics in mires reclaimed for forestry in eastern Finland. *Annales Botanici Fennici* 23(1), 59–67.

Silvola, J., Alm, J., Ahlholm, U., Nykänen, H., and Martikainen, P. J. (1996) CO_2 fluxes from peat in boreal mires under varying temperature and moisture conditions. *Journal of Ecology* 84, 219–228.

Simmons, J. A., Fernandez, I. J., Briggs, R. D., and Delaney, M. T. (1996) Forest floor carbon pools and fluxes along a regional climate gradient in Maine, USA. *Forest Ecology and Management* 84, 81–95.

Šimůnek, J., and Suarez, D. L. (1993) Modeling of carbon dioxide transport and production in soil: 1, Model development. *Water Resources Research* 29(2), 487–497.

Singh, J. S., and Gupta, S. R. (1977) Plant decomposition and soil respiration in terrestrial ecosystems. *Botanical Review* 43, 449–528.

Sinsabaugh, R. L., Carreiro, M. M., and Repert, D. A. (2002) Allocation of extracellular enzymatic activity in relation to litter composition, N deposition, and mass loss. *Biogeochemistry* 60, 1–24.

Sirotnak, J. M., and Huntly, N. J. (2000) Direct and indirect effects of herbivores on nitrogen dynamics, voles in riparian areas. *Ecology* 81, 78–87.

Sitaula, B. K., Bakken, L. R., and Abrahamsen, G. (1995) N-fertilization and soil acidification effects on N_2O and CO_2 emission from temperate pine forest soil. *Soil Biology and Biochemistry* 27(11): 1401–1408.

Six, J., Elliott, E. T., Paustian, K., and Doran, J. W. (1998) Aggregation and soil organic matter accumulation in cultivated and native grassland soils. *Soil Science Society of America Journal* 62, 1367–1377.

Six, J., Conant, R. T., Paul, E. A., and Paustian, K. (2002) Stabilization mechanisms of soil organic matter: Implications for C-Saturation of soils. *Plant and Soil* 241, 155–176.

Skopp, J., Jawson, M. D., and Doran, J. W. (1990) Steady-state aerobic microbial activity as a function of soil water content. *Soil Science Society of America Journal* 54, 1619–1625.

Smith, F. B., and Brown, P. E. (1933) The diffusion of carbon dioxide through soils. *Soil Science* 35(6), 413–423.

Smith, S. E., and Read, D. J. (1997) *Mycorrhizal symbiosis*. 2nd edition, Academic Press, London.

Smith, D. L., and Johnson, L. (2004) Vegetation-mediated changes in microclimate reduce soil respiration as woodlands expand into grasslands. *Ecology* 85, 3348–3361.

Spano, D., Snyder, R. L., Duce, P., and Paw, U. K. T. (1997) Surface renewal analysis for sensible heat flux density using structure functions. *Agricultural and Forest Meteorology* 86(3–4), 259–271.

Spano, D., Snyder, R. L., Duce, P., *et al.* (2000) Estimating sensible and latent heat flux densities from grapevine canopies using surface renewal. *Agricultural and Forest Meteorology* 104(3), 171–183.

Sparling, G. P. (1992) Ratio of microbial biomass carbon to soil organic carbon as a sensitive indicator of changes in soil organic matter. *Australian Journal of Soil Research* 30, 195–207.

Stark, J. M., and Firestone, M. K. (1995) Mechanisms for soil moisture effects on activity of nitrifying bacteria. *Applied Environmental Microbiology* 61, 218–221.

Stark, S., Strömmer, R., and Tuomi, J. (2002) Reindeer grazing and soil microbial processes in two suboceanic and two subcontinental tundra heaths. *Oikos* 97, 69–78.

Stark, S., Tuomi, J., Strömmer, R., and Helle, T. (2003) Non-parallel changes in soil microbial carbon and nitrogen dynamics due to reindeer grazing in northern boreal forests. *Ecography* 26, 51–59.

Startsev, N. A., Mcnabb, D. H., and Startsev, A. D. (1997) Soil biology in recent clear-cuts in west-central Alberta. *Canadian Journal of Soil Science* 78, 69–76.

Steele, S. J., Gower, S. T., Vogel, J. G., and Norman, J. M. (1997) Root mass, net primary production and turnover in aspen, jack pine and black spruce forests in Saskatchewan and Manitoba, Canada. *Tree Physiology* 17(8–9), 577–587.

Stevenson, I. L. (1956) Some observations on the microbial activity in a remoistened air-dried soil. *Plant and Soil* 8, 170–182.

Stoklasa, J., and Ernest, A. (1905) Uber den Ursprung, die Menge und die Bedeutung des Kohlendioxyds in Boden. *Centro Bakteriol* 14, 723–736.

Stolzy, L. H. (1974) Soil atmosphere. In *The plant root and its environment* (E. W. Carson, ed.), pp. 335–362, University Press of Virginia, Charlottesville, VA.

Striegl, R. G., and Wickland, K. P. (1998) Effects of a clear-cut harvest on soil respiration in a jack pine-lichen woodland. *Canadian Journal of Forest Research* 28, 534–539.

Striegl, R. G., and Wickland, K. P. (2001) Soil respiration and photosynthetic uptake of carbon dioxide by ground-cover plants in four ages of jack pine forest. *Canadian Journal of Forest Research* 31, 1540–1550.

Strömgren, M. (2001) Soil-surface CO_2 flux and growth in a boreal Norway spruce stand: Effects of soil warming and nutrition. Doctoral thesis, Acta Universitatia Agriculturae Sueciae, Silvestria 220, Swedish University of Agricultural Sciences, Uppsala, ISBN 91-576-6304-1.

Stull, R. B. (1997) *An introduction to boundary layer meteorology.* Kluwer Academic Publishers, Dordrecht, the Netherlands.

Styles, J. M., Raupach, M. R., Farquhar, G. D., Kolle, O., Lawton, K. A., Brand, W. A., Werner, R. A., Jordan, A., Schulze, E. D., Shibistova, O., and Lloyd, J. (2002) Soil and canopy CO_2: $^{13}CO_2$, H_2O and sensible heat flux partitions in a forest canopy inferred from concentration measurements. *Tellus* 54B, 655–676.

Su, B. (2005) Interactions between ecosystem carbon, nitrogen and water cycles under global change: Results from field and mesocosm experiments. Ph.D. Dissertation, University of Oklahoma, Norman, OK.

Subke, J. A., Reichstein, M., and Tenhunen, J. D. (2003) Explaining temporal variation in soil CO_2 efflux in a mature spruce forest in southern Germany. *Soil Biology and Biochemistry* 34, 1467–1483.

Subke, J. A., Haln, V., Battipaglia, G., Linder, S., Buchmann, N., and Cotrufo, M. F. (2004) Feedback interactions between needle litter decomposition and rhizosphere activity. *Oecologia* 139, 551–559.

Subke, J. A., and Tenhunen, J. D. (2004) Direct measurements of CO_2 flux below a spruce forest canopy. *Agricultural and Forest Meteorology* 126(1–2), 157–168.

Sulzman, E. W., Brant, J. B., Bowden, R. D., and Lajtha, K. (2005) Contribution of aboveground litter, belowground litter, and rhizosphere respiration to total soil CO_2 efflux in an old growth coniferous forest. *Biogeochemistry* 73(1), 231–256.

Svensson, A. S., Johansson, F. I., Moller, I. M., and Rasmusson, A. G. (2002) Cold stress decreases the capacity for respiratory NADH oxidation in potato leaves. *FEBS Letters* 517, 79–82.

Swift, M., Heal, O. W., and Anderson, J. M., (1979) *Decomposition in terrestrial ecosystems.* University of California Press, Berkeley, CA.

Takagi, K., Miyata, A., Harazono, Y., Ota, N., Komine, M., and Yoshimoto, M. (2003) An alternative approach to determining zero-plane displacement, and its application to a lotus paddy field. *Agricultural and Forest Meteorology* 115, 173–181.

Tang, J., Baldocchi, D. D., and Xu, L. (2005a) Tree photosynthesis modulates soil respiration on a diurnal time scale. *Global Change Biology* 11, 1298–1304.

Tang, J., Qi, Y., Xu, M., Misson, L., and Goldstein, A. H. (2005b) Forest thinning and soil respiration in a ponderosa pine plantation in the Sierra Nevada. *Tree Physiology* 25, 57–66.

Tang, J., and Baldocchi, D. D. (2005) Spatial-temporal variation in soil respiration in an oak-grass savanna ecosystem in California and its partitioning into autotrophic and heterotrophic components. *Biogeochemistry* 73(1), 183–207.

Tans, P. P., Fung, I. Y., and Takahashi, T. (1990) Observational constraints on the global atmospheric CO_2 budget. *Science* 247, 1431–1438.

Tans, P. P. (1998) Oxygen isotopic equilibration between carbon dioxide and water in soils. *Tellus* 50B, 163–178.

Tate, C. M., and Striegl, R. G. (1993) Methane consumption and carbon-dioxide emission in tallgrass prairie: Effects of biomass burning and conversion to agriculture. *Global Biogeochemical Cycles* 7(4), 735–748.

Taylor, B., and Parkinson, D. (1988) Aspen and pine leaf litter decomposition in laboratory microcosms: 2, Interactions of temperature and moisture level. *Canadian Journal of Botany* 66(10), 1966–1973.

Taylor, J., and Lloyd, J. (1992) Sources and sinks of CO_2. *Australian Journal of Botany* 40, 407–418.

Tedla, A. B. (2004) Carbon and nitrogen dynamics and microbial community structure of a tall grass prairie soil subjected to simulated global warming and clipping. M.S. Thesis, University of Oklahoma, Norman, OK.

Thierron, V., and Laudelout, H. (1996) Contribution of root respiration to total CO_2 efflux from the soil of a deciduous forest. *Canadian Journal of Forest Research* 26, 1142–1148.

Thomas, S. M., Cook, F. J., Whitehead, D., and Adams, J. A. (2000) Seasonal soil-surface carbon fluxes from the root systems of young *Pinus radiata* trees growing at ambient and elevated CO_2 concentration. *Global Change Biology* 6(4), 393–406.

Thompson, M. V., and Randerson, J. T. (1999) Impulse response functions of terrestrial carbon cycle models: Method and application. *Global Change Biology* 5, 371–394.

Thornley, J. H. M. (1970) Respiration, growth and maintenance in plants. *Nature* 227, 304–305.

Thuille, A., Buchmann, N., and Schulze, E. D. (2000) Carbon stocks and soil respiration rates during deforestation, grassland use and subsequent Norway spruce afforestation in southern Alps, Italy. *Tree Physiology* 20, 849–857.

Tien, M., and Myer, S. B. (1990) Selection and characterization of mutants of *Phanerochaete chrysosporium* exhibiting ligninolytic activity under nutrient rich conditions. *Applied and Evironmental Microbiology* 56, 2540–2544.

Tingey, D. T., Phillips, D. L., and Johnson, M. G. (2000) Elevated CO_2 and conifer roots: Effects on growth, life span and turnover. *New Phytologist* 147, 87–103.

Toland, D. E., and Zak, D. R. (1994) Seasonal patterns of soil respiration in intact and clear-cut northern hardwood forests. *Canadian Journal of Forest Research* 24, 1711–1716.

Torbert, H. A., Rogers, H. H., Prior, S. A., Schlesinger, W. H., and Runion, G. B. (1997) Effects of elevated atmospheric CO_2 in agro-ecosystems on soil carbon storage. *Global Change Biology* 7, 513–521.

Torbert, H. A., Prior, S. A., Rogers, H. H., and Wood, C. W. (2000) Review of elevated atmospheric CO_2 effects on agro-ecosystems: Residue decomposition processes and soil C storage. *Plant and Soil* 224, 59–73.

Trettin, C. C., Song, B., Jurgensen, M. F., and Li, C. (2001) *Existing soil carbon models do not apply to forested wetlands*. USDA Forest Service GTR SRS-46.

Trettin, C. C., and Jurgensen, M. F. (2003) Carbon cycling in wetland forest soils. In *The potential of U.S. forest soil to sequester carbon and mitigate the greenhouse effect* (J. M. Kimble, L. S. Heath, R. A. Birdsey, and R. Lal, eds.), pp. 311–331, CRC Press, Boca Raton, FL.

Trumbore, S. E., Chadwick, O. A., and Amundson, R. (1996) Rapid exchange between soil carbon and atmospheric carbon dioxide driven by temperature change. *Science* 272, 393–396.

Trumbore, S. (2000) Age of soil organic matter and soil respiration: Radiocarbon constraints on belowground C dynamics. *Ecological Applications* 10(2), 399–411.

Tryon, P. R., and Chapin, F. S. (1983) Temperature control over root growth and root biomass in taiga forest trees. *Canadian Journal of Forest Research* 13, 827–833.

Tu, K., and Dawson, T. (2005) Partitioning ecosystem respiration using stable carbon isotope analyses of CO_2. In *Stable isotopes and biosphere-atmosphere interactions: Processes and biology controls* (L. B. Flanagan, R. Ehleringer, and D. E. Pataki, eds.), pp. 125–153, Academic Press, London.

Tufekcioglu, A., Raich, J. W., Isenhart, T. M., and Schulz, R. C. (2001) Soil respiration within riparian buffers and adjacent crop fields. *Plant and Soil* 229, 117–124.

Tulaphitak, T., Pairintra, C., and Kyuma, K. (1983) Soil fertility and tilth. In *Shifting cultivation: An experiment at Nam Phrom, northeast Thailand, and its implications for upland farming in the monsoon tropics* (K. Kyuma, and C. Pairintra, eds.), pp. 62–83, Faculty of Agriculture, Kyoto University, Japan.

Tulaphitak, T., Pairintra, C., and Kyuma, K. (1985) Changes in soil fertility and tilth under shifting cultivation. *Soil Science and Plant Nutrition* 31, 251–261.

Turner, M. G. (1989) Landscape ecology: The effect of pattern on process. *Annual Review of Ecology and Systematics* 20, 191–190.

Turpin, H. W. (1920) The carbon dioxide of the soil air. *Cornell University Agricultural Experiment Station Memoir* 32, 319–362.

Valentini, R., Matteucchi, G., Dolman, H., Schulze, E.-D., Rebmann, C., Moors, E. J., Granier, A., Gross, P., Jensen, N. O., Pilgaard, K., Lindroth, A., Grelle, A., Bernhofer, C., Grünwald, T., Aubinet, M., Ceulemans, R., Kowalski, A. S., Vesala, T., Rannik, Ü., Berbigier, P., Lousteau, D., Gudmundsson, J., Thorgairsson, H., Ibrom, A., Morgenstern, K., Clement, R., Moncrieff, J., Montagnani, L., Minerbi, S., and Jarvis, P. G. (2000) Respiration as the main determinant of carbon balance in European forests. *Nature* 404, 861–865.

Vance, E. D., and Chapin III, F. S. (2001) Substrate limitations to microbial activity in taiga forest floors. *Soil Biology and Biochemistry* 33(2), 173–188.

Van der Peijl, M. J., and Verhoeven, J. T. A. (1999) A model of carbon, nitrogen and phosphorus dynamics and their interactions in river marginal wetlands. *Ecological Modelling* 118, 95–130.

Van der Werf, A., Welschen, R., and Lambers, H. (1992). Respiratory losses increase with decreasing inherent growth rate of a species and with decreasing nitrate supply: A search for explanations for these observations. In *Molecular, biochemical and physiological aspects of plant respiration* (H. Lambers, and L. H. W. Van der Plas, eds.), pp. 421–432, SPB Academic Publishing, the Hague, the Netherlands.

Van Gestel, M., Ladd, J. N., and Amato, M. (1991) Carbon and nitrogen mineralization from two soils of contrasting texture and microaggregate stability: Influence of sequential fumigation, drying and storage. *Soil Biology and Biochemistry* 23, 313–322.

Van Kessel, C., Nitschelm, J., Horwath, W. R., Harris, D., Walley, F., Luscher, A., and Hartwig, U. (2000) Carbon-13 input and turn-over in a pasture soil exposed to long-term elevated atmospheric CO_2. *Global Change Biology* 6, 123–135.

Vanhala, P. (2002) Seasonal variation in the soil respiration rate in coniferous forest soils. *Soil Biology and Biochemistry* 34, 1375–1379.

van't Hoff, J. H. (1884) *Études de dynamique chimique* (Studies of chemical dynamics). Frederik Muller and Co., Amsterdam, the Netherlands.

Van Vuuren, M. M. I., and Van der Eerden, L. J. (1992) Effects of three rates of atmospheric nitrogen deposition enriched with ^{15}N on litter decomposition in a heathland. *Soil Biology and Biochemistry* 24, 527–532.

Verburg, P. J., Arnone III, J. A., Obrist, D., Schorran, D., Evans, R. D., Leroux-Swarthout, D., Johnson, D. W., Luo, Y., and Coleman, J. S. (2004) Net ecosystem carbon exchange in two experimental grassland ecosystems. *Global Change Biology* 10, 498–508.

Vermes, J. F., and Myrold, D. D. (1992) Denitrification in forest soil of Oregon. *Canadian Journal of Forest Research* 22, 504–512.

Vitousek, P. M., and Sanford, R. L. (1986) Nutrient cycling in moist tropical forest. *Annual Review of Ecology and Systematics* 17, 137–167.

Vitousek, P. M., and Howarth, R. W. (1991) Nitrogen limitation on land and in the sea, How can it occur? *Biogeochemistry* 13, 87–115.

Vose, J. M., Elliott, K. J., and Johnson, D. W. (1995) Soil CO_2 flux in response to elevated atmospheric CO_2 and nitrogen fertilization, patterns and methods. In *Soil and global change* (R. Lal, J. M. Kimble, E. Levine, and B. A. Stewart, eds.), CRC Press, Boca Raton, FL.

Vose, J. M., and Ryan, M. G. (2002) Seasonal respiration of foliage, fine roots, and woody tissue in relation to growth, tissue N, and photosynthesis. *Global Change Biology* 8, 182–193.

Wadman, W. P., and de Haan, S. (1997) Decomposition of organic matter from 36 soils in a long-term pot experiment. *Plant and Soil* 189(2), 289–301.

Waksman, S. A., and Starkey, R. L. (1924) Microbiological analysis of soil as an index of soil fertility: VII, Carbon dioxide evolution. *Soil Science* 17(2), 141–161.

Wallander, H., Arnebrant, K., and Dahlberg, A. (1999) Relationships between fungal uptake of ammonium, fungal growth and nitrogen availability in ectomycorrhizal *Pinus sylvestris* seedlings. *Mycorrhiza* 8(4), 215–223.

Walter, H. (1952) Eine einfache Methode zur ökologischen Erfassung des CO_2-Faktors am Standort. *Berichte der Deutschen Botanischen Gesellschaft* 65, 175–182.

Wan, S., Luo, Y., and Wallace, L. (2002) Changes in microclimate induced by experimental warming and clipping in tallgrass prairie. *Global Change Biology* 8, 754–768,

Wan, S., and Luo, Y. (2003) Substrate regulation of soil respiration in a tallgrass prairie: Results of a clipping and shading experiment. *Global biogeochemical cycles* 17, 1054, doi: 10.1029/2002GB001971.

Wan, S., Hui, D., Wallace, L. L., and Luo, Y. (2005) Direct and indirect warming effects on ecosystem carbon processes in a tallgrass prairie. *Global Biogeochemical Cycles* 19, GB2014, doi: 10.1029/2004GB002315.

Wang, Y., Amundson, R., and Trumbore, S. (1999) The impact of land use change on C turnover in soils. *Global Biogeochemical Cycles* 13(1), 47–57.

Wang, H., Curtin, D., Jame, Y. W., McConkey, B. G., and Zhou, H. F. (2002) Simulation of soil carbon dioxide flux during plant residue decomposition. *Soil Science Society of America Journal* 66(4), 1304–1310.

Wang, X. Z., and Curtis, P. (2002) A meta-analytical test of elevated CO_2 effects on plant respiration. *Plant Ecology* 161, 251–261.

Warembourg, F. R., and Kummerow, J. (1991) Photosynthesis/translocation studies in terrestrial ecosystems. pp. 11–38. In *Carbon isotope techniques* (D. C. Coleman, and B. Fry, eds.), Academic Press, New York.

Watson, R. T., Noble, I. R., Bolin, B., Ravindranath, N. H., Verardo, D. J., and Dokken, D. J. (2000) *Land use, land-use change and forestry.* Cambridge University Press, Cambridge, UK.

Weart, S. (2003) The discovery of rapid climate change. *Physics Today* 56(8), 30–36.

Weber, M. G. (1985) Forest soil respiration in eastern Ontario jack pine ecosystems. *Canadian Journal of Forest Research* 15, 1069–1073.

Weber, M. G. (1990) Forest soil respiration after cutting and burning in immature aspen ecosystems. *Forest Ecology and Management* 31, 1–14.

Wedin, D. A., Tieszen, L. L., Dewey, B., and Pastor, J. (1995) Carbon isotope dynamics during grass decomposition and soil organic matter formation. *Ecology* 76, 1383–1392.

Welles, J. M., Demetriades-Shah, T. H., and McDermitt, D. K. (2001) Considerations for measuring ground CO_2 effluxes with chambers. *Chemical Geology* 177(1–2), 3–13.

Weltzin, J. F., Pastor, J., Harth, C., Bridgham, S. D., Updegraff, K., and Chapin, C. T. (2000) Response of bog and fen plant communities to warming and water table manipulations. *Ecology* 81, 3464–3478.

Weltzin, J. F., Bridgham, S. D., Pastor, J., Chen, J. Q., and Harth, C. (2003) Potential effects of warming and drying on peatland plant community composition. *Global Change Biology* 9 (2), 141–151.

West, T. O., and Marland, G. (2002) Net carbon flux from agricultural ecosystems: Methodology for full carbon cycle analyses. *Environmental Pollution* 116(3), 439–444.

West, T. O., and Post, W. M. (2002) Soil organic carbon sequestration rates by tillage and crop rotation: A global data analysis. *Soil Science Society of America Journal* 66, 1930–1946.

Western, A. W., Bloschl, G., and Grayson, R. G. (1998) Geostatistical characterization of soil moisture patterns in the Tarrawarra catchment. *Journal of Hydrology* 250, 20–37.

Whipps, J. M., and Lynch, J. M. (1983) Substrate flow and utilization in the rhizosphere of cereals. *New Phytologist* 95, 605–623.

White, J. D., Running, S. W., Thornton, P. E., Keane, R. E., Ryan, K. C., Fagre, D. B., and Key, C. H. (1998) Assessing simulated ecosystem processes for climate variability research at Glacier National Park, USA. *Ecological Applications* 8, 805–823.

Whitman, W. B., Bowen, T. L., and Boone, D. R. (1992) The methanogenic bacteria. In *The prokaryotes* (A. Balows, H. G. Truper, M. Dwarkin, W. Harder, and K. L. Schleifer, eds.), pp. 719–767, 2nd edition, Springer-Verlag, New York.

Wiant, H. V. (1967) Influence of temperature on the rate of soil respiration. *Journal of Foresty* 65, 489–490.

Wichern, F., Luedeling, E., Muller, T., Joergensen, R. G., and Buerkert, A. (2004) Field measurements of the CO_2 evolution rate under different crops during an irrigation cycle in a mountain oasis of Oman. *Applied Soil Ecology* 25(1), 85–91.

Wieser, G. (2004) Seasonal variation of soil respiration in a *Pinus cembra* forest at the upper timberline in the central Austrian Alps. *Tree Physiology* 24, 475–480.

Wildung, R. E., Garland, T. R., and Buschbom, R. L. (1975) The interdependent effects of soil temperature and water content on soil respiration rate and plant root decomposition in arid grassland soils. *Soil Biology and Biochemistry* 7, 373–378.

Williams, J. H. H., and Farrar, J. F. (1990) Control of barley root respiration. *Physiologia Plantarum* 79, 259–266.

Wilsey, B. J., Parent, G., Roulet, N. T., Moore, T. R., and Potvin, C. (2002) Tropical pasture carbon cycling, relationships between C source/sink strength, above-ground biomass and grazing. *Ecology Letters* 5, 367–376.

Winston, G. C., and Sundquist, E. T. (1997) Winter CO_2 fluxes in a boreal forest. *Journal of Geophysical Research* 102(D24), 28795–28804.

Wirth, S. J. (2001) Regional-scale analysis of soil microbial biomass and soil basal CO_2-respiration in northeastern Germany. In *Sustaining the global farm: Selected papers from the 10th International Soil Conservation Organization meeting, May 24–29, 1999, West Lafayette, IN.* (D. E. Stott, R. H. Mohtar, and G. C. Steinhardt, eds.), pp. 486–493, ISCO in cooperation with the USDA and Purdue University, West Lafayette, IN. Available online: http.//topsoil. nserl.purdue.edu/nserlweb/isco99/pdf/isco99pdf.htm.

Witkamp, M. (1966) Rates of carbon dioxide evolution from forest floor. *Ecology* 47(3), 492.

Witkamp, M., and Frank, M. L. (1969) Evolution of CO_2 from litter, humus, and subsoil of a pine stand. *Pedobiologia* 9, 358–365.

Wohlfahrt, G., Anfang, C., Bahn, M., Haslwanter, A., Newesely, C., Schmitt, M., Drösler, M., Pfadenhauer, J., and Cernusca, A. (2005) Quantifying nighttime ecosystem respiration of a meadow using eddy covariance, chamber and modelling. *Agricultural and Forest Meteorology* 128, 141–162.

Wojick, D. E. (1999) Carbon storage in soils: The ultimate no-regrets policy? A report to Greening Earth Society. Available online: http.//www.greeningearthsociety.org/pdf/carbon.pdf.

Wollny, E. (1831) Untersuchungen über den Einfluss der physikalischen Eigenschaften des Bodens auf dessen Gehalt an freier Kohlensaure. *Forschende Gebiete–Agricultural Physics* 4, 1–28.

Wood, B. D., Keller, C. K., and Johnstone, D. L. (1993) In situ measurement of microbial activity and controls on microbial CO_2 production in the untreated zone. *Water Resource Research* 29, 647–659.

Wyngaard, J. C. (1990) Scalar fluxes in the planetary boundary layer: Theory, modeling, and measurement. *Boundary-Layer Meteorology* 50, 49–75.

Xu, M., and Qi, Y. (2001a) Soil-surface CO_2 efflux and its spatial and temporal variations in a young ponderosa pine plantation in northern California. *Global Change Biology* 7, 667–677.

Xu, M., and Qi, Y. (2001b) Spatial and seasonal variations of Q_{10} determined by soil respiration measurements at a Sierra Nevadan forest. *Global Biogeochemical Cycles* 15, 687–696

Xu, L., Baldocchi, D. D., and Tang, J. (2004) How soil moisture, rain pulses, and growth alter the response of ecosystem respiration to temperature. *Global Biogeochemical Cycles* 18, GB4002, doi: 10.1029/2004GB002281.

Yavitt, J. B., and Fahey, T. J. (1986) Long-term litter decay and forest floor leaching in *Pinus contorta* ecosystems, southeastern Wyoming. *Journal of Ecology* 74, 525–545.

Yim, M. H., Joo, S. J., and Nakane, K. (2002) Comparison of field methods for measuring soil respiration, a static alkali absorption method and two dynamic closed chamber methods. *Forest Ecology and Management* 170(1–3), 189–197.

Zak, D. R., Grigal, D. F., and Ohmann, L. F. (1993) Kinetics of microbial respiration and nitrogen mineralization in Great Lakes forests. *Soil Science Society of America Journal* 57(4), 1100–1106.

Zak, D. R., Pregitzer, K. S., King, J. S., and Holmes, W. E. (2000) Elevated atmospheric CO_2, fine roots and the response of soil microorganisms: A review and hypothesis. *New Phytologist* 147, 201–222.

Zak, D. R., Holmes, W. E., Finzi, A. C., Norby, R. J., and Schlesinger, W. H. (2003) Soil nitrogen cycling under elevated CO_2: A synthesis of forest FACE experiments. *Ecological Applications* 13, 1508–1514.

Zepp, R. G., Miller, W. L., Burke, R. A., Parsons, D. A. B., and Scholes, M. C. (1996) Effects of moisture and burning on soil-atmosphere exchange of trace carbon gases in a southern African savanna. *Journal of Geophysical Research-Atmosphere* D101, 23699–23706.

Zhang, W., Parker, K., Luo, Y., Wallace, L., and Hu, S. (2005) Soil microbial responses to experimental atmospheric warming and clipping in a tallgrass prairie. *Global Change Biology* 11, 266–277.

Zheng, D., Chen, J., Noormets, A., Le Moine, J., and Euskirchen, E. (2005) Effects of climate and land use on landscape soil respiration in northern Wisconsin, 1972–2001. *Climate Research* 28, 163–173.

Zhou, X., Sherry, B., An, Y., Wallace, L. L., and Luo, Y. (2006) Main and interactive effects of warming, clipping, and doubled precipitation on soil CO_2 efflux in a grassland ecosystem. *Global Biogeochemical Cycles* (in press).

Zhuang, Q., McGuire, J. M., Clein, J. S., Clein, J. S., Dargaville, R. J., Kicklighter, D. W., Myneni, R. B., Dong, J., Romanovsky, V. E., Harden, J., and Hobbie, J. E. (2003) Carbon cycling in extratropical terrestrial ecosystems of the northern hemisphere during the 20th century, a modeling analysis of the influences of soil thermal dynamics. *Tellus* 55B, 751–776.

Zibilske, L. M. (1994) Carbon mineralization. In *Methods of soil analysis, Part* 2 (R. W. Weaver *et al.*, eds.), pp. 835–863, SSSA Book Ser. 5 SSSA, Madison, WI.

Zimov, S. A., Davidov, S. P., Voropaev, Y. V., Prosiannikov, S. F., Semiletov, I. P., Chapin, M. C., and Chapin III. F. S. (1996) Siberian CO_2 efflux in winter as a CO_2 source and cause of seasonality in atmospheric CO_2. *Climate Change* 33, 111–120.

Index

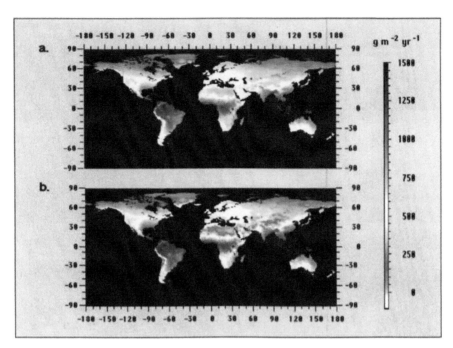

FIGURE 10.12 Global annual soil CO_2 emissions as predicted by the (a) log-transformed model, $\log R_s = F + (QT \dfrac{P}{K+P})$ $R_s = e^{\log R_s} - 1.0$ and (b) untransformed model $R_s = F_e^{QT} \dfrac{P}{K+P}$, where R_s (g C m^{-2}d^{-1}) is soil CO_2 efflux, F (g C m^{-2}d^{-1}) is the efflux rate when temperature is zero, Q(°C^{-1}) is temperature coefficient, T(°C) is mean monthly air temperature, P(cm) is mean monthly precipitation, and K(cm month^{-1}) defines the half-saturation coefficient of the precipitation function (Provided by J. W. Raich with permission form Global Biogeochemical Cycles: Raich and Potter 1995).

FIGURE A Schematic showing path of air flow between 6400-09 and LI-6400 console (left) and the measurement of soil CO_2 efflux in the field (right).

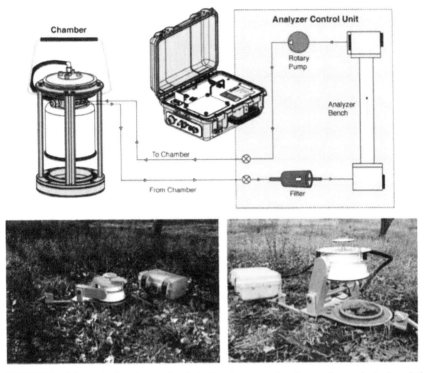

FIGURE B Schematic showing path of air flow between chamber and console (up) and the continuous measurement of soil CO_2 efflux with closed (left at bottom) and open (right at bottom) chamber in the field.

FIGURE C Soil respiration system that SRC-1 connected with EGM (left) and CIRAS (right) from PP Systems.

FIGURE D   CFX-2 from PP Systems.

FIGURE E   The measurement of soil respiration in the field by SRS1000 Ultra compact soil flux system (left) and SRS2000 Intelligent portable soil flux system (right).

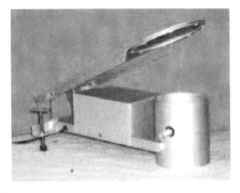

FIGURE F   SRC-MV5 system.

FIGURE G   Automated soil respiration chamber in the open position.

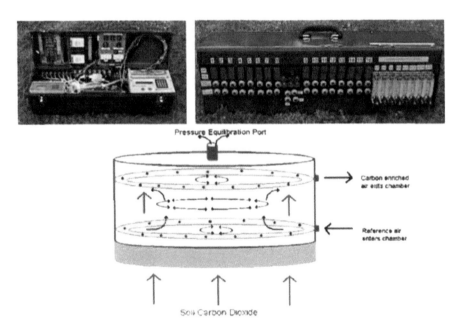

FIGURE H   The console of 16-port Automatic Carbon Efflux System (up) and soil respiration chamber showing air flow (bottom).

Printed and bound by CPI Group (UK) Ltd, Croydon, CR0 4YY

03/10/2024

01040412-0006